AF556876

STRUCTURAL BIOLOGY SERIES

JONATHAN KING, *Editor*

Ribosome Structure
and
Protein Biosynthesis

STRUCTURAL BIOLOGY SERIES

Squire *Muscle: Design, Diversity, and Disease*

Ribosome Structure and Protein Biosynthesis

Alexander S. Spirin

Institute of Protein Research
U.S.S.R. Academy of Sciences
and
Moscow State University

1986

THE BENJAMIN/CUMMINGS PUBLISHING COMPANY, INC.
Advanced Book Program
Menlo Park, California

Reading, Massachusetts • Don Mills, Ontario • Wokingham, U.K. • Amsterdam
Sydney • Singapore • Tokyo • Mexico City • Bogota • Santiago • San Juan

Sponsoring editor: Paul Elias
Interior design: Richard Kharibian
Cover design: Daved Garza

Library of Congress Cataloging-in-Publication Data

Spirin, A. S. (Aleksandr Sergeevich)
Ribosome structure and protein biosynthesis.

(Structural biology series)
Bibliography: p.
Includes index.
1. Protein biosynthesis. 2 Ribosomes.
3. Genetic translation. I. Title. II. Series.
QP551.S72 1986 574.19'296 85-20953
ISBN 0-8053-8390-5
ABCDEFGHIJ– MA –89876
The Benjamin/Cummings Publishing Company
2727 Sand Hill Road
Menlo Park, California 94025

Contents

PART III

FUNCTIONING OF THE RIBOSOME

Preface

This book is based on an advanced course of lectures on ribosome structure and protein biosynthesis that I offer at Moscow State University. These lectures have been part of a general course on molecular biology for almost 20 years. It is quite natural that they have undergone considerable evolution as knowledge has progressed in this field. This evolution continues, and readers should be prepared that some of the statements and ideas that I have ventured to put in this book may not stand the test of time. Nevertheless, since I decided to keep a lecture style as far as possible, I took these risks in writing this book. After all, this is a textbook which, on the one hand, should provide a background of knowledge and current ideas in the field and, on the other hand, should stimulate students to think, discuss, and formulate problems. The presentation of some of the topics should perhaps be regarded just as examples of today's interpretations or generalizations. Anyway, in a number of cases I have preferred not to leave white spots but to give a tentative interpretation of disputable and controversial results as I do in my lectures.

The literature sources in this book are cited mainly for teaching purposes. Each of the chapters is not intended as a review, and the reference lists are far from complete. To give an insight into the histories of discoveries I have tried to cite pioneer studies in the fields discussed. To provide information on the present state of knowledge, I have referred the reader to some of the recent papers. As this is an advanced textbook, I thought it would be helpful to give several dozen references in each chapter rather than limiting them to the very minimum. Many chapters are also supplemented with a list of books, reviews and particularly fundamental papers for further reading; these include titles where the point of view and interpretations may differ from my own.

In any textbook the choice of illustrations is the most complex and difficult part. When lecturing and using the blackboard, it is quite easy to draw a diagram, whereas providing a corresponding drawing for a written textbook is quite another matter. I decided in the main not to reproduce standard diagrams used in other books or textbooks but to create my own. In making the final illustrations I was greatly assisted by the Department of Scientific Information of the Institute of Protein Research, and, personally, by its Head, Mr Ariel Raiher; I am particularly grateful to Pavlina Zavozina and Vera Nevskaya for doing these drawings.

The illustrations to some of the chapters could not have appeared without the personal assistance of Dr. Valery Lim and Dr. Victor Vasiliev from the Institute of Protein Research, who not only supplied the pictures of stereochemical models and the majority of the original electron micrographs for this book, but were my constant consultants during the writing of the corresponding sections. Their associates Andrey Kayava, Dmitry Lyakhov, Olga Selivanova, and Sergei Ryazantsev also helped with the drawings and photographs of models.

I am grateful to the scientists who have read parts of the manuscript and made their comments. Particularly valuable comments were made by Dr. Vadim Agol, Dr. Alexey Bogdanov, Dr. Alexander Chetverin, Dr. Alexander Efimov, Dr. Alexander Girshovich, Dr. Anatoly Gudkov, Dr. Lev Kiselev, Dr. Dmitry Knorre, Dr. Alexander Krayevsky, Dr. Marina Kukhanova, Dr. Oleg Ptitsyn, Dr Evgeny Shakhnovich, Dr. Valery Schwartz and Dr. Leonid Voronin. Dr. Georgyi Gause who has translated this book into English made several valuable suggestions which contributed to clearer formulations.

Finally, I would particularly like to thank Ruma Makarova and Larisa Rozhanskaya for their invaluable help in preparing the manuscript.

Unfortunately, I had no real opportunity to familiarize many of my colleagues with this manuscript, particularly in other countries. I very much hope that they will respond to the publication of the book and send their critical comments which will be greatly appreciated.

Pushchino and Moscow **Alexander Spirin**
January, 1984

PART I

INTRODUCTION

Chapter 1

General Scheme of Protein Biosynthesis

The proteins of all living cells are synthesized by *ribosomes*. The ribosome is a large macromolecule consisting of ribonucleic acids (*ribosomal RNAs*) and proteins; it has a complex asymmetric quaternary structure. In order to synthesize protein, the ribosome must be supplied with (1) a program determining the sequence of amino acid residues in the polypeptide chain, (2) the amino acid substrate from which the protein is to be made, and (3) chemical energy. The ribosome itself plays a catalytic role and is responsible for forming peptide bonds, i.e. for the polymerization of amino acid residues into the polypeptide chain.

The program that sets the sequence of amino acid residues in a polypeptide chain comes from *deoxyribonucleic acid* (DNA), i.e. from the cell genome. Sections of the double-stranded DNA, which are called genes, serve as templates for synthesizing single-stranded RNA molecules. The synthesized RNA species are complementary replicas of just one of the DNA chains and therefore are faithful copies of the nucleotide sequence of the other DNA chain. This process of gene copying, accomplished by the enzyme RNA polymerase, is called *transcription*. In eucaryotic cells, and to a lesser extent in procaryotic

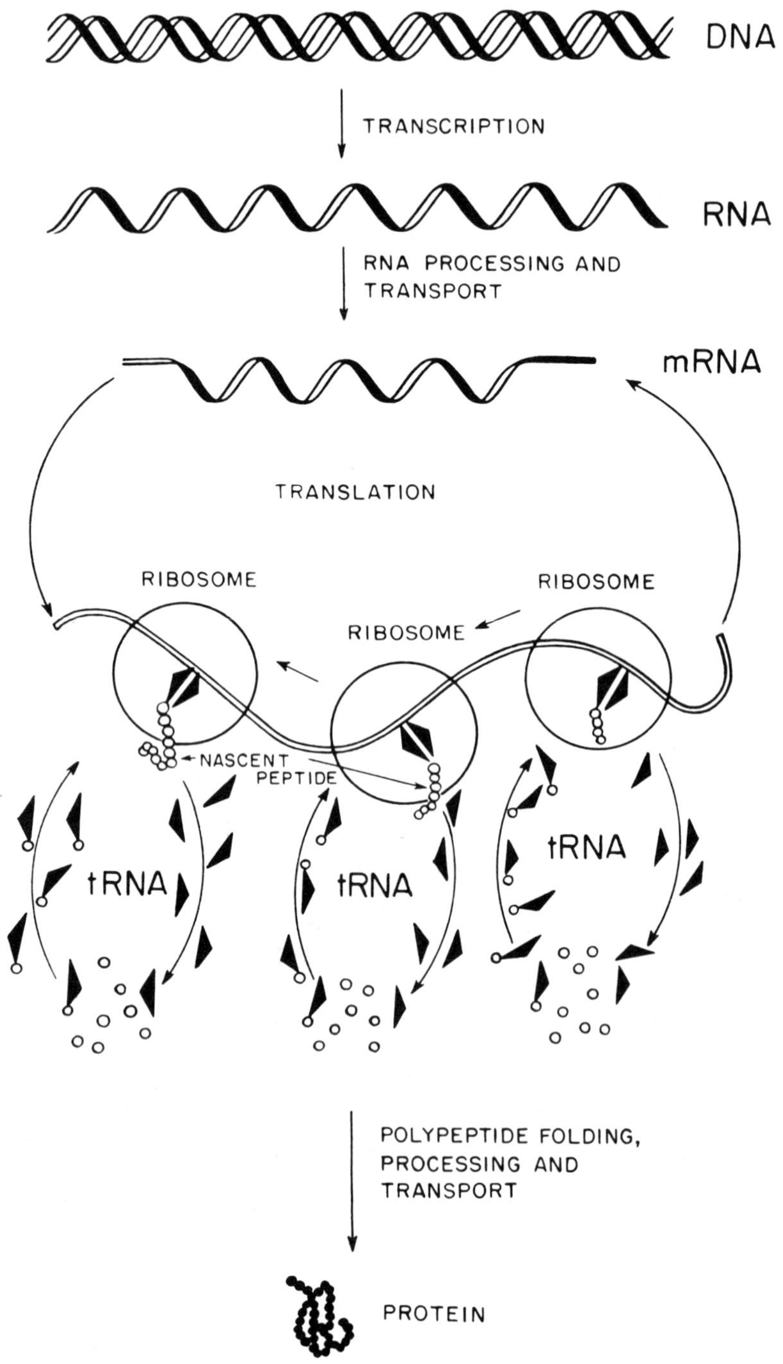

Figure 1 General scheme of protein biosynthesis (DNA → RNA → protein).

cells, nascent RNA may undergo a number of additional changes called *processing*; as a result, certain parts of the nucleotide sequence may be excised from RNA. The mature RNA becomes associated with the ribosomes and serves as a program, or template, which determines the amino acid sequence in the synthesized protein. This template RNA is usually called *messenger RNA* (mRNA). In other words, the flow of information from DNA to ribosomes is mediated by gene transcription, resulting in the formation of mRNA.

Proteins consist of amino acids. Free amino acids, however, are not used in the synthetic machinery of the ribosome. To become a substrate for protein synthesis, an amino acid must be *activated* by coupling with the adenylic moiety of ATP and then *accepted* by (covalently linked to) a special RNA molecule called *transfer RNA* (tRNA); this process is performed by the enzyme aminoacyl-tRNA synthetase. The resulting aminoacyl-tRNA is used by the ribosome as a substrate for protein synthesis, and the energy of the chemical bond between the amino acid residue and tRNA is used for forming a peptide bond. Thus, the activation of amino acids and formation of aminoacyl-tRNAs provide both material and energy to protein synthesis.

Using mRNA as a program and aminoacyl-tRNAs as energy-rich substrates, the ribosome *translates* genetic information from the nucleotide language of mRNA into the amino acid language of polypeptide chains. In molecular terms this implies that while moving along the mRNA, the ribosome consecutively selects appropriate aminoacyl-tRNA species from the medium. The specificity of the aminoacyl residue of a corresponding aminoacyl-tRNA selected by the ribosome is defined by the combination of nucleotides in a corresponding stretch of mRNA associated with the ribosome. This brings us to the problem of *genetic coding*, i.e. the question of nucleotide combinations that determine, or code, each of the 20 natural amino acids.

Hence, the movement of the ribosome along the mRNA chain (or, in other words, the passing of mRNA through the ribosome) establishes a temporal order of entering the various aminoacyl-tRNA species into the ribosome. This order depends on the sequence of coding nucleotide combinations along the mRNA. The aminoacyl residue of each selected aminoacyl-tRNA is being attached covalently to a growing polypeptide chain by the ribosomal machinery. Deacylated tRNA is released by the ribosome into solution. This results in the step-by-step formation of the polypeptide chain.

The general scheme of protein biosynthesis outlined above is presented in fig. 1.

Further reading

Lewin, B. (1983). *Genes*. New York: Wiley.

Watson, J. D. (1976). *Molecular biology of the gene*. 3rd ed. Menlo Park, Calif.: W. A. Benjamin.

Chapter 2

Messenger RNA and the Genetic Code

2-1 Discovery of mRNA

After the discovery and final recognition of the genetic function of DNA (1944–1953)[1–4], it rapidly became clear that DNA itself does not serve as a direct template for protein synthesis. In addition, a number of early observations suggested that ribonucleic acid was closely connected to cellular protein synthesis[5,6]. These ideas were developed and resulted in the concept that RNA is the intermediate responsible for the transfer of genetic information from DNA to proteins; in particular it has been suggested that RNA serves as a template upon which amino acid residues are polymerized (DNA → RNA → → protein)[7–10].

This conceptual advance coincided with the discovery of protein-synthesizing ribonucleoprotein particles of the cell which were later called ribosomes (see chapter 4). It had also been established that the RNA of these particles accounted for the main bulk of cellular RNA. Hence, it was naturally assumed that genes are transcribed into ribosomal RNA species, which in turn serve as templates for protein synthesis. This led to a "one gene-one ribosome-one protein" hypothesis.

During 1956 and 1957, in order to test this hypothesis, a comparative analysis of DNA and RNA base composition in a large number of microorganisms was conducted[11]. DNA base compositions may be rather different in different groups of microorganisms, and it was hypothesized that if the above formulation of a "DNA-RNA-protein" scheme was correct, the base composition of total RNA would strongly correlate with the DNA base composition in bacteria. The experimental results, however, were unexpected. Despite great differences of DNA base composition in various bacterial species, the composition of total RNA was found to be similar in all of the studied bacteria, and did not imitate DNA base composition. These results implied that the bulk of cellular RNA, i.e. most likely ribosomal RNA, could not serve as a direct informational intermediate between DNA and proteins.

At the same time, RNA base composition was shown to vary slightly for different bacterial species, and to be positively correlated with the base composition of DNA. The conclusion based on this correlation was that cells may contain a special *minor RNA fraction* which imitates DNA base composition and could possibly serve as an intermediate between genes and protein-synthesizing particles[11,12].

On the other hand, Volkin and Astrachan[13] have studied RNA synthesis in bacteria infected by DNA-containing T2 bacteriophages. Bacterial protein synthesis ceases soon after infection, and the entire cellular protein-synthesizing machinery is switched over to producing phage proteins. Most of the cellular RNA does not undergo any change during this process, but the cell begins to synthesize a small fraction of metabolically unstable short-lived RNA, the nucleotide composition of which is similar to the base composition of phage DNA.

Several years later, in 1961, the minor RNA fraction, termed *DNA-like RNA*, was separated from the total cellular RNA. Its function as messenger, carrying information from the DNA to the ribosomes, was demonstrated in the direct experiments of Brenner, Jacob, and Meselson[14], and those of Gros, Watson, and co-workers[15]; similar observations have been made by Spiegelman and associates[16]. It has been demonstrated that DNA-like RNA formed after the T4 phage infection binds to the preexisting host ribosomes (incidentally, no new ribosomes are synthesized after phage infection), and the ribosomes associated with the phage-specific RNA synthesize the phage proteins. This RNA could be detached easily from the ribosomes *in vitro* without destroying the particles. It has been shown that this RNA is indeed complementary to one of the phage DNA chains.

On the basis of their results on genetic regulation in bacteria, Jacob and Monod (1961) advanced the idea that a special short-lived RNA transfers information from genes to ribosomes and serves as a direct template for protein synthesis[17]. The term *messenger RNA* was accepted in all subsequent studies.

2-2 Deciphering the code

The first step after the discovery of mRNA (1956–1961) was to elucidate the code by which amino acid sequences of proteins are written in the nucleotide sequences of mRNA and correspondingly in the nucleotide sequence of one of the two DNA chains[18]. Even before the discovery of mRNA, theoretical considerations led to the assumption that each amino acid had to be coded by a combination of at least three nucleotides. Indeed, proteins are comprised of 20 natural amino acids (fig. 2), whereas nucleic acids contain only 4 types of nucleotide residues [the nitrogeneous bases of nucleic acids are adenine (A), guanine (G), cytosine (C), and either uracil (U) for RNA or thymine (T) for DNA]. It was obvious that one nucleotide could not code for one amino acid (4 vs. 20). There could be 16 dinucleotide combinations, or doublets, a number again insufficient to code for 20 amino acids. Thus, the minimal number of nucleotide residues in the combination coding for one amino acid had to be three; in other words, amino acids most probably had to be coded by the *nucleotide triplets*. The number of possible triplets is 64, more than enough for the coding of 20 amino acids.

There are two possible explanations for excessive triplets: either only 20 triplets are "meaningful," i.e. may code for one or another amino acid, while the other 44 are nonsense ones, or amino acids may be coded by more than one triplet, in which case the code would be *degenerate*.

Furthermore, the triplet code could be overlapping when a given nucleotide is part of three strongly overlapping or two less overlapping coding triplets; alternatively, it could be nonoverlapping when independent coding triplets are adjacent to each other in the template nucleic acid or are even separated by noncoding nucleotides. The observation that point mutations (i.e. changes of a single nucleotide in the nucleic acid molecule) usually lead to a change of only one amino acid in the corresponding protein provided evidence against the idea of an overlapping code. Moreover, the overlapping code would

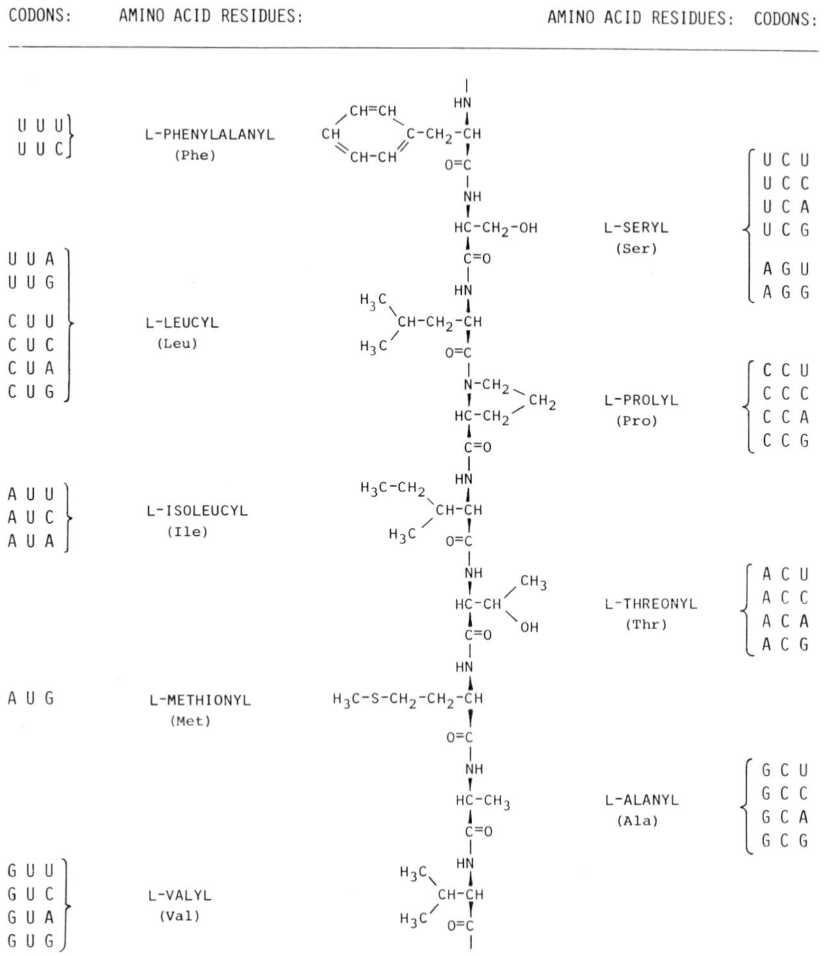

CODONS:	AMINO ACID RESIDUES:	AMINO ACID RESIDUES:	CODONS:
UUU, UUC	L-PHENYLALANYL (Phe)		
		L-SERYL (Ser)	UCU, UCC, UCA, UCG, AGU, AGG
UUA, UUG, CUU, CUC, CUA, CUG	L-LEUCYL (Leu)		
		L-PROLYL (Pro)	CCU, CCC, CCA, CCG
AUU, AUC, AUA	L-ISOLEUCYL (Ile)		
		L-THREONYL (Thr)	ACU, ACC, ACA, ACG
AUG	L-METHIONYL (Met)		
		L-ALANYL (Ala)	GCU, GCC, GCA, GCG
GUU, GUC, GUA, GUG	L-VALYL (Val)		

Figure 2 Amino acid residues present in proteins and the corresponding codons.

inevitably result in the possible neighbors of a given amino acid residue being restricted, a situation that has never been observed in actual protein sequences. Therefore a *nonoverlapping* code appeared more likely.

Finally, it had to be demonstrated whether the coding triplets were

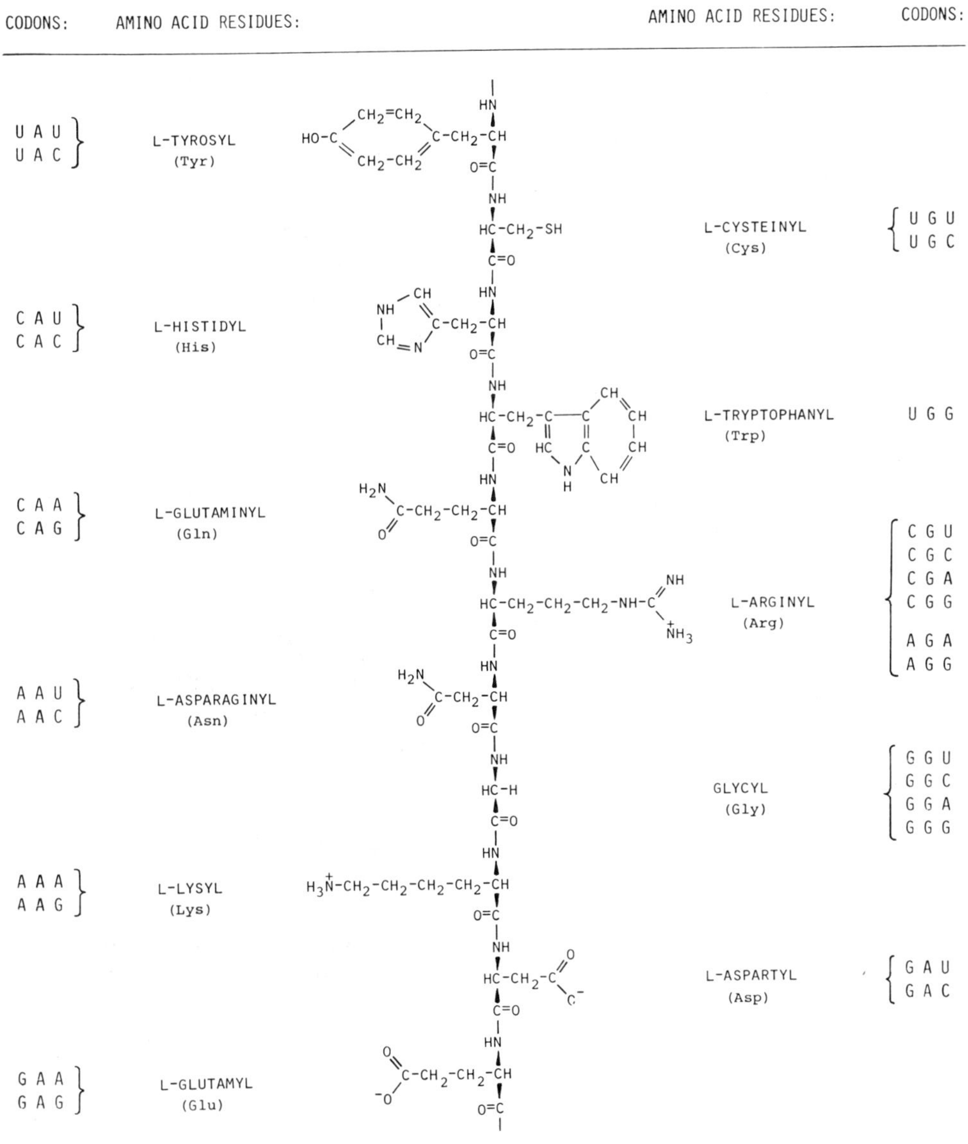

CODONS:	AMINO ACID RESIDUES:
U A U U A C	L-TYROSYL (Tyr)
U G U U G C	L-CYSTEINYL (Cys)
C A U C A C	L-HISTIDYL (His)
U G G	L-TRYPTOPHANYL (Trp)
C A A C A G	L-GLUTAMINYL (Gln)
C G U C G C C G A C G G A G A A G G	L-ARGINYL (Arg)
A A U A A C	L-ASPARAGINYL (Asn)
G G U G G C G G A G G G	GLYCYL (Gly)
A A A A A G	L-LYSYL (Lys)
G A U G A C	L-ASPARTYL (Asp)
G A A G A G	L-GLUTAMYL (Glu)

Figure 2 continued.

separated by noncoding residues, or commas, or whether they were read along the chain without any punctuation; in other words, whether the code was *comma free* or not. The comma free case leads to the problem of the reading frame of the template nucleic acid: only a strict triplet-by-triplet readout from a fixed point on the polynucleotide chain could result in an unambiguous amino acid sequence.

The classic experiments of Crick and Brenner[19] published at the end of 1961 established that the code was triplet, degenerate, nonoverlapping, and comma free. In these experiments, numerous mutants were obtained in the rII region of the T4 bacteriophage gene B using chemical agents which produced either insertions or deletions of one nucleotide residue during DNA replication. Proflavine and other acridine dyes were used for this purpose. Nucleotide insertions or deletions close to the gene origin resulted in a loss of gene expression. By recombining different mutant phages in *Escherichia coli* cells, phenotypic revertants showing normal gene expression were obtained. An analysis of the revertants demonstrated that gene expression was restored if the region with the deletion was located near the region with the insertion, or vice versa. Gene expression could also be restored if two additional insertions (or deletions) were introduced near the region with the initial insertion (or, respectively, deletion). The following conclusions were drawn: (1) Insertion or deletion of a single nucleotide at the beginning of the coding region appeared to result in a loss of all the coding potential of the corresponding gene instead of simply a point mutation; the inactivation could be the result of a shift of the reading frame. (2) Deletion or insertion located close to the initial insertion or deletion, respectively, restored the coding potential of the sequence because the original reading frame was restored. (3) Three, but no fewer, closely located insertions or deletions also restored the initial coding potential of the nucleotide sequence. From the results of these experiments, it follows that the code is triplet, and that triplets are read sequentially without commas from a strictly fixed point in the same frame. These experiments also provided additional evidence that the code is degenerate: if many of the 64 possible triplets were nonsense ones, it was highly probable that at least one nonsense triplet appeared in the region between the insertion and deletion or between the three insertions where the readout occurs with a shift of frame; this would lead to an interruption of the polypeptide chain synthesis.

Deciphering the nucleotide triplets also began in 1961 when Nirenberg and Matthaei discovered the coding properties of synthetic polyribonucleotides in cell-free translation systems[20]. The possibility of preparing synthetic polyribonucleotides of various compositions using a special enzyme, polynucleotide phosphorylase, was first demonstrated by Grunberg-Manago and Ochoa several years earlier[21]. The composition of polynucleotides synthesized in the system that they described depended only on the selection of ribonucleoside diphosphates supplied as substrates; homopolynucleotides such as polyuridylic acid, polyadenylic acid, and polycytidylic acid prepared from UDP, ADP, and CDP, respectively, were the simplest polyribonucleotides synthesized. Using poly(U) as a template polynucleotide for *E. coli* ribosomes, Nirenberg and Matthaei demonstrated that this template directs synthesis of polyphenylalanine. It has been concluded that the triplet UUU codes for phenylalanine. Similarly, experiments with polyadenylic and polycytidylic acid have shown that AAA codes for lysine, and CCC for proline.

Further elucidation of the genetic code was based on the use of synthetic statistical heteropolynucleotides of a different composition, which was set by the number and ratio of substrate nucleoside diphosphates in the polynucleotide phosphorylase reaction[22,23]. Thus, it has been demonstrated that the statistical poly(U,C) copolymer directed the incorporation of four amino acids into the polypeptide chain; these were phenylalanine, leucine, serine, and proline. If the U-to-C ratio in the polynucleotide was 1:1, then all four amino acids were incorporated into the polypeptide with equal probabilities. If the U-to-C ratio was 5:1, the probabilities of amino acid incorporation were as follows: Phe > Leu = Ser > Pro. Thus phenylalanine should be coded by triplets consisting of three U or of two U and one C. Leucine and serine are coded by triplets consisting of two U and one C or of two C and one U. Proline is coded by triplets consisting of three C or of two C and one U. Unfortunately, this approach could provide only the composition of the coding triplets, not their nucleotide sequence, since the nucleotide sequence of the template polynucleotide used was statistical.

Due to the invention of a new technique by Nirenberg and Leder[24], the nucleotide sequences of the coding triplets were soon determined. These investigators found that individual trinucleotides possessed coding properties: after an association with the ribosome they supported the selective binding of aminoacyl-tRNA species with the ribosome. For example, UUU and UUC triplets stimulated the binding of phenylalanyl-tRNA, UCU and UCC the binding of seryl-tRNA, CUU and CUC the binding of leucyl-tRNA, and CCU and CCC the binding of prolyl-tRNA. By 1964, methods for synthesizing trinucleotides with the desired sequence were available. In the subsequent two years a wide variety of trinucleotides were tested and, as a result, virtually the whole code was deciphered[25] (fig. 3).

The end of the story was marked by the use of synthetic polynucleotides with a regular nucleotide sequence as templates in the cell-free ribosomal systems of polypeptide synthesis. Methods allowing regular polynucleotides to be synthesized have been developed by Khorana, who has also verified the genetic code by directly using these polynucleotides as templates[26]. In complete agreement with the previously established code dictionary, the use of poly(UC)$_n$ as a template resulted in the synthesis of a polypeptide consisting of

First letter	Second letter: U	C	A	G	Third letter
U	UUU } Phe UUC UUA } Leu UUG	UCU UCC } Ser UCA UCG	UAU } Tyr UAC UAA Ochre UAG Amber	UGU } Cys UGC UGA Opal UGG Trp	U C A G
C	CUU CUC } Leu CUA CUG	CCU CCC } Pro CCA CCG	CAU } His CAC CAA } Gln CAG	CGU CGC } Arg CGA CGG	U C A G
A	AUU AUC } Ile AUA AUG Met	ACU ACC } Thr ACA ACG	AAU } Asn AAC AAA } Lys AAG	AGU } Ser AGC AGA } Arg AGG	U C A G
G	GUU GUC } Val GUA GUG	GCU GCC } Ala GCA GCG	GAU } Asp GAC GAA } Glu GAG	GGU GGC } Gly GGA GGG	U C A G

Figure 3 Codon dictionary[27].

alternating serine and leucine residues, while poly(UG)$_n$ directed synthesis of the regular copolymer with alternating valine and cysteine residues. Poly(AAG)$_n$ directed the synthesis of three homopolymers: polylysine, polyarginine, and polyglutamic acid.

2-3 Some features of the code dictionary

The complete code dictionary[27] is given in fig. 3. Of the 64 triplets termed *codons*, 61 are meaningful or sense ones: they code for amino acids. Only 3 codons—UAG ("amber"), UAA ("ochre"), and UGA ("opal")—do not code for amino acids and therefore are sometimes called *nonsense codons*. The nonsense triplets play an important part in translation, since in mRNA these codons serve as signals for the termination of polypeptide chain synthesis; at present they are usually referred to as *termination* or *stop codons*.

The degeneracy of the code does not extend to all amino acids. Two amino acids, methionine and tryptophan, are coded by one codon each, i.e. by AUG and UGG, respectively. On the contrary, three amino acids, specifically leucine, serine, and arginine, have six codons each.

The remaining amino acids, with the exception of isoleucine, are coded either by two or by four codons; only isoleucine is coded by three codons.

It should be emphasized that the triplets coding for a given amino acid differ in most cases only in the third base. Only when the amino acid is coded by more than four codons do differences occur in the first and second positions of the triplet as well. A group of four codons differing only in the third nucleotide and coding for one and the same amino acid is often called the *codon family*. The code dictionary contains eight such codon families: for leucine, valine, serine, proline, threonine, alanine, arginine, and glycine (fig. 3).

The code illustrated in fig. 3 is universal for the protein-synthesizing systems of bacteria and for the cytoplasmic extraorganellar protein-synthesizing systems of all eucaryotes, i.e. animals, fungi, and plants. There are some exceptions, however. For example, the protein-synthesizing system of mammalian and fungal mitochondria shows a number of deviations from this universal code[28]. Tryptophan in mitochondria is coded by both UGG and UGA; UGA is therefore not used as a termination codon. In mammalian (human) mitochondria AGA and AGG do not code for arginine but play the part of termination codons. In yeast mitochondria the whole codon family CUU, CUC, CUA, and CUG codes for threonine but not for leucine, although in another fungus, *Neurospora*, these codons correspond to leucine as given by the universal code.

2-4 Structure of mRNA

Primary structure

In contrast to DNA, messenger RNA, as well as other cellular RNA species, is a *single-stranded* polynucleotide. It consists of four kinds of linearly arranged ribonucleoside residues—adenosine (A), guanosine (G), cytidine (C), and uridine (U)—sequentially connected by phosphodiester bonds between the 3′-position of the ribose of one nucleoside and the 5′-position of the adjacent one (fig. 4). The terminal nucleoside, the 5′-position of which does not participate in forming the internucleotide bond, is referred to as the 5′-end of RNA. The terminal nucleoside with a free 3′-hydroxyl is referred to as the 3′-end. It is accepted practice to read and write RNA nucleotide

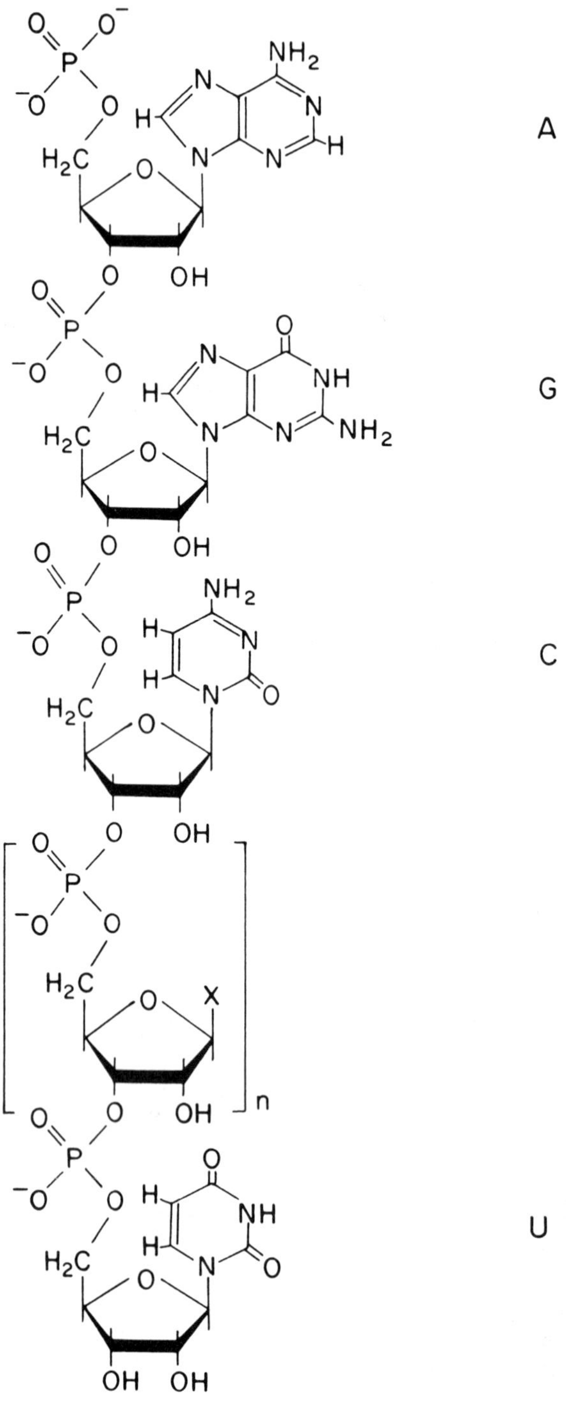

A

G

C

U

Figure 4 Nucleotide residues in RNA.

sequences from the 5′- to the 3′-end, i.e. in the direction of the internucleotide phosphodiester bond from the 3′-position to the 5′-position of the neighbor (3′—Ⓟ—5′ bond direction). This direction corresponds to the polarity of mRNA readout by the ribosome.

The terminal 5′-position in natural mRNAs is always substituted. In procaryotic organisms this end is either simply phosphorylated (fig. 4) or carries the triphosphate group. Eucaryotic mRNAs generally have a special group, the so-called *cap*, at the terminal 5′-position. The cap is a N^7-methylated residue of guanosine—5′—triphosphate linked with the 5′-terminal nucleoside by the unusual 5′—5′ pyrophosphate bond[29–31] (fig. 5). Eucaryotic cells possess a special system including

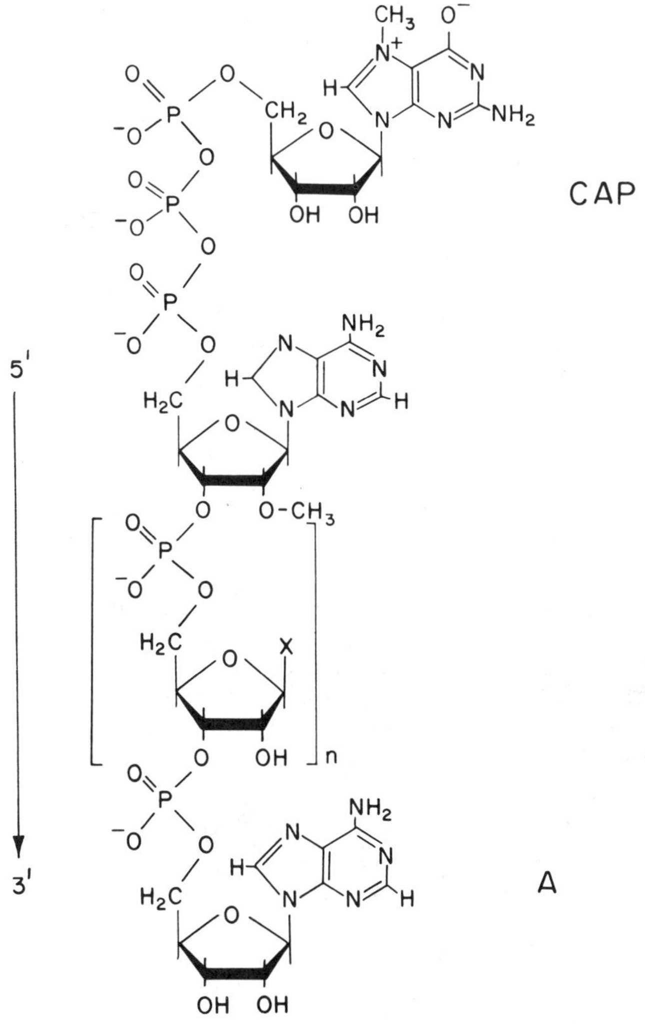

Figure 5 Cap at the 5′-end of eucaryotic mRNA.

guanylyl transferase and methyl transferase, enzymes that are responsible for mRNA capping. In addition, the capping is usually accompanied by methylation of the 2′-hydroxyl group of ribose and the base in the 5′-terminal nucleoside adjacent to the cap. Often the 5′-terminal residue in mRNA is a purine nucleoside, either G or A.

The 3′-terminal hydroxyl of natural mRNA remains unsubstituted. Thus, this end possesses two hydroxyl groups in *cis*-position (*cis*-glycol group) (see fig. 4).

Functional regions

The physical length of the mRNA chain is always greater than the length of its coding sequence. The coding sequence includes only part of the total mRNA length. The first codon is preceded by a noncoding 5′-terminal sequence, the length of which varies for different mRNAs. Furthermore, the terminal codon is never located at the 3′-end of an mRNA chain, but is always followed by a noncoding 3′-terminal sequence. In addition, many eucaryotic mRNAs contain a long noncoding sequence of adenylic acid residues at their 3′-end[32]. This poly(A) tract is added to mRNA after the end of transcription by a special enzyme, polyadenylate polymerase.

Identifying the factors that determine the starting point of the coding nucleotide sequence within an mRNA chain is an important problem. Each polypeptide is known to begin with a N-terminal methionine residue, and therefore the first codon in the coding sequence should be that of methionine[33]. In most cases AUG, rarely GUG or UUG, plays the role of the *initiation codon*. The codon AUG codes for methionine both when it is the first codon of the mRNA coding sequence and when it occurs in internal positions. The codon GUG, however, codes for valine in internal positions and for the initiator methionine only if it occupies the first position in the coding sequence. The same is true for codon UUG coding for leucine in internal positions. In some exceptional cases, AUU and AUA may also serve as initiation codons for the first methionine in the chain.

The identification of the initiation codons, however, does not solve the starting point problem of the coding sequence. The difficulty is that by no means every AUG (the more so GUG or UUG) triplet becomes an initiation codon. Generally, translation cannot be initiated from internal AUG, GUG, or UUG triplets. If an mRNA chain is scanned from its 5′-end, AUG as well as GUG and UUG triplets may be found repeatedly both in frame with the subsequent coding sequence and out of frame, but they cannot initiate translation. Finally, many AUG, GUG, and UUG triplets located within the coding sequence but out of the reading frame fortunately do not initiate synthesis of erroneous polypeptides. Thus, in contrast to all other codons, both sense and nonsense ones, the choice of a given codon as an initiation point depends not only on the codon structure, i.e. its nucleotide composition and sequence, but also on the position of the codon in the mRNA. It remains to be seen which structural elements confer the capacity to serve as initiation codon to a given AUG (or GUG or UUG). There is evidence that the nucleotide sequence preceding the initiation codon plays an important part (see chapter 16). It may well be that the particular secondary and tertiary structures of this mRNA region are vital for the corresponding triplet to be exposed as an initiation codon (see below).

A given mRNA polynucleotide chain does not necessarily contain just one coding sequence. In procaryotic mRNAs it is common for one polynucleotide chain to contain coding sequences for several proteins. Such mRNAs are usually called *polycistronic mRNAs*. (This term comes from the word *cistron*, which Benzer introduced as an equivalent of a gene). Different coding sequences (*cistrons*) within a given mRNA chain are usually separated by internal noncoding sequences. Such an internal noncoding sequence begins from the termination codon of the preceding cistron. Sometimes an additional termination codon is found at the beginning of such a spacer sequence; it appears to double for the termination signal present in the coding sequence in order to make termination fail-safe. The next cistron begins from an initiation codon such as AUG (or GUG).

The RNA present in the small bacteriophage MS2 is a good example of polycistronic mRNA[34]. This phage has a spherical shape; its diameter is 250 Å, its molecular mass 3.6×10^6 daltons. The phage particle contains 180 coat protein subunits, each with a molecular mass of 14,700 daltons, one molecule of the so-called A-protein having a molecular mass of 38,000 daltons, and one molecule of RNA with a molecular mass of about 10^6 daltons. After infecting the *E. coli* cells or in a cell-free translation system, the phage RNA serves as a template for the synthesis of coat protein, A-protein, and a subunit of RNA replicase with a molecular mass of 62,000 daltons (this subunit is not a component of the phage particle). The location of corresponding cistrons along the MS2 RNA chain is shown schematically in fig. 6. The chain starts from G, bearing triphosphate at its 5′-position. This is followed by a long noncoding sequence, with a total length of 129 nucleotide residues; this sequence contains AUG and GUG triplets which, however, do not serve as initiation codons. The first initiation codon, GUG, starts the coding sequence of the A-cistron corresponding to the A-protein. The A-cistron is 1179 nucleotide residues in length and ends with a UAG termination triplet. It is followed by a

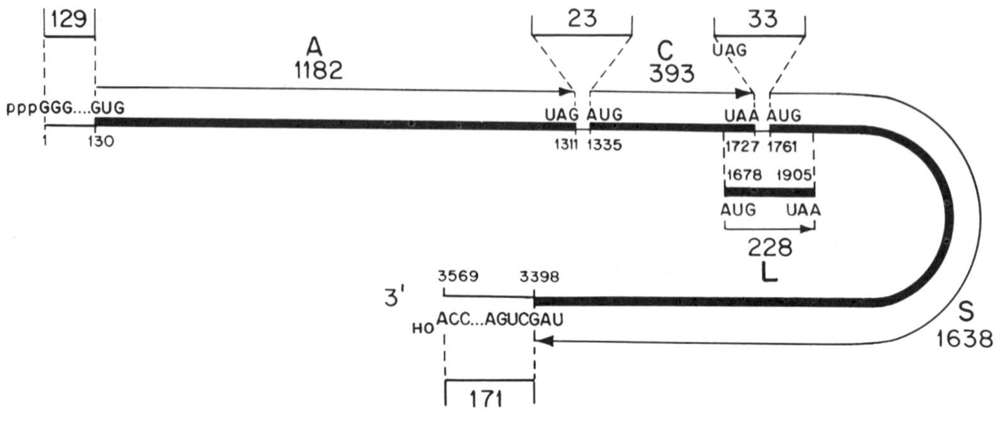

Figure 6 Scheme illustrating the arrangement of the coding and noncoding nucleotide sequences in the MS2 bacteriophage RNA chain. A, C, L, and S are cistrons coding for "maturation," coat, lysis, and replicase (synthetase) proteins, respectively. The larger figures are the number of nucleotide residues in a given section; the smaller numbers represent the nucleotide residue positions.

noncoding region 26 residues long. The next coding sequence starts from AUG and is 390 nucleotides in length; this is the C-cistron coding for coat protein. This cistron is terminated by UAA and followed by the second termination codon UAG. The C-cistron is separated from the S-cistron, which codes for RNA replacase subunit, by 36 nucleotides. The S-cistron begins from AUG, is 1635 nucleotides in length, and ends with a UAG. At a distance of one nucleotide from its termination signal, i.e. out of the reading frame, is found yet another termination triplet, UGA. The 3′-terminal noncoding sequence has a total length of 174 nucleotide residues and ends with an adenosine having a *cis*-glycol group. The complete primary structure of MS2 RNA was published by Fiers and co-workers[34].

An analysis of the MS2 phage RNA also illustrates another way of accommodating different coding sequences in one mRNA chain. In addition to the three proteins already mentioned, MS2 phage RNA codes for a fourth protein, which is known as the lysis protein, or the L-protein[35,36]. This protein appears to be involved in host cell lysis at the late stage of infection. It is coded by an RNA region that begins within the end section of the C-cistron, contains the entire 36-nucleotide region between the C-cistron and the S-cistron, and terminates within the S-cistron; the reading frame of the L-cistron is shifted to the right by one residue (+1 shift), so that this cistron is not translated during S-protein or C-protein synthesis. The L-cistron has its own initiation codon AUG, which is out of frame with C-cistron codons, and its own termination codon UAA, which is out of frame with the codons belonging to the S-cistron (fig. 7). The presence of such overlapping coding sequences within one mRNA is rare and is largely characteristic of viral systems where the economy of space for accommodating cistrons in a nucleic acid molecule plays a particularly important role.

In contrast to procaryotes, in eucaryotic organisms mRNAs are as a rule *monocistronic*, i.e. they code for just one polypeptide chain. The eucaryotic mRNA coding sequence is flanked both at the 5′-end and at the 3′-end by noncoding sequences. As has already been mentioned, the 5′-end is usually modified by the cap (fig. 5), which appears to be essential for the association between the mRNA and the ribosome prior to initiation.

It is appropriate to emphasize here that the mechanisms responsible for searching for the initiation codon in procaryotic and eucaryotic translation systems are different. Procaryotic ribosomes form a complex with the mRNA and recognize the initiation codon independently of the 5′-end; it is for this reason that they can initiate from internal sites in the polycistronic mRNA. In contrast, eucaryotic ribosomes usually need the mRNA 5′-end to form the association complex; the cap appears to contribute to such an association (see chapter 17). With eucaryotic mRNA it is generally thought that the first AUG from the 5′-end in most cases serves as an initiation codon, although there are exceptions to this rule. The 3′-end of the eucaryotic mRNA often shows some heterogeneity: the length of the 3′-terminal noncoding sequence can vary for different molecules of a given mRNA. Moreover, it has already been mentioned that the vast majority of eucaryotic mRNAs have poly(A) tracts of various length at the 3′-end.

Folding

The three-dimensional structure of mRNA has yet to be determined. Measurements of various physical parameters of several mRNAs have demonstrated that these molecules possess extensively folded structures with a large number of intrachain interactions due to the Watson-Crick complementary base pair formation[37–42]. Although mRNAs are not double helices of the DNA type, they do have a well developed secondary structure because of the complementary pairing of different regions of the same chain with each other; this results in a

```
                 MET GLU THR ARG PHE PRO GLN GLN SER GLN GLN THR PRO ALA SER
CAAGGUCUCCUAAAAGAUGGAAACCCGAUUCCCUCAGCAAUCGCAGCAAACUCCGGCAUC
GLN GLY LEU LEU LYS ASP GLY ASN PRO ILE PRO SER ALA ILE ALA ALA ASN SER GLY ILE
                1678

THR ASN ARG ARG ARG PRO PHE LYS HIS GLU ASP TYR PRO CYS ARG ARG GLN GLN ARG SER
UACUAAUAGACGCCGGCCAUUCAAACAUGAGGAUUACCCAUGUCGAAGACAACAAAGAAG
TYR                                                   MET SER LYS THR THR LYS LYS

SER THR LEU TYR VAL LEU ILE PHE LEU ALA ILE PHE LEU SER LYS PHE THR ASN GLN LEU
UUCAACUCUUUAUGUAUUGAUCUUCCUCGCGAUCUUUCUCUCGAAAUUUACCAAUCAAUU
PHE ASN SER LEU CYS ILE ASP LEU PRO ARG ASP LEU SER LEU GLU ILE TYR GLN SER ILE

LEU LEU SER LEU LEU GLU ALA VAL ILE ARG THR VAL THR THR LEU GLN GLN LEU LEU THR
GCUUCUGUCGCUACUGGAAGCGGUGAUCCGCACAGUGACGACUUUACAGCAAUUGCUUACUUAA
ALA SER VAL ALA THR GLY SER GLY ASP PRO HIS SER ASP ASP PHE THR ALA ILE ALA TYR LEU
                                                                             1902
```

Figure 7 Nucleotide sequence coding for the L-protein of the MS2 bacteriophage (beginning with nucleotide residue 1678)[35,36]. The amino acid sequence of the L-protein is written above the nucleotide sequence, whereas the sequences of the end of the C-protein and the start of the S-protein are given below the nucleotide sequence.

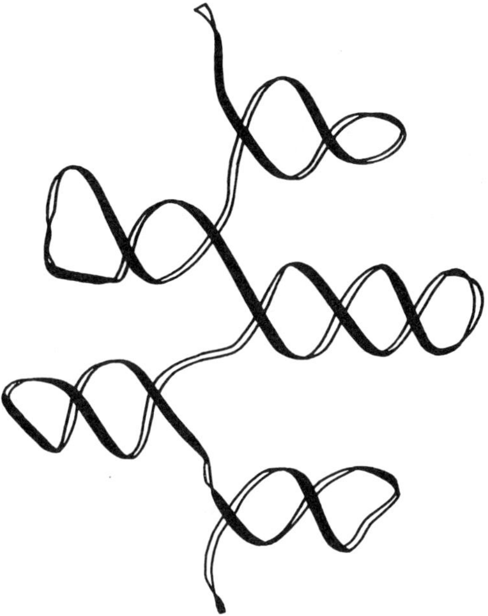

Figure 8 Scheme illustrating the formation of a secondary structure (double-helical hairpins) by pairing of the adjacent sections of the RNA polynucleotide chain.

large number of relatively short double-helical regions being formed. About 70% of all the nucleotide residues in the chain participate in the complementary pairing and, correspondingly, in the formation of intramolecular helices. Most of the double-helical regions appear to be formed by the complementary pairing of adjacent sections in the polynucleotide chain; the scheme of the formation of such short helices is given in fig. 8. The complementary pairing of distant chain sections should result in the additional and extensive folding of the structure, leading to compact domains being formed in mRNA, as is schematically illustrated in fig. 9[43,44]. These interactions, which are responsible for the mRNA secondary structure, are based mainly on A·U and G·C pairing (Watson-Crick pairs), as well as on G·U pairing (see chapter 3, fig. 19).

The forces responsible for forming the mRNA tertiary structure are as unknown as the mRNA tertiary structure itself.

Meanwhile, there is evidence suggesting that the secondary and tertiary structures of mRNA may play a role in translation[45–47]. It has been pointed out that the three-dimensional structure of the initiation

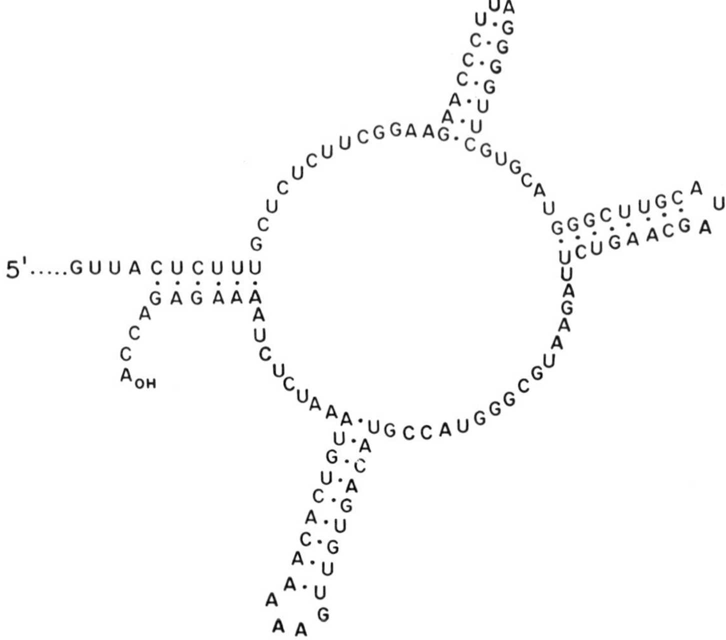

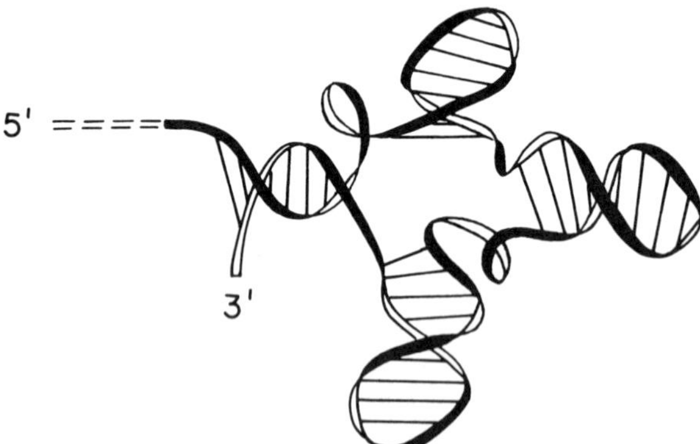

Figure 9 Possible secondary structure for the 3′-terminal domain of brome mosaic virus RNA. *Top*, Nucleotide sequence and the complementarity of the paired regions of the chain[43,44]. *Bottom*, Same structure shown as double-helical regions connected by single-stranded sections.

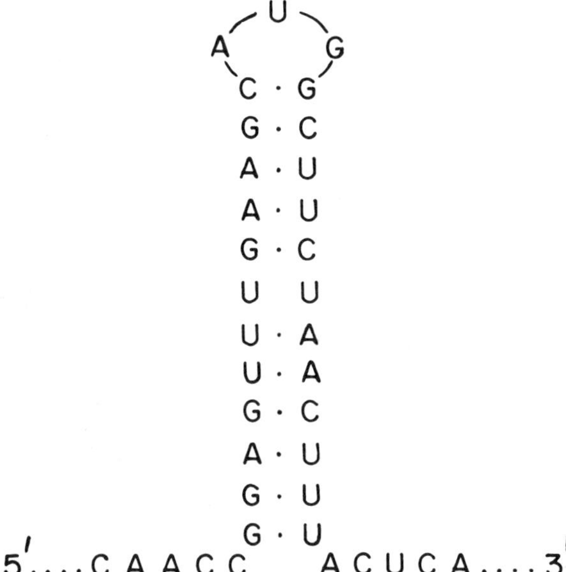

Figure 10 Possible secondary structure (double-helical hairpin) for the region of bacteriophage MS2 RNA containing the initiation AUG codon of the coat protein cistron. Redrawn from J. A. Steitz (1969), *Nature* 224:957–964, with permission.

region may be important or even decisive for the recognition of AUG (or GUG) triplet as initiation codon. Apparently either the initiation triplet should not be involved in the Watson-Crick pairing with other mRNA nucleotides, or such a pairing should be weak (unstable). In other words, the initiation codon either has to be exposed or should be subject to easy exposure for interaction with the initiator tRNA on the ribosome. Such an unpaired AUG triplet was indeed found at the beginning of the coat protein cistron in the predicted secondary structure of the MS2 phage RNA (fig. 10); this triplet appears to be located at the apex loop of the double-helical hairpin structure[46]. It is known to serve as a preferential site for protein synthesis initiation on MS2 RNA. It is interesting that initiation on the initiation codons of other MS2 RNA cistrons, e.g. the RNA replicase cistron is far less effective. This is in line with observations that the AUG initiation triplet of the S-cistron is part of a helical region which, however, is not very stable because it is short and imperfect. The presence of the secondary structure in the region containing the termination codon of the coat protein cistron and the initiation codon of the RNA replicase cistron has been demonstrated directly in experiments with the

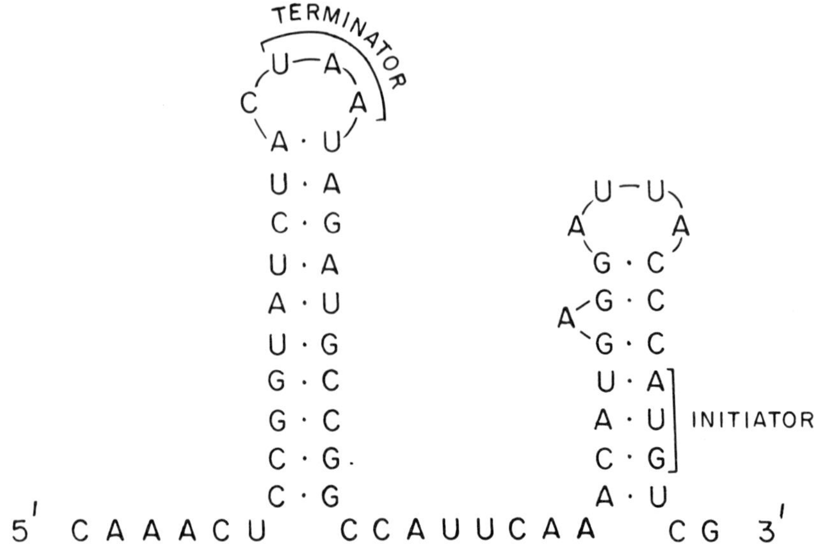

Figure 11 Secondary structure for the R17 bacteriophage RNA fragment, including the end of the coat protein cistron and the beginning of the replicase cistron. Redrawn from J. Gralla et al. (1974) *Nature* 248:204–208, with permission.

corresponding RNA fragment of the related bacteriophage R17[47] (fig. 11).

It is conceivable that the stabilization of a double-helical region carrying the initiation triplet, which is achieved either by RNA tertiary folding or by the specific attachment of RNA-binding protein, may completely inhibit initiation at a given site. Thus it is quite likely that the initiation triplets of both the A-cistron and the S-cistron of the MS2 RNA, as well as of RNAs of the related phages R17, f2, etc., are made inaccessible by the tertiary structure[45]. Initiation of A-cistron translation appears to take place in the course of RNA synthesis only when the complete three-dimensional structure has not yet been formed. Initiation of S-cistron translation occurs in the course of the translation of the preceding C-cistron: ribosomes unfold RNA during C-cistron translation and thus release the region carrying the initiation triplet of S-cistron from its buried state. After the complete coat protein molecules have appeared, the initiation of the S-cistron is turned off: the phage coat protein has a specific affinity to the unstable hairpin containing the initiation AUG triplet (see fig. 11) and, when bound to this hairpin, results in its stabilization.

After initiation the ribosomes may perform a readout more or less independently on the secondary and tertiary structure of mRNA. It is likely that they sequentially unfold the mRNA chain while moving along (of course, the chain sections refold after the ribosomes have moved away). At the moment, very little is known about the part played by the secondary and tertiary mRNA structures in the rate with which ribosomes move along the RNA chain, i.e. the rate of polypeptide elongation. It is known that this rate is nonuniform[48,49] and it may well be that it depends on the stability of the secondary and tertiary structure in different mRNA regions[49–51].

Particular attention should be paid to noncoding mRNA sequences. Specifically, their function may be to create the specialized three-dimensional secondary and tertiary structures that control initiation, elongation, and the transition of ribosomes from one cistron to another; they may also play a part in the binding to mRNA of special recognition proteins which may affect translation.

References

1. O. T. Avery, C. M. MacLeod, and M. McCarty (1944), "Studies in the chemical nature of the substance inducing transformation of pneumococcal types," *J. Exptl. Med.* 78:137–158.
2. A. D. Hershey and M. Chase (1952), "Independent functions of viral protein and nucleic acid in growth of bacteriophage," *J. Gen. Physiol.* 36:39–56.
3. J. D. Watson and F. H. C. Crick (1953), "Molecular structure of nucleic acids," *Nature* 171:738–740.
4. J. D. Watson and F. H. C. Crick (1953), "Genetical implications of the structure of deoxyribose nucleic acid," *Nature* 171:964–967.
5. T. Caspersson, H. Landström-Hyden, and L. Aquilonius (1941), Cytoplasmanukleotide in eiweissproduzierenden Drüsenzellen," *Chromosoma* 2:111–131.
6. J. Brachet (1941–1942), "La détection histochimique et le microdosage des acides pentosenucléiques," *Enzymologia* 10:87–96.
7. A. Boivin and R. Vendrely (1947), "Sur le rôle possible des deux acides nucléique dans la cellule vivante," *Experientia* 3:32–34.
8. A. L. Dounce (1952), "Duplicating mechanism for peptide chain and nucleic acid synthesis," *Enzymologia* 15:251–258.
9. F. H. Crick (1958), "On protein synthesis," *Symp. Soc. Exptl. Biol.* 12:138–163.
10. F. H. C. Crick (1959), "The present position of the coding problem," *Brookhaven Symp. Biol.* 12:35–38.

11. A. S. Spirin, A. N. Belozersky, N. V. Shugaeva, and B. F. Vanushin (1957), "Studies on species specificity of nucleic acids in bacteria," *Biokhimiya* 22:744–754.

12. A. N. Belozersky and A. S. Spirin (1958), "A correlation between the compositions of deoxyribonucleic and ribonucleic acids," *Nature* 182:111–112.

13. E. Volkin and L. Astrachan (1956)," Phosphorus incorporation in *Escherichia coli* ribonucleic acid after infection with bacteriophage T2," *Virology* 2:149–161.

14. S. Brenner, F. Jacob, and M. Meselson (1961), "An unstable intermediate carrying information from genes to ribosomes for protein synthesis," *Nature* 190:576–581.

15. F. Gros, H. Hiatt, W. Gilbert, C. G. Kurland, R. W. Riseborough, and J. D. Watson (1961), "Unstable ribonucleic acid revealed by pulse labeling of *Escherichia coli*," *Nature* 190:581–585.

16. S. Spiegelman (1961), "The relation of informational RNA to DNA," *Cold Spring Harbor Symp. Quant. Biol.* 26:75–90.

17. F. Jacob and J. Monod (1961), "Genetic regulatory mechanisms in the synthesis of proteins," *J. Mol. Biol.* 3:318–356.

18. G. Gamov, A. Rich, and M. Yčas (1956), "The problem of information transfer from the nucleic acids to proteins," *Advan. Biol. Med. Phys.* 4:23–68.

19. F. H. C. Crick, L. Barnett, S. Brenner, and R. J. Watts-Tobin (1961), "General nature of the genetic code for proteins," *Nature* 192:1227–1232.

20. M. W. Nirenberg and J. H. Matthaei (1961), "The dependence of cell-free protein synthesis in *E. coli* upon naturally occurring or synthetic polyribonucleotides," *Proc. Nat. Acad. Sci. U.S.A.* 47:1588–1602.

21. M. Grunberg-Manago and S. Ochoa (1955), "Enzymatic synthesis and breakdown of polynucleotides: Polynucleotide phosphorylase," *J. Amer. Chem. Soc.* 77:3165–3166.

22. M. W. Nirenberg, O. W. Jones, P. Leder, B. F. C. Clark, W. S. Sly, and S. Pestka (1963), "On the coding of genetic information," *Cold Spring Harbor Symp. Quant. Biol.* 28:549–557.

23. J. F. Speyer, P. Lengyel, C. Basilio, A. J. Wahba, R. S. Gardner, and S. Ochoa (1963), "Synthetic polynucleotides and the amino acid code," *Cold Spring Harbor Symp. Quant. Biol.* 28:559–567.

24. M. W. Nirenberg and P. Leder (1964), "RNA codewords and protein synthesis: The effect of trinucleotides upon the binding of sRNA to ribosomes," *Science* 145:1399–1407.

25. M. Nirenberg, T. Caskey, R. Marshall, R. Brimacombe, D. Kellogg, B. Doctor, D. Hatfield, J. Levin, F. Rottman, S. Pestka, M. Wilcox, and F. Anderson (1966); "The RNA code and protein synthesis," *Cold Spring Harbor Symp. Quant. Biol.* 31:11–24.

26. H. G. Khorana, H. Büchi, H. Ghosh, N. Gupta, T. M. Jacob, H. Kössel, R. Morgan, S. A. Narang, E. Ohtsuka, and R. D. Wells (1966), "Polynucleotide synthesis and the genetic code," *Cold Spring Harbor Symp. Quant. Biol.* 31:39–49.

27. F. H. C. Crick (1966), "The genetic code: Yesterday, today, and tomorrow," *Cold Spring Harbor Symp. Quant. Biol.* 31:1–9.

28. P. Slonimski, P. Borst, and G. Attardi, eds. (1982), *Mitochondrical genes* (Cold Spring Harbor, N.Y.: Cold Spring Harbor Laboratory).

29. Y. Furuichi and K.-I. Miura (1975), "A blocked structure at the 5′-terminus of mRNA from cytoplasmic polyhedrosis virus," *Nature* 253:374–375.

30. Y. Furuichi, M. Morgan, S. Muthukrishnan, and A. J. Shatkin (1975)," Reovirus messenger RNA cont ..ns a methylated blocked 5′-terminus structure: $m^7G(5')ppp(5')G^mpCp$," *Proc. Nat. Acad. Sci. U.S.A.* 72:362–366.

31. A. J. Shatkin (1976), "Capping of eucaryotic mRNAs," *Cell* 9:645–653.

32. G. Brawerman (1976), "Characteristics and significance of the polyadenylate sequence in mammalian messenger RNA," Progress in nucleic acid research and molecular biology, ed. W. E. Cohn, vol. 17, pp. 117–148 (New York: Academic Press).

33. B. F. C. Clark and K. A. Marcher (1966), "The role of N-formylmethionyl-sRNA in protein biosynthesis," *J. Mol. Biol.* 17:394–406.

34. W. Fiers, R. Contreras, F. Duerinck, G. Haegeman, D. Iserentant, J. Merregaert, W. Min You, F. Molemans, A. Raeymaekers, A. Van den Berghe, G. Volckaert, and M. Ysebaert (1976), "Complete nucleotide sequence of bacteriophage MS2 RNA: Primary and secondary structure of the replicase gene," *Nature* 260:500–507.

35. J. F. Atkins, J. A. Steitz, C. W. Anderson, and P. Model (1979), "Binding of mammalian ribosomes to MS2 phage RNA reveals an overlapping gene encoding a lysis function," *Cell* 18:247–256.

36. M. N. Beremand and T. Blumenthal (1979), "Overlapping genes in RNA Phage: A new protein implicated in lysis," *Cell* 18:257–266.

37. P. Doty, H. Boedtker, J. R. Fresco, R. Haselkorn, and M. Litt (1959), "Secondary structure in ribonucleic acids," *Proc. Nat. Acad. Sci. U.S.A.* 45:482–499.

38. W. B. Gratzer and E. G. Richards (1971), "Evaluation of RNA conformation from circular dichroism and optical rotatory dispersion data," *Biopolymers* 10:2607–2614.

39. A. M. Bobst, Y.-C. E. Pan, and D. J. Phillips (1974), "Comparative optical property studies on polycistronic R17 phage ribonucleic acid and rabbit globin messenger ribonucleic acid," *Biochemistry* 13:2129–2133.

40. J. W. Holder and J. B. Lingrel (1975), Determination of secondary structure in rabbit globin messenger RNA by thermal denaturation," *Biochemistry* 14:4209–4212.

41. J. N. Vournakis, M. S. Flashner, M. A. Katopes, G. A. Kitas, N. C. Vamvakopoulos, M. S. Sell, and R. M. Wurst (1976), "Structural studies on intact and deadenylated rabbit globin mRNA," Progress in nucleic acid research and molecular biology, ed. W. E. Cohn and E. Volkin, vol. 19, pp. 233–252 (New York: Academic Press).

42. N. T. Van, J. W. Holder, S. L. C. Woo, A. R. Means, and B. W. O'Malley (1976), "Secondary structure of ovalbumin messenger RNA," *Biochemistry* 15:2054–2062.

43. R. Dasgupta and P. Kaesberg (1977), "Sequence of an oligonucleotide derived from the 3′-end of each of the four brome mosaic viral RNAs," *Proc. Nat. Acad. Sci. U.S.A.* 74:4900–4904.

44. M. R. Gunn and R. H. Symons (1980), "The RNAs of bromoviruses: 3′-Terminal sequences of the four brome mosaic virus RNAs and comparison with cowpea chloratic mottle virus RNa 4," *FEBS Letters* 115:77–82.

45. H. F. Lodish and H. D. Robertson (1969), "Regulation of *in vitro* translation of bacteriophage f2 RNA," *Cold Spring Harbor Symp. Quant. Biol.* 34:655–673.

46. J. A. Steitz (1969), "Polypeptide chain initiation: Nucleotide sequences of the three ribosomal binding sites in bacteriophage R17 RNA," *Nature* 224:957–964.

47. J. Gralla, J. A. Steitz, and D. M. Crothers (1974), "Direct physical evidence for secondary structure in an isolated fragment of R17 bacteriophage mRNA," *Nature* 248:204–208.

48. A. Protzel, and A. J. Morris (1974), "Gel chromatographic analysis of nascent globin chains. Evidence of nonuniform size distribution", *J. Biol. Chem.* 249:4594–4600.

49. W. G. Chaney, and A. J. Morris (1979), "Nonuniform size distribution of nascent peptides. The effect of messenger RNA structure upon the rate of translation", *Arch. Biochem. Biophys.* 194:283–291.

50. S. Varenne, M. Knibiehler, D. Cavard, J. Morlon, and C. Lazdunski (1982), "Variable rate of polypeptide chain elongation for colicins A, E2 and E3", *J. Mol. Biol.* 159:57–70.

51. Y. V. Svitkin and V. I. Agol (1983), "Translational barrier in central region of encephalomyocarditis virus genome: Modulation by elongation factor 2 (eFF-2)," *Eur. J. Biochem.* 133:145–154.

Further reading

Bloemendal, H. (1972). Mammalian messenger RNA. In *The mechanism of protein synthesis and its regulation* (L. Bosch, ed.), pp. 487–514. Amsterdam and London: North-Holland.

Bosch, L. and Voorma, H. O. (1972). Translation of viral RNA. In *The mechanism of protein synthesis and its regulation* (L. Bosch, ed.), pp. 395–440. Amsterdam and London: North-Holland.

Cellular regulatory mechanisms (1961). *Cold Spring Harbor Symp. Quant. Biol.*, vol. 26.

Crick, F. H. C. (1963). The recent excitement in the coding problem. Progress in nucleic acid research (J. N. Davidson and W. E. Cohn, eds.), vol. 1, pp. 163–217. New York: Academic Press.

——— (1968). The origin of the genetic code. *J. Mol. Biol.* 38:367–379.

The genetic code (1966). *Cold Spring Harbor Symp. Quant. Biol.*, vol. 31.

Lipmann, F. (1963). Messenger ribonucleic acid. Progress in nucleic acid research (J. N. Davidson and W. E. Cohn, eds.), vol. 1, pp. 135–161. New York: Academic Press.

The mechanism of protein synthesis (1969). *Cold Spring Harbor Symp. Quant. Biol.*, vol. 34.

mRNA: The relation of structure to function (1976). Progress in nucleic acid research (W. E. Cohn and E. Volkin, eds.), vol. 19. New York: Academic Press.

Orgel, L. E. (1968). Evolution of the genetic apparatus. *J. Mol. Biol.* 38:381–393.

Schlessinger, D. (1972). Longevity and translation yield of mRNA. In *The mechanism of protein synthesis and its regulation* (L. Bosch, ed.), pp. 441–464. Amsterdam and London: North-Holland.

Shafritz, D. A. (1977). Messenger RNA and its translation. In *Molecular mechanisms of protein biosynthesis* (H. Weissbach and S. Pestka, eds.), pp. 555–601. New York: Academic Press.

Synthesis and structure of macromolecules (1963). *Cold Spring Harbor Symp. Quant. Biol.*, vol. 28.

Woese, C. R. (1967*a*). *The genetic code: The molecular basis for genetic expression.* New York: Harper and Row.

——— (1967*b*). The present status of the genetic code. Progress in nucleic acid research and molecular biology (J. N. Davidson and W. E. Cohn, eds.), vol. 7, pp. 107–172. New York: Academic Press.

Yčas, M. (1969). *The biological code.* Amsterdam and London: North-Holland.

Chapter 3

Transfer RNA and Aminoacyl-tRNA Synthetases

3-1 Discovery

Information on the amino acid sequences of proteins is written down as nucleotide sequences of the messenger RNA. The template triplet (*codon*) should determine unambiguously the position of a corresponding amino acid. However, there is no apparent steric fit between the structure of amino acids and their respective codons. In other words, codons cannot serve as direct template surfaces for amino acids. In order to solve this problem, in 1955 Francis Crick put forward his "adaptor hypothesis" in which he proposed the existence of special small adaptor RNA species and of specialized enzymes covalently attaching the amino acid residues to these RNA[1,2]. According to this hypothesis each of the amino acids has its own species of adaptor RNA, and the corresponding enzyme attaches this amino acid only to a given adaptor. On the other hand, the adaptor RNA possesses a nucleotide triplet (subsequently termed the *anticodon*) that is complementary to the appropriate codon of the template RNA. Hence, the recognition of a codon by the amino acid is indirect and is mediated through a system consisting of the adaptor RNA and the enzyme: a specific enzyme concomitantly recognizes an amino acid and the corresponding adaptor molecule, so that they become ligated to each other; in its turn, the adaptor recognizes an mRNA codon, and thus the amino acid attached becomes assigned specifically to this codon. In addition, this mechanism implied the energy supply for amino acid polymerization at the expense of chemical bond energy between the amino acid residues and the adaptor molecules.

It is a striking fact that this model was soon confirmed experimentally. In 1957 Hoagland, Zamecnik, and Stephenson, and simultaneously Ogata and Nohara, reported the discovery of a relatively low-molecular-weight RNA ("soluble RNA") and a special enzyme fraction ("pH 5 enzyme") that attached amino acids to this RNA[3,4]. It was demonstrated that the aminoacyl-tRNA formed was indeed an intermediate in the transfer of amino acids into a polypeptide chain. Subsequently, this RNA was termed *transfer RNA* (tRNA); the enzymes were called *aminoacyl-tRNA synthetases*.

The cell contains a specific aminoacyl-tRNA synthetase for each of the 20 amino acids participating in protein synthesis. Therefore, procaryotic cells contain 20 different aminoacyl-tRNA synthetases. The situation with eucaryotic cells is more complex, particularly because, in addition to the main cytoplasmic synthetases, there are special sets of aminoacyl-tRNA synthetases for chloroplasts and mitochondria.

The number of different tRNA species is always greater than the number of amino acids and aminoacyl-tRNA synthetases. Bacteria, such as *E. coli*, possess at least 40 tRNA species coded by different genes. This implies that several different tRNAs may be recognized by the same aminoacyl-tRNA synthetase and, correspondingly, can be ligated to the same amino acid; such tRNAs are called *isoacceptor* tRNAs. In some cases, different isoacceptor tRNA species recognize different codons for a given amino acid. For example, *E. coli* has 5 different leucine tRNA species, with anticodons CAG, GAG, NAG (N is an unidentified derivative of one of the bases), CAA, and A*AA, recognizing 6 leucine codons; among them, $tRNA_1^{Leu}$ recognizes the leucine codon CUG (anticodon CAG), and $tRNA_5^{Leu}$ recognizes the leucine codons UUA and UUG (the anticodon is A*AA, where A* is the unidentified derivative of A). Quite often, however, isoacceptor tRNA species differing in their primary structure recognize identical codons and have similar anticodons. This is, for example, true for $tRNA_4^{Leu}$ and $tRNA_5^{Leu}$, both of which recognize codons UUA and UUG. The situation is similar in the cytoplasm of eucaryotic cells.

Mitochondria contain a different set of tRNA species which is markedly simpler than that of the cytoplasm: only 23 or 24 tRNA species can be found in mitochondria, and they are sufficient to recognize all 61 or 62 sense codons of mitochondrial mRNA[5].

3-2 Structure of tRNA

Primary structure

In 1965 Holley and co-workers reported the nucleotide sequence of the first tRNA molecule[6]. This molecule was yeast alanine tRNA (fig. 12)[7]. Since then, hundreds of sequences of different tRNA from various sources have been determined[8]. All of these structures have several common features.

The length of tRNA chains varies from 74 to 95 nucleotide residues. At the 3′-end all tRNA species contain a universal trinucleotide sequence, CCA_{OH}; it is the terminal invariant adenosine that accepts the amino acid residue when the aminoacyl-tRNA is being formed.

The anticodon triplet is located approximately in the middle of the tRNA chain (IGC in positions 34 to 36 in fig. 12). As a rule, the 5′-side from the anticodon contains two pyrimidine residues, whereas the 3′-side most often contains two purine residues, although the second residue on the 3′-side may be a pyrimidine, as in the case of $tRNA^{Ala}$ (fig. 12). These seven nucleotide residues together form the so-called anticodon loop (AC loop) which interacts with the mRNA and possesses a characteristic three-dimensional structure (see below).

Approximately one-third of the way along the tRNA chain from its 3′-end there is a region common to most tRNA species; this region contains a sequence GTψC or, much less frequently, GUψC (or

1 10 20
pG–G–G–C–G–U–G–U–m¹G–G–C–G–U–A–G–hU–C–G–G–hU–

30 40
–A–G–C–G–C–m²₂G–C–U–C–C–C–U–U–I–G–C–m¹I–Ψ–G–G–

50 60
–G–A–G–A–G–G–U–C–U–C–C–G–G–T–Ψ–C–G–A–U–U–

70 76
–C–C–G–G–A–C–U–C–G–U–C–C–A–C–C–A_{OH}

Figure 12 Nucleotide sequence of the alanine tRNA of yeast ($tRNA_1^{Ala}$)[7]. The anticodon triplet is shown by the wavy line. Conservative (invariant) nucleotide residues and sequences are underscored by the continuous line. Positions where either purine or pyrimidine bases are conserved are underscored by the broken line.

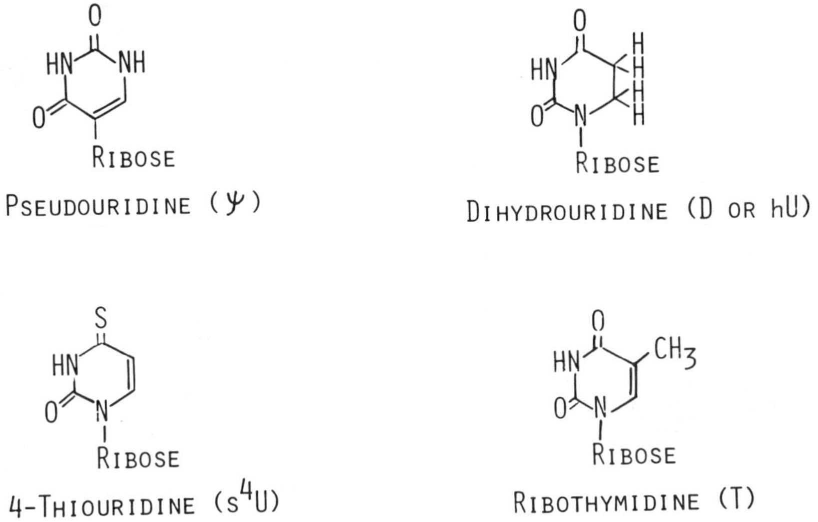

PSEUDOURIDINE (ψ)

DIHYDROURIDINE (D OR hU)

4-THIOURIDINE (s^4U)

RIBOTHYMIDINE (T)

Figure 13 Modified uridine derivatives widely occurring in tRNAs.

$Gm^1\psi\psi C$ in archaebacteria), and is flanked on both sides by purine residues. In the eucaryotic initiator $tRNA_F^{Met}$, this sequence is substituted for by GAψC or GAUC. This sequence is the principal conservative sequence of tRNA. (In mitochondrial tRNAs, however, the corresponding sequence region varies strongly.)

Some other conservative parts of the sequence in the region of nucleotide residues 8 to 25 should be mentioned. Several invariants and semiinvariants are present here: U or its thio-derivative (s^4U) in position 8, G or its methyl-derivative (m^2G) in position 10, AG or AA in positions 14 or 15, GG in positions 17 to 21, and AG in positions 21 to 24 of different tRNAs.

In addition to the four main types of nucleotide residues (i.e. A, G, C, and U), the tRNA polynucleotide chain is characterized by a variety of modified nucleosides frequently referred to as "minor" nucleosides. These nucleosides are the result of posttranscriptional enzymatic modification of the usual nucleotide residues at specific positions of the tRNA polynucleotide chain. Up to now, several dozen various modified nucleosides have been identified. Ribothymidine (5-methyluridine, abbreviated T or m^5U) and pseudouridine (5-ribofuranosyl-uracil, ψ) are found in nearly all tRNAs and are particularly characteristic of the universal sequence GTψC (fig. 13). 5,6-Dihydrouridine (D or hU) is also an almost universal minor residue, especially in the region of residues 15 to 24. Bacterial tRNAs typically contain 4-thiouridine (s^4U) in position 8. The most common

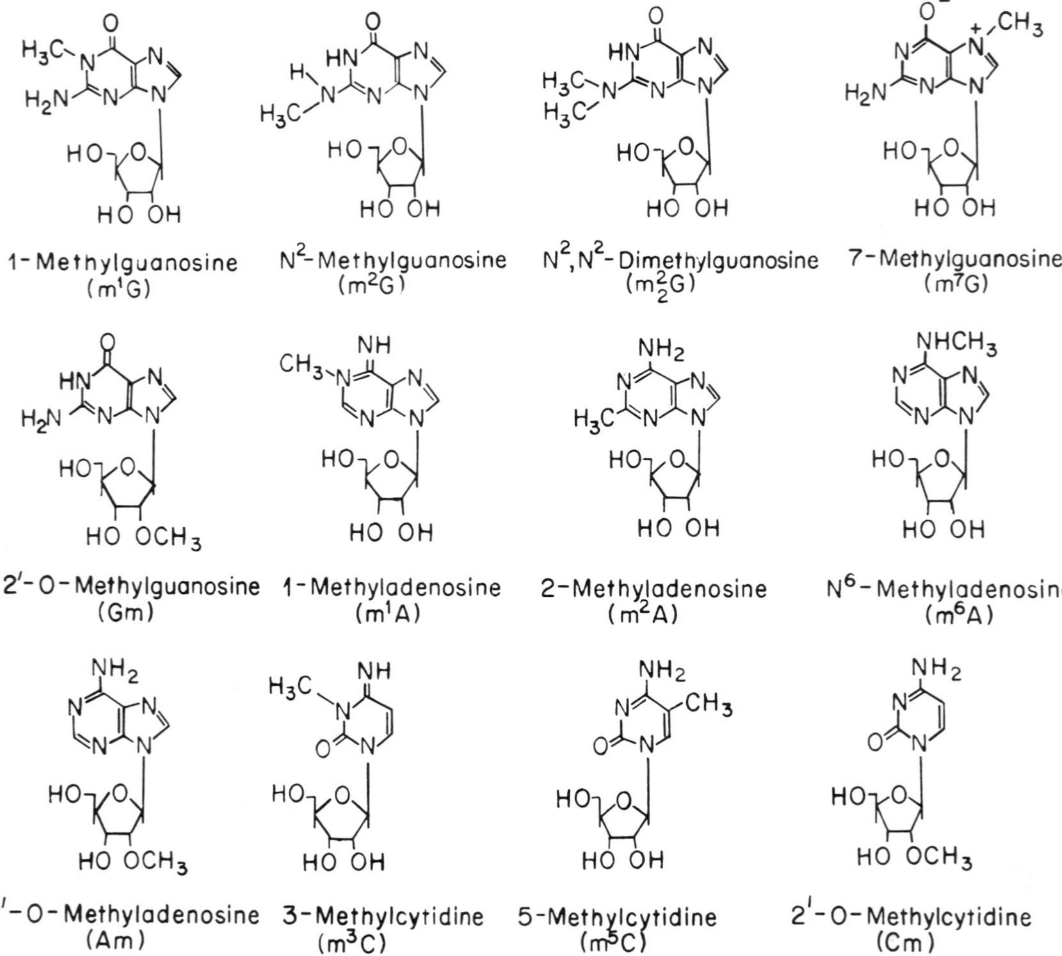

Figure 14 Methylated derivatives of guanosine, adenosine, and cytidine occurring in tRNAs.

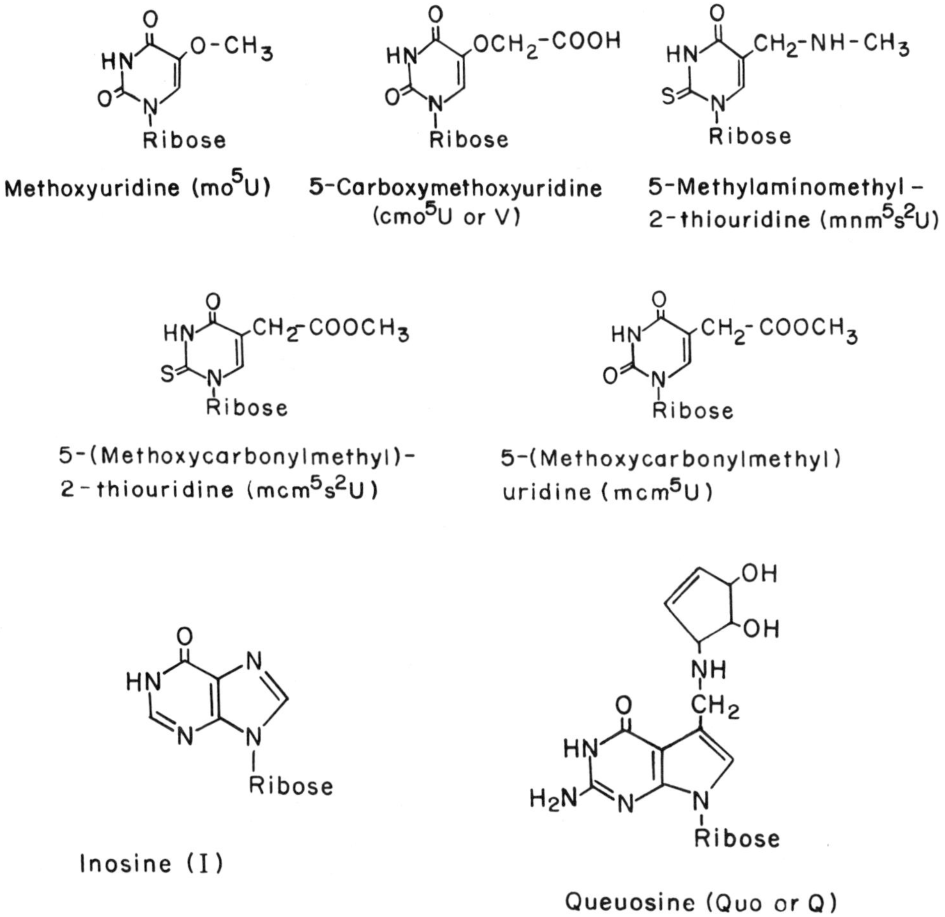

Methoxyuridine (mo^5U)

5-Carboxymethoxyuridine (cmo^5U or V)

5-Methylaminomethyl-2-thiouridine (mnm^5s^2U)

5-(Methoxycarbonylmethyl)-2-thiouridine (mcm^5s^2U)

5-(Methoxycarbonylmethyl) uridine (mcm^5U)

Inosine (I)

Queuosine (Quo or Q)

Figure 15 Modified nucleosides occurring in the first position of the tRNA anticodon.

minor residues are various methylated derivatives of the usual nucleosides, such as 1-methylguanosine (m^1G), N^2-methylguanosine (m^2G), N^2,N^2-dimethylguanosine (m^2_2G), 7-methylguanosine (m^7G), 2′-O-methylguanosine (Gm), 1-methyladenosine (m^1A), 2-methyladenosine (m^2A), N^6-methyladenosine (m^6A), 2′-O-methyladenosine (Am), 3-methylcytidine (m^3C), 5-methylcytidine (m^5C), 2′-O-methylcytidine (Cm), etc. (fig. 14).

The first position of the anticodon may contain nonmodified G and C but A and U are always derivativized, except the cases of U in mitochondrial and a few special tRNA species (see below). The A in the first position of the anticodon is usually deaminated into inosine (I) (fig. 15). I in this position is particularly characteristic of eucaryotic tRNAs, such as $tRNA^{Ile}$, $tRNA^{Val}$, $tRNA^{Ser}$, $tRNA^{Pro}$, $tRNA^{Thr}$, $tRNA^{Ala}$, and $tRNA^{Arg}$. The U derivatives present in the first anticodon position are 5-methoxyuridine (mo^5U) or 5-carboxymethoxyuridine (cmo^5U or V) in $tRNA^{Ala}$, $tRNA^{Ser}$, and $tRNA^{Val}$ of bacteria; 5-methylaminomethyl-2-thiouridine (mnm^5s^2U) in bacterial $tRNA^{Glu}$ and $tRNA^{Lys}$; 5-(methoxycarbonylmethyl)-2-thiouridine (mcm^5s^2U) in $tRNA^{Glu}$ and $tRNA^{Lys}$ of fungi; or 5-(methoxycarbonylmethyl)uridine (mcm^5U) in $tRNA^{Arg}$ of fungi (fig. 15). The presence of unmodified U has been demonstrated for one species of $tRNA^{Gly}$ in several bacteria and for one of the yeast $tRNA^{Leu}$; it is, however, typical of mitochondrial tRNAs.

In some tRNAs, such as $tRNA^{Asp}$, $tRNA^{Asn}$, $tRNA^{His}$, and $tRNA^{Tyr}$ of bacteria and animals, the first position of the anticodon contains a hypermodified G derivative, the so-called queuosine (Quo or Q), the chemical name of which is 7-{[(*cis*-4,5-dioxy-2-cyclopenten-1-yl)amino]methyl}-7-deazaguanosine (see fig. 15).

One special type of modification, termed *hypermodification*, can be found in the anticodon loop at the position of the purine nucleoside adjacent to the anticodon on the 3′-side. For example, the residue flanking the anticodon at the 3′-side is N^6-isopentenyl adenosine (i^6A)

N^6-Isopentenyladenosine (i^6A)

2-Methylthio-N^6-isopentenyladenosine (ms^2i^6A)

N^6-(threoninocarbonyl) adenosine (t^6A)

Wybutosine (yW or Y)

Peroxywybutosine (oyW)

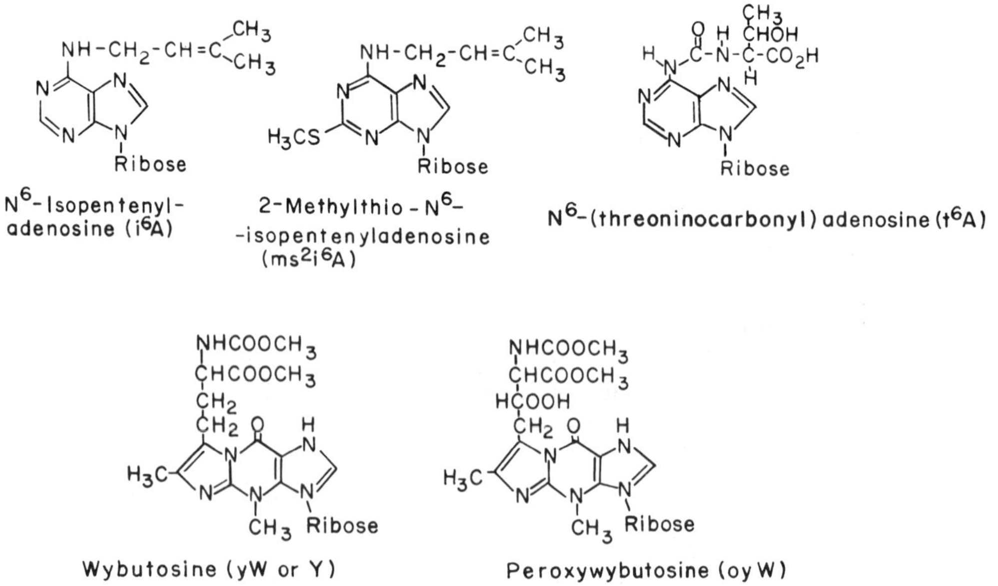

Figure 16 Hypermodified nucleosides occurring in the position adjacent to the anticodon at its 3′-side.

in eucaryotic $tRNA^{Cys}$, $tRNA^{Ser}$, and $tRNA^{Tyr}$; 2-methylthio-N^6-isopentenyladenosine (N^6-isopentenyl-2-methylthioadenosine, ms^2i^6A) in the analogous bacterial tRNAs; and N^6-(threoninocarbonyl)-adenosine (t^6A) in $tRNA^{Ile}$, $tRNA^{Thr}$, $tRNA^{Lys}$, and $tRNA^{Met}$ of both eucaryotes and bacteria (fig. 16). This position is even more hypermodified in the $tRNA^{Phe}$ of all eucaryotes, where it is represented by the so-called wybutosine (yW or Y) or its hydroxy-derivative (oyW) (see fig. 16).

Secondary structure

An analysis of even the first tRNA primary structure (i.e. $tRNA^{Ala}$ of yeast, see fig. 12) revealed a number of interesting features concerning possible chain folding into the secondary structure[6]. First of all, the 5′-terminal section (positions 1 to 7) has a marked complementarity with the 3′-end-adjacent section (positions 66 to 72) if the sections are arranged in an antiparallel fashion. In addition, three inner sections of the tRNA chain display self-complementarity when folded upon themselves; because of this they are capable of forming hairpinlike structures. These self-complementary regions are the parts of the molecule between residues 10 and 25, 27 and 43, and 49 and 65 where the two terminal four- or five-nucleotide-long sequences of each section are complementary to each other. Pairing these complementary sequences results in the structure schematically presented in fig. 17, commonly called a *cloverleaf* structure. It is remarkable that without exception the nucleotide sequences of all the tRNA species studied so far reveal similar self-complementarity features and correspondingly can be folded into very similar cloverleaves.

The parts of the cloverleaf structure have been designated as follows: the *acceptor stem* (AA stem), with the universal 3′-terminal sequence CCA which accepts an amino acid residue; the *dihydrouridylic arm* (D arm), with the corresponding loop varying somewhat in length and containing, as a rule, between one and five dihydrouridylic acid residues; the *anticodon arm* (AC arm), with an anticodon loop of constant length equal to seven nucleotides; and the *thymidyl-pseudouridylic arm* (Tψ arm), which has a loop with the universal $GT\psi C{}^{G}_{A}A$ sequence. In addition, the cloverleaf contains a *variable loop* (V loop) between the anticodon and Tψ arms; in $tRNA^{Ala}$ this loop is only five nucleotides long whereas in other tRNA species it may reach 15 to 20 nucleotide residues in length (the latter is the case for $tRNA^{Leu}$, $tRNA^{Ser}$, and bacterial $tRNA^{Tyr}$).

YEAST tRNA^{Ala}

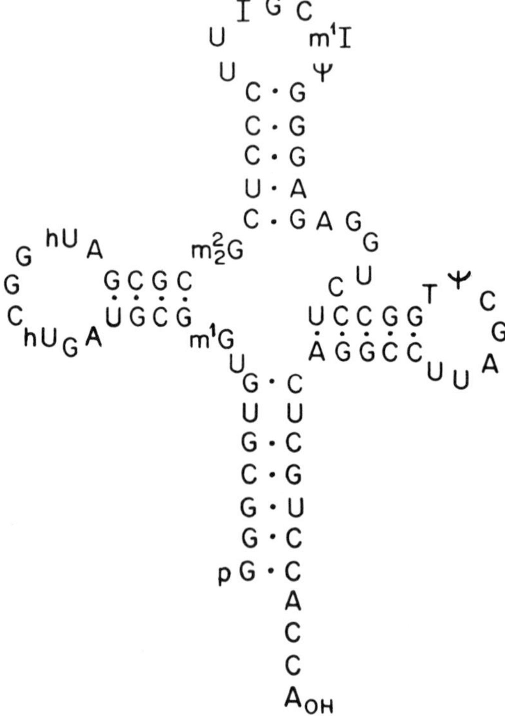

Figure 17 Yeast alanine tRNA secondary structure (the "cloverleaf)[6].

Structurally, the paired (double-stranded) part of each arm of tRNA is a double helix. The RNA double helix contains 11 pairs of nucleotide residues per turn. The parameters of this helix are similar to those of the A-form of DNA. The double helix is the main element of tRNA secondary structure[9,10] (fig. 18). In addition to the canonical Watson–Crick base pairs G·C and A·U, the double-stranded regions of tRNA often contain the G·U pair, which is close by its steric parameters to the canonic pairs (fig. 19).

The secondary structure of unpaired regions, such as loops and the acceptor $^{A}_{G}$CCA-terminus, is of a different type. A single-helical arrangement of several residues maintained by base-stacking interactions can occur here. The structure of the anticodon loop is particularly interesting[9,10] (fig. 20): three anticodon bases and two subsequent

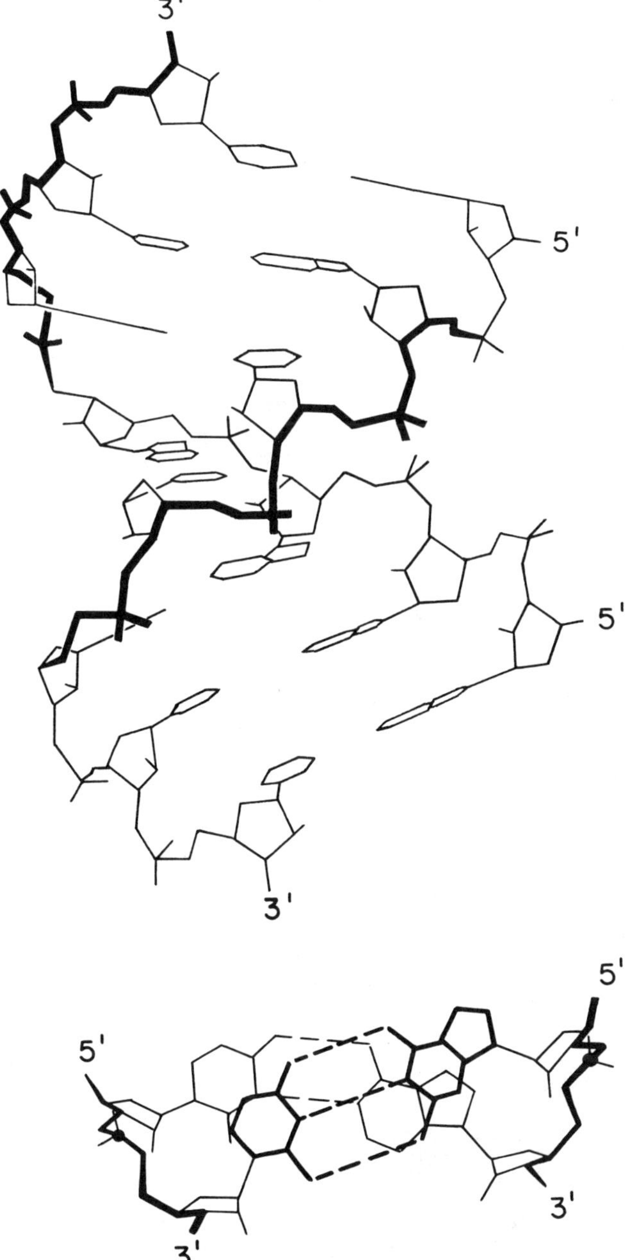

Figure 18 Double-helical fragment of RNA (A-form): skeletal model showing only covalent bonds. *Top*, Side view. Seven pairs of nucleotide residues are given. The backbone sections facing the viewer are drawn with thicker lines. *Bottom*, View at the helix top illustrating the base-stacking pattern. For the sake of clarity only two stacked nucleotide pairs are shown. The pair facing the viewer is drawn with thicker lines. The broken lines are hydrogen bonds.

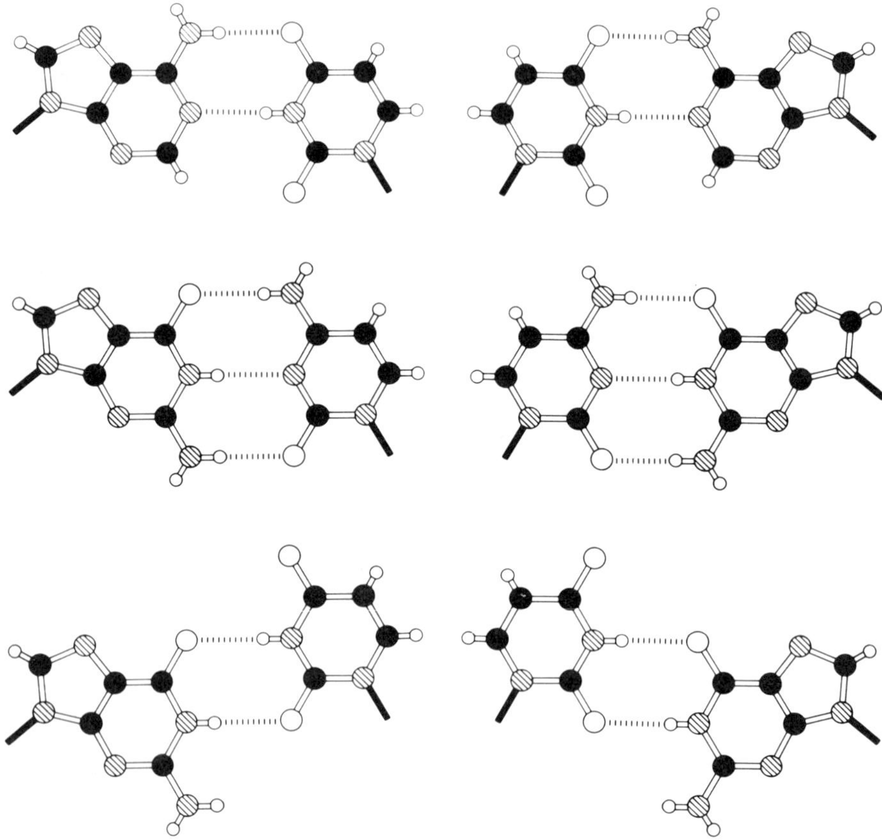

Figure 19 Base pairing in RNA double helices: ball-and-stick drawings. *Top to bottom*, A·U and U·A; G·C and C·G; G·U and U·G. Solid circles = carbons, hatched circles = nitrogens, large open circles = oxygens, and small open circles = hydrogens; solid sticks are N-glycosidic bonds between the base and ribose.

bases adjacent to the anticodon from the 3′-side are stacked with each other and form a single-stranded, right-handed helix; the first base of the anticodon is located at the top of the helix, and the groups capable of forming hydrogen bonds of all three anticodon bases are exposed outward. Such an orientation of the anticodon bases is extremely important to interaction with the mRNA codon. The features of the primary structure of the anticodon loop contribute specifically to the maintenance of the spatial arrangement described: the hypermodified purine base directly adjacent to the anticodon from the 3′-side as well

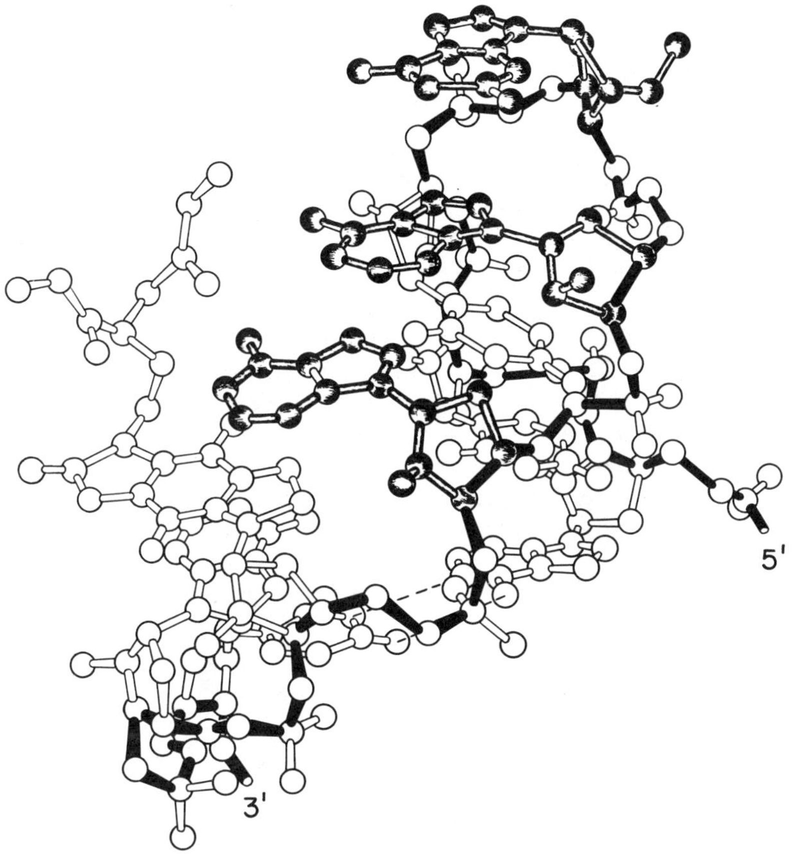

Figure 20 Anticodon loop of yeast phenylalanine tRNA[9,10]: ball-and-stick skeletal model (without hydrogens). The path of the backbone is given in solid black; three anticodon residues are shaded.

as the next base, usually also a purine, provides for stable stacking interactions in the single-stranded helix, while the two "small" pyrimidine bases at the 5′-side of the anticodon, and particularly the adjacent invariant U, make a sharp bend in the chain (between the anticodon and U) and maintain the loop conformation, particularly at the expense of a hydrogen bond between the invariant U and the phosphate group of the third residue of the anticodon.

It is noteworthy that the Tψ loop has a similar single-stranded conformation with an analogous "uridine turn" (in this case the role of invariant U is played by ψ).

Tertiary structure

The three-dimensional structure of tRNA was first reported for yeast tRNA^{Phe}. This structure was determined independently by the groups of Alexander Rich[9] and Aaron Klug[10] in 1974 through the use of X-ray analysis of tRNA^{Phe} crystals. A great deal of indirect evidence as well as direct determinations of the three-dimensional structures of several other tRNA species has demonstrated that the main pattern of tRNA chain folding into the tertiary structure is universal. Schematically, this folding may be represented as follows. The acceptor stem and the T-arm are arranged along a common axis, forming a continuous double helix 12 nucleotide pairs in length; the anticodon arm and the dihydrouridylic arm are also arranged along a common axis and yield another double helix, this one 9 nucleotide pairs long. These two helices are oriented toward each other at approximately a right angle so that the dihydrouridylic loop is brought into proximity with the T-loop, and the interaction between the GG invariant and the ψC invariant fastens them together (fig. 21). As a result, the structure looks like the letter *L* with the tops of its two limbs corresponding to

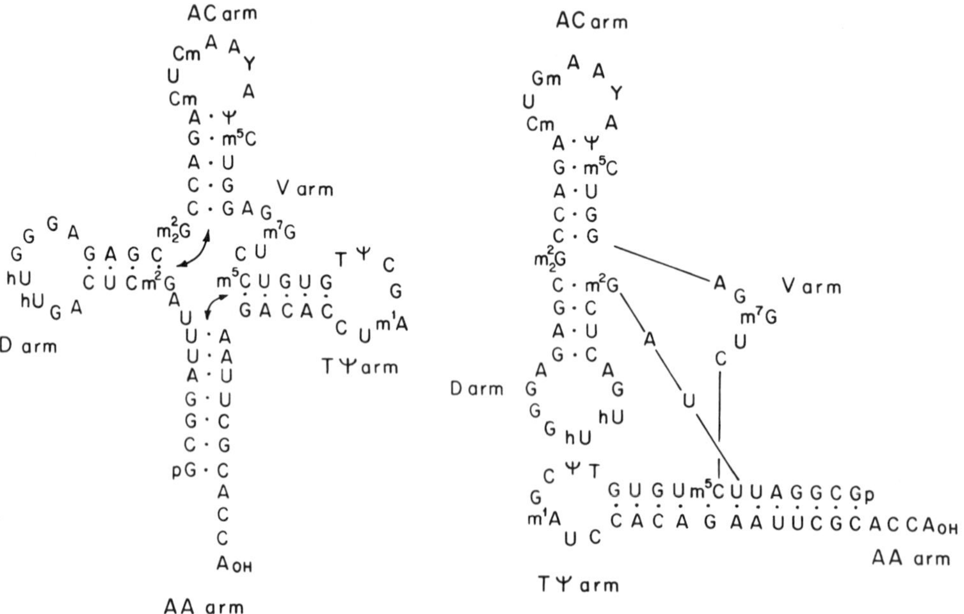

Figure 21 Scheme illustrating the folding of the tRNA helical regions into the tertiary structure (yeast tRNA^{Phe}). (After S.-H. Kim et al. (1973), *Science* 179:285–288; (1974) *Science* 185:435–440)

the anticodon and the acceptor 3′-end. The short, single-stranded bridge between the acceptor stem and dihydrouridylic helix (residues 8 and 9), part of the dihydrouridylic loop, and the additional variable loop are superimposed on the dihydrouridylic helix in the region of the inner corner of the L-shaped molecule, resulting in the formation of the so-called core of the molecule with a number of tertiary interactions. In a schematical drawing of the model of the yeast

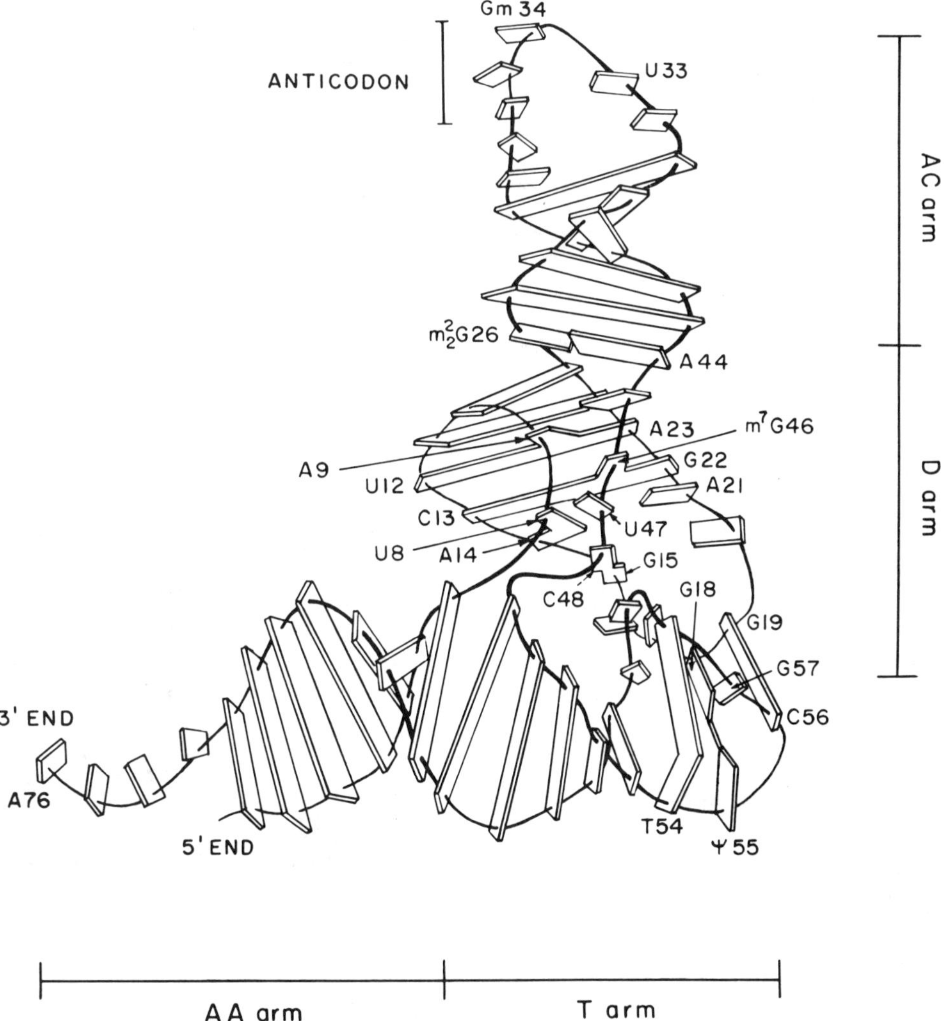

Figure 22 The tRNA tertiary structure: schematic drawing of the three-dimensional structure of yeast $tRNA^{Phe}$. (Redrawn from S.-H. Kim (1975) *Nature* 256:679–681, with permission). (See also Ref. 11)

tRNA^{Phe}, the core can be seen as a concentration and intertwining of the chain sections in the region of the corner, especially at its inner side (fig. 22).

Each limb of the L-shaped tRNA molecule is about 70 Å long, and the molecule has a "thickness" of around 20 Å. The distance between the anticodon and the acceptor end is 76 to 78 Å. All three bases of the anticodon on the top of one of the limbs are turned toward the inner side of the corner in the L-shaped molecule.

There are a large number of non-canonical (non-Watson–Crick) interactions between chain bases in the tRNA tertiary structure[11]. First, the corner of the L-shaped molecule is stabilized by both the stacking interactions and the hydrogen bonding between the dihydrouridylic loop and the T loop. The interaction between the invariant G19 and C56 is of the Watson–Crick type (see fig. 19), whereas the interaction between the invariant G18 and ψ55 is unusual, including the hydrogen bonding of O at C4 of the ψ pyrimidine ring both with N1 and with the nitrogen atom at C2 of the purine ring of the G (fig. 23 (bottom right)). In addition, there is an unusually strong stacking interaction between three guanosine residues in the same corner: G57 is found to be intercalated between G18

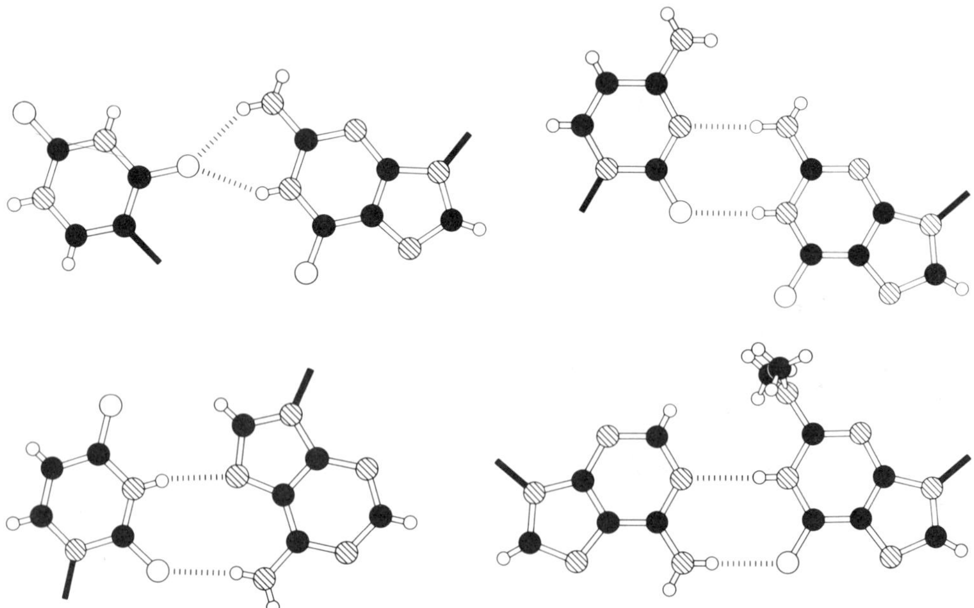

Figure 23 Unusual base pairing in the tRNA tertiary structure (yeast tRNA^{Phe}): ball-and-stick drawings. *Top left*, m_2^2G26·A44. *Top right*, A14·U8. *Bottom left*, G15·C48. *Bottom right*, G18·ψ55. Atom and bond designations as in fig. 19.

and G19. Moreover, G57 through N at C2 seems to form hydrogen bonds with the ribose residues of G18 and G19, whereas through its N7 it forms a hydrogen bond with the ribose of ψ55.

Even more complex tertiary interactions are observed in the core. As has already been mentioned, different sections of the polynucleotide chain are interbound here. A characteristic feature is the non-canonical purine-purine G·A or A·G pairing (depending on the RNA species) between residues 26 and 44 (fig. 23 (top left)). G·C pairing, or A·U pairing in other tRNA species, between residues 15 and 48, is unusual for double helices: in this case the orientation of chains is parallel (fig. 23 (bottom left)). Even more unusual is A·U pairing between residues 14 and 8 where N7 of the pruine ring participates in

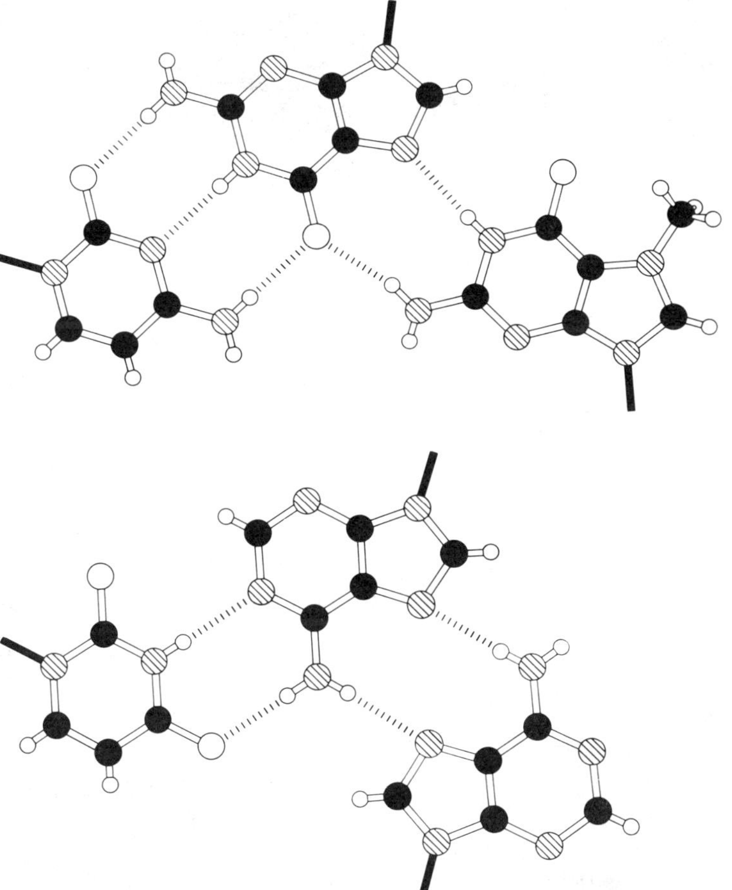

Figure 24 Base triple interactions through hydrogen bonds typical of the tRNA tertiary structure: ball-and-stick drawings. *Top*, C13·G22·m^{7}G46. *Bottom*, U12·A23·A9. Atom and bond designations as in fig. 19.

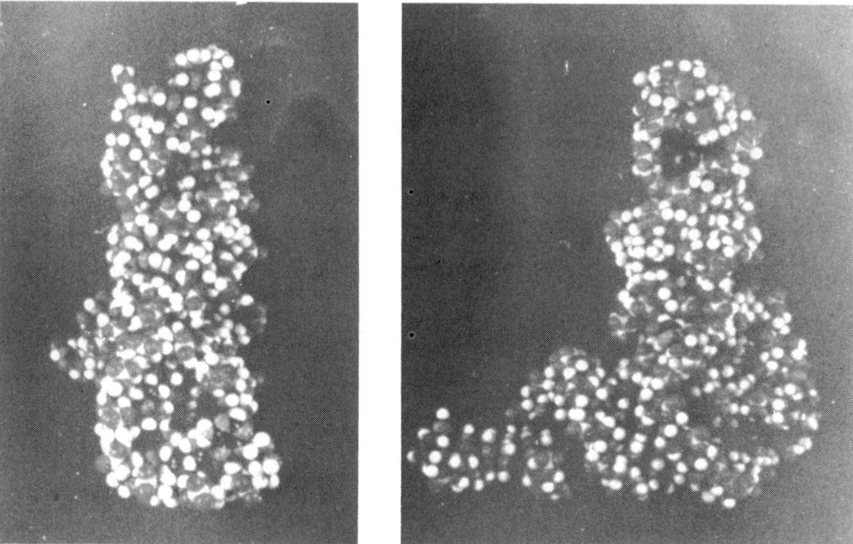

Figure 25 Space-filling model of yeast phenylalanine tRNA; two perpendicular views. (Courtesy of Dr. S.-H. Kim, University of California, Berkeley)

the formation of the hydrogen bond (fig. 23 (top right)). Triple hydrogen bond interactions, such as U·A·A, or the equivalent G·C·G or U·A·G in other tRNAs, between residues 12, 23, and 9, respectively, are characteristic of this part of the molecule (fig. 24 (bottom)). The triple hydrogen bond interaction is found also for C·G·G, or the equivalent U·A·A in other tRNAs, between residues 13, 22, and 46, respectively (fig. 24 (top)).

A photograph of the atomic space-filling model of tRNA is given in fig. 25.

3-3 Aminoacyl-tRNA synthetases

Despite the universality of the main features of the three-dimensional structure of tRNAs, aminoacyl-tRNA synthetases show marked differences depending on their amino acid specificity (see review by Kiselev and Favorova (1974)). In most cases, they are large proteins with a molecular mass ranging from 100,000 to 400,000 daltons, possessing a subunit structure, although monomeric enzymes can also be found among them. Usually the synthetase molecule consists of two or four protein subunits which may or may not be identical. For example, bacterial tyrosyl-tRNA synthetase consists of two identical subunits with a molecular mass of 47,000 daltons each, and bacterial methionyl-tRNA synthetase consists of two identical subunits although each has a molecular mass of about 80,000 to 90,000 daltons (α_2 type). Phenylalanyl-tRNA synthetase of bacteria, as well as of yeasts, consists of four subunits of two types with molecular masses of about 50,000 and about 60,000 daltons ($\alpha_2\beta_2$ type). Such bacterial enzymes as valyl-, leucyl-, and isoleucyl-tRNA synthetases are monomeric (α_1 type) consisting of a single polypeptide chain with a molecular mass of about 110,000 daltons. It appears, however, that in the latter case the polypeptide chain consists of two homologous regions forming two similar domains, each with a molecular mass of about 50,000 to 60,000 daltons. At the same time, cysteinyl- and glutaminyl-tRNA synthetases of bacteria consist of a single polypeptide chain (also α_1 type) with a molecular mass of 40,000 to 60,000 daltons, and do not appear to be subdivided into two homologous regions. Arginyl-tRNA synthetase, with a molecular mass of about 70,000 daltons, is also monomeric.

As a rule, synthetases with the same amino acid specificity which are isolated from different biological sources have similar quaternary structures and their polypeptide chains are of a similar size. However, compared with procaryotic enzymes, the synthetases of eucaryotic cells are characterized by somewhat larger subunits.

Despite the diversity of aminoacyl-tRNA synthetases, a generalized scheme of the subunit or domain structure may be proposed. Indeed, the synthetases that have a molecular mass of about 100,000 daltons often consist either of two subunits or of two similar domains. Synthetases with greater molecular masses (about 200,000 daltons) appear to be "duplicated" enzymes of the previous type. Therefore, if the principal unique building unit (i.e. domain or subunit) has a molecular mass ranging between 35,000 and 60,000 daltons, then many of the aminoacyl-tRNA synthetases are built according to the following scheme: $(35{,}000 - 60{,}000)_2 \times 1$ or 2.

The non-repeating unit in some cases may, however, be markedly larger; bacterial alanyl-tRNA synthetase, for example, consists of either two or four identical subunits, each with a molecular mass of 100,000 daltons, and shows no evidence of any repeats in its amino acid sequence. On the other hand, as has been already mentioned, cysteinyl-tRNA synthetase does not belong to the two-subunit- or the two-domain-type enzymes. Arginyl-tRNA synthetase also does not display features of the two-domain organization.

In any case, according to functional tests the molecules of most aminoacyl-tRNA synthetases (but not arginyl-tRNA synthetase!) possesses two sets of substrate-binding sites; in other words, they are dimers in the functional sense as well. The active sites, however, are not independent and can markedly affect each other in the complete dimeric or two-domain enzyme, thus displaying a certain cooperativity (see below).

Several characteristic features of eucaryotic aminoacyl-tRNA synthetases, primarily of animal aminoacyl-tRNA synthetases, should be discussed. Like many other proteins that play a role in the protein-synthesizing machinery of the eucaryotic cell (see, for example, chapter 12, section 12.2), the eucaryotic, and particularly mammalian, synthetases are organized in large multienzyme complexes or aggregates. In cell extracts they are usually present as a material possessing sedimentation coefficients between 12 and 30S corresponding to molecular masses from 3×10^5 to 10^6 and more daltons[12]. A given synthetase may exist in complexes of different sizes; for example, rat liver lysyl-tRNA synthetase is found in 24S, 18S, and 12S components. The complexes may be isolated and purified. For instance, a complex with a molecular mass of about 10^6 daltons and containing seven aminoacyl-tRNA synthetases specific to lysine, arginine, glutamine, glutamic acid, leucine, isoleucine, and methionine can be isolated from mammalian cells[13]. The existence of these enzymes in the form of a complex is not, however, vital to their activity: on one hand, the enzymes may be present in cells individually and not in the aggregate; on the other hand, they appear to function independently on each other in the complex. Besides synthetases, the complexes may contain other proteins of the cellular protein-synthesizing machinery.

In addition, a large proportion of eucaryotic aminoacyl-tRNA synthetases are associated directly with polyribosomes[14–16]. Although the association is quite loose and reversible, at each given moment a large proportion of cellular aminoacyl-tRNA synthetases appear to be in a labile association with the functional ribosomes. These characteristic features of eucaryotic systems appear to be due to the molecular attributes of the enzymes. It has been demonstrated that in contrast to procaryotic aminoacyl-tRNA synthetases, the eucaryotic analogs may possess a rather high nonspecific affinity to high-molecular-mass RNA, including mRNA and ribosomal RNA[17]. This affinity (i.e. RNA-binding capacity) of eucaryotic synthetases may be responsible for their concentration and partial compartmentation on the protein-synthesizing particles.

3-4 Aminoacylation of tRNA

The ligation of an amino acid to the tRNA 3′-end catalyzed by aminoacyl-tRNA synthetase is coupled with ATP cleavage. The overall equation of the process may be written as follows:

$$\text{Aa} + \text{ATP} + \text{tRNA} \overset{E}{\rightleftharpoons} \text{Aa-tRNA} + \text{AMP} + \text{PP}_i,$$

where Aa is the amino acid, Aa-tRNA is the aminoacyl-tRNA, and PP_i is the inorganic pyrophosphate. It has been demonstrated that the enzyme catalyzes two different reactions comprising two consecutive steps of the above process[2].

A reaction proceeding at the first stage that is catalyzed by aminoacyl-tRNA synthetase is the so-called amino acid *activation*, where the carboxyl group of the amino acid attacks the bond between the α- and β-phosphates of ATP, resulting in the formation of a mixed anhydride aminoacyl adenylate and inorganic pyrophosphate (fig. 26(1)):

$$\text{Aa} + \text{ATP} \overset{E}{\rightleftharpoons} \text{Aa-AMP} + \text{PP}_i. \tag{3.1}$$

This reaction is reversible and may be conveniently traced by pyrophosphate exchange: if [^{32}P]-pyrophosphate is added to the reaction mixture, the label is soon detected in [^{32}P]-ATP. Aminoacyl adenylate formed in the reaction given by (3.1) remains bound to the enzyme and is not released into solution.

The reaction, and consequently the overall reaction, is markedly shifted in the direction of aminoacyl adenylate and aminoacyl-tRNA formation due to the hydrolysis of the inorganic pyrophosphate which is catalyzed by pyrophosphatase. Therefore, the production of pyrophosphate in the amino acid activation step and the subsequent hydrolysis of the pyrophosphate to the inorganic orthophosphate play an important part in providing the energy that ensures the direction of the entire process.

A reaction catalyzed at the second stage by the same aminoacyl-tRNA synthetase involves the so-called *accepting* of the amino acid where the 2′- or 3′-hydroxyl of the ribose residue of the tRNA 3′-terminal adenosine attacks the anhydride group of the aminoacyl adenylate, resulting in the formation of an ester bond between the aminoacyl residue and the tRNA, with the accompanying release of AMP (fig. 26(2)):

$$\text{Aa-AMP} + \text{tRNA} \overset{E}{\rightleftharpoons} \text{Aa-tRNA} + \text{AMP}. \tag{3.2}$$

The case of arginyl-tRNA synthetase seems to be an exception: the

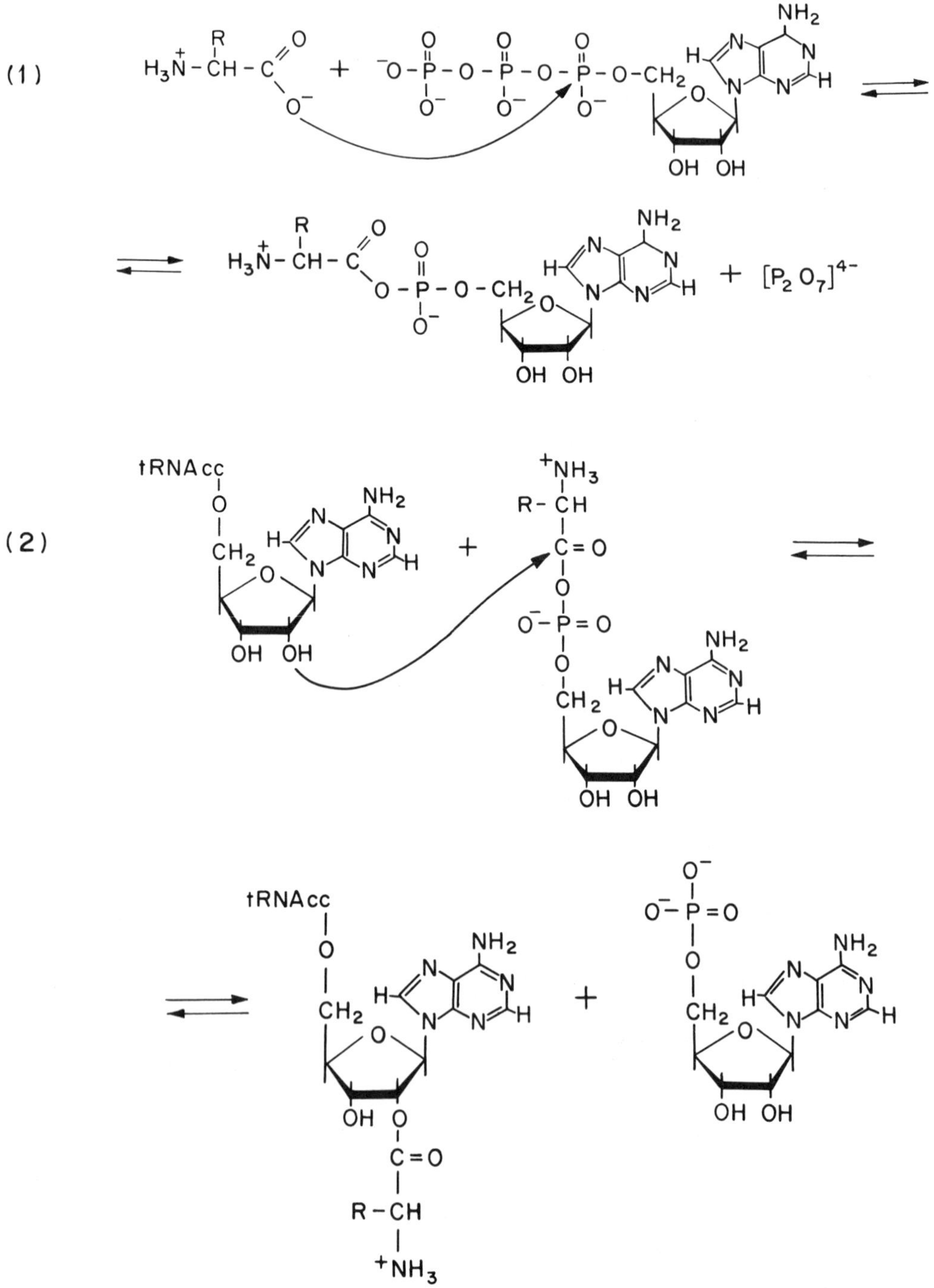

(1)

(2)

Figure 26 Reactions of amino acid activation (1) and acceptance of the aminoacyl residue by the tRNA molecule (2), catalyzed by aminoacyl-tRNA synthetase.

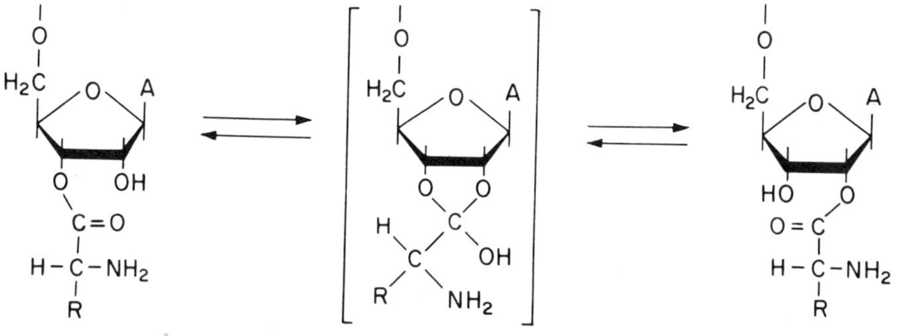

Figure 27 Spontaneous migration of the aminoacyl residue between the 2′- and 3′-positions of the ribose in the terminal tRNA adenosine through a short-lived intermediate.

intermediate formation of the aminoacyl adenylate is not detected here[18].

It is noteworthy that different aminoacyl-tRNA synthetases possess different specificity with regard to the position of the ribose hydroxyl participating in the transacylation reaction[19,20]. For example, phenylalanyl-, leucyl-, isoleucyl-, valyl-, methionyl-, and arginyl-tRNA synthetases couple the amino acids to the 2′-position, whereas seryl-, glycyl-, threonyl-, and histidyl-tRNA synthetases ligate the amino acid to the 3′-position of tRNA. Tyrosyl- and cysteinyl-tRNA synthetases catalyze the reaction with both the 2′- and the 3′-hydroxyl groups. This is of no great importance to the subsequent fate of the aminoacyl-tRNA formed because in an aqueous solution the aminoacyl residue spontaneously migrates between the 2′- and 3′-positions (through the formation of 2′,3′-triester), and eventually the two forms are in equilibrium (fig. 27).

Thus, an aminoacyl-tRNA synthetase uses three substrates of a different chemical nature: ATP, an amino acid, and tRNA. Correspondingly, it must possess three different substrate-binding sites. ATP is the universal substrate for all aminoacyl-tRNA synthetases, whereas for the amino acid and tRNA, each aminoacyl-tRNA synthetase displays high specificity.

As has already been mentioned, in many cases aminoacyl-tRNA synthetases are dimers or pseudodimers, and, correspondingly, they possess two sets of substrate-binding sites. The substrate-binding sites both within each subunit (or the equivalent domain) and on different subunits (or domains) are interdependent. Frequently synergism is observed: the binding of one substrate molecule facilitates the binding of the other. On the other hand, there is a negative

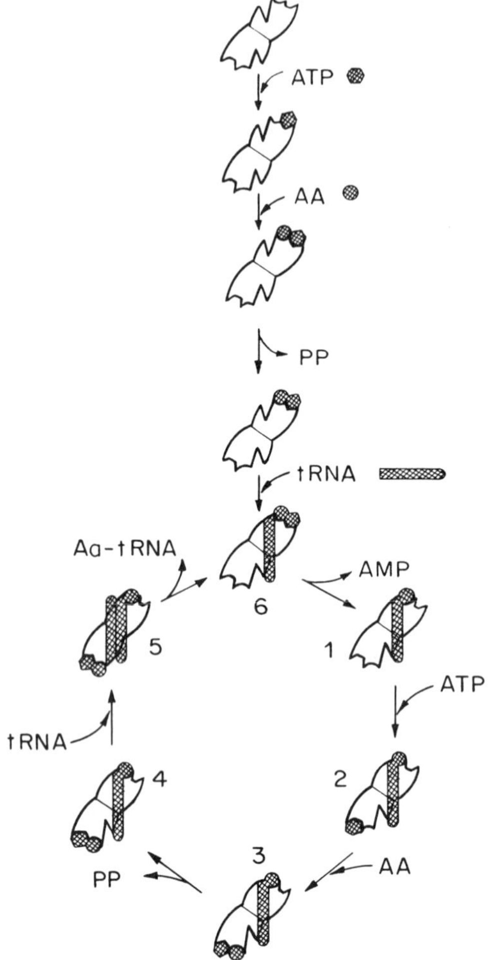

Figure 28 Possible sequence of events in the functioning of the two-domain (or dimeric) aminoacyl-tRNA synthetase.

cooperativity in the binding of two tRNA molecules: the binding of one tRNA molecule makes the binding of the other one less tight.

An example sequence of substrate addition to the dimeric enzyme is schematically presented in fig. 28[21–23]. Beginning with the enzyme free of substrates (upper part of the scheme), the first stages are often found to involve the binding of small substrates, such as ATP and the amino acid; and the binding of one of these may stimulate the binding of the other (synergism). The substrates bound to the enzyme interact to yield aminoacyl adenylate, and the resulting pyrophosphate is released into solution. The binding of the small substrates and the formation of aminoacyl adenylate stimulate tRNA binding, resulting in the aminoacylation of tRNA by the enzyme and the release of AMP into solution. The aminoacyl-tRNA, when present in a single copy per enzyme molecule, dissociates from the enzyme rather slowly, but the binding of the second tRNA molecule stimulates the dissociation. This leads to the cycle shown in the lower part of fig. 28, where one of the tRNA-binding sites is permanently occupied and the enzyme displays the reactivity of only half of its substrate-binding sites ("half-of-the-sites-reactivity").

In reality, under conditions where the enzyme works in substrate excess, the pathway shown in the lower part of fig. 28 is the route that appears to occur. State 1 is exhibited when the active site of one subunit (or domain) is occupied by aminoacyl-tRNA while the other one is vacant. Therefore, only the substrate-binding sites of the other active center of the enzyme are capable of binding substrate ligands. The consecutive or independent binding of the small substrates, ATP, and the amino acid (states 2 and 3) results in the formation of the enzyme-bound aminoacyl adenylate (state 4), which in turn stimulates tRNA association with the second active center of the enzyme (state 5). Because of the negative cooperativity mentioned above, the binding of tRNA with the second active center weakens the holding of aminoacyl-tRNA in the first binding center; as a result, this aminoacyl-tRNA dissociates into solution, leaving the enzyme with one active center occupied and the other vacant (state 6). Thus, the two active centers of a dimeric (or two-domain) enzyme appear to work alternately. The final product aminoacyl-tRNA is not released into solution immediately after its synthesis has been completed, but "waits" until the second substrate tRNA enters its binding site.

(It should be pointed out again that the above scheme is just an example of a possible reaction pathway and cannot be regarded as general or generally accepted one).

3-5 Specificity of tRNA aminoacylation

Specificity for amino acids

To provide unambiguous mRNA decoding during translation, aminoacyl-tRNA synthetases should possess an extremely high specificity when selecting amino acids and tRNAs as substrates. In the case of amino acid selection the enzyme has to discriminate between substrates, which sometimes possess very similar structures, such as isoleucine and valine. The error rate in tRNA aminoacylation is indeed extremely low, and even for related amino acids, e.g. isoleucine and valine, it does not appear to exceed one per 10,000.

However, analysis of the stages of amino acid binding and the subsequent reversible formation of aminoacyl adenylate measured by ATP-pyrophosphate exchange has shown that the enzyme cannot provide such high specificity in the discrimination of related amino acids at these stages. For example, isoleucyl-tRNA synthetase can effectively bind valine and form valyl adenylate. Similarly, valyl-tRNA synthetase can bind and activate isoleucine as well as a number of other amino acids, e.g. alanine, serine, cysteine, and threonine. The phenylalanine enzyme activates methionine, leucine, and tyrosine. Nevertheless, none of the listed misactivated amino acids becomes accepted by tRNA.

In addition to amino acid discrimination at the binding stage, the enzyme may possess a special error-correcting mechanism which acts *after* the aminoacyl adenylate has been formed[24]. Basically, the binding of the cognate tRNA by the enzyme results in a hydrolytic release of the free amino acid if the amino acid residue was noncognate for the enzyme. It seems that in most cases misactivated amino acid bound to the enzyme as aminoacyl adenylate is normally transferred to tRNA, but the ester bond between the noncognate amino acid and tRNA is hydrolyzed immediately by the enzyme:

$$\text{Val} + \text{ATP} + \text{E}^{\text{Ile}} \rightleftharpoons \text{Val-AMP}\cdot\text{E}^{\text{Ile}} + \text{PP}_i;$$
$$\text{Val-AMP}\cdot\text{E}^{\text{Ile}} + \text{tRNA}^{\text{Ile}} \rightleftharpoons \text{Val-tRNA}^{\text{Ile}}\cdot\text{E}^{\text{Ile}} + \text{AMP};$$
$$\text{Val-tRNA}^{\text{Ile}}\cdot\text{E}^{\text{Ile}} + \text{H}_2\text{O} \rightarrow \text{Val} + \text{tRNA}^{\text{Ile}} + \text{E}^{\text{Ile}}.$$

This implies that the enzyme has a second chance to discriminate between the aminoacyl residues, now in the form of their ester derivatives; if the residue is noncognate, then the water molecule is activated and the ester bond is attacked (see review by Sprinzl and Cramer (1979)). In the case of the hydrolysis of valyl-tRNA^{Ile} by isoleucyl-tRNA synthetase, the free hydroxyl of the tRNA terminal ribose has been seen to play an important part[25].

It may well be that in some other cases a different correction (proofreading) mechanism is in operation, and that the noncognate aminoacyl adenylate may be hydrolyzed by the enzyme prior to the transfer of the aminoacyl residue to tRNA[26].

Specificity for tRNA

It has already been stated that the binding of tRNA with aminoacyl-tRNA synthetase is a multistep process. The initial tRNA binding is not very specific, and so the enzyme may interact with a number of noncognate tRNAs. The isoleucyl-tRNA synthetase, for example, can bind $tRNA^{Val}$, and its binding is only one-fifth the strength of the binding of the cognate $tRNA^{Ile}$. The enzyme interacts with $tRNA^{Glu}$ as well; this latter binding, however, is 10,000 times weaker than the binding of the cognate $tRNA^{Ile}$. Generally, very different affinities are found for various combinations of aminoacyl-tRNA synthetases with noncognate tRNA species, from an almost total absence of affinity to an affinity close to that of the cognate tRNA. The degree of affinity of the enzymes to tRNA usually increases with the decrease in pH and ionic strength and is stimulated by organic solvents; this suggests that ionic interactions contribute considerably toward binding. Correspondingly, the same factors stimulate the nonspecific binding of tRNA by aminoacyl-tRNA synthetases. Magnesium ions, however, frequently have the opposite effect: they may decrease the binding of noncognate tRNA species to an aminoacyl-tRNA synthetase, i.e. increase binding specificity. The latter effect is usually considered to be the result of the action of magnesium ions upon the conformation of both the enzyme and tRNA.

The initial binding of tRNA to the enzyme is a fast step, i.e. the rates of both forward and reverse reaction (association and dissociation) are high. This fast step of initial recombination may be followed by a slower step, when the complex somehow rearranges. Such a rearrangement takes place only if the enzyme has bound the cognate tRNA. This is the recognition stage during which the main discrimination between the cognate and noncognate tRNA species is accomplished. Thus, the first binding step involves only a rough selection of tRNAs, and the main function of this step is the rapid scanning of the various tRNA species (see, e.g., Ref. 27). If the bound tRNA is noncognate, it will be inactive in the induction of the structural rearrangement of the enzyme complex and, hence, is incapable of entering the next stage; as a result, it will be dissociated easily from the fast reversible initial complex. Only if the cognate tRNA is bound, is the next phase, involving the *rearrangement* of the complex and proper *fitting* of tRNA and the enzyme, required for the subsequent aminoacylation reaction initiated:

$$\text{E·Aa-AMP} + \text{tRNA} \xrightleftharpoons[]{\substack{\text{fast,}\\ \text{not very}\\ \text{specific}}} (\text{E·Aa-AMP·tRNA})' \xrightarrow{\substack{\text{slow,}\\ \text{specific}}}$$

$$(\text{E·Aa-AMP·tRNA})'' \rightarrow \text{E·Aa-tRNA} + \text{AMP}.$$

It is likely that this mechanism is not universal for all aminoacyl-tRNA synthetases. For example, the tyrosyl-tRNA synthetase from *E. coli* and the serine enzyme from yeasts, as well as the arginyl-tRNA synthetase, show a very high specificity even at the stage of the initial complex; they bind little of the noncognate tRNA species.

Regardless of which mechanism is realized, the final result is a very high specificity of the selection of tRNA by the enzyme. This raises the problem of the specific tRNA-protein recognition. It is apparent that certain specific regions of the tRNA molecule are involved in this recognition. The available evidence indicates that tRNA regions recognized by the enzyme may be located in several, often distant, parts of the three-dimensional structure. Evidently, different aminoacyl-tRNA synthetases use different tRNA regions for recognition. The binding and recognition of $tRNA_F^{Met}$ by methionyl-tRNA synthetase, for example, involve the anticodon and the acceptor stem, as well as the variable loop located in the core region of the molecule. It is likely that simultaneous interaction with all three tRNA regions is needed for the specific recognition and eventual fitting to the enzyme. In other pairs of aminoacyl-tRNA synthetases and tRNAs, the anticodon arm and the acceptor stem also often prove to be regions important to recognition; in addition, the dihydrouridylic arm and the additional variable loop sometimes are essential[28]. Usually, all aminoacyl-tRNA synthetases recognize regions generally located on, or close to, the inner part of the corner of the L-shaped tRNA molecule[29]. It is quite likely that the inner side of the corner of the L-shaped tRNA and the adjacent regions of the two flat sides of the tRNA molecule interact with the surfaces of the enzyme molecule.

References

1. F. H. C. Crick (1957). Discussion in The structure of nucleic acids and their role in protein synthesis, Biochemical Society Symposium no. 14, pp. 25–26 (Cambridge: At the University Press).

2. M. B. Hoagland (1960), "The relationship of nucleic acid and protein synthesis as revealed by studies in cell-free systems," Nucleic acids, ed. E. Chargaff and J. N. Davidson, vol. 3, pp. 349–408 (New York: Academic Press).

3. M. B. Hoagland, P. C. Zamecnik, and M. L. Stephenson (1957), "Intermediate reactions in protein biosynthesis," *Biochim. Biophys. Acta* 24:215–216.

4. K. Ogata and H. Nohara (1957), "The possible role of the ribonucleic acid (RNA) of the pH 5 enzyme in amino acid activation," *Biochim. Biophys. Acta* 25:659–660.

5. P. Slonimski, P. Borst, and G. Attardi, eds. (1982), *Mitochondrial genes* (Cold Spring Harbor, N.Y.: Cold Spring Harbor Laboratory).

6. R. W. Holley, J. Apgar, G. A. Everett, J. T. Madison, M. Marquisee, S. H. Merrill, J. R. Penswick, and A. Zamir (1965), "Structure of a ribonucleic acid," *Science* 147:1462–1465.

7. J. R. Penswick, R. Martin, and G. Dirheimer (1975), "Evidence supporting a revised sequence for yeast alanine tRNA," *FEBS Letters* 50:28–31.

8. D. H. Gauss and M. Sprinzl (1983), "Compilation of tRNA sequences," *Nucleic Acids Res.* 11:r1–r53.

9. S. H. Kim, F. L. Suddath, G. J. Quigley, A. McPherson, J. L. Sussman, A. H.-J. Wang, N. C. Seeman, and A. Rich (1974), "Three-dimensional tertiary structure of yeast phenylalanine transfer RNA," *Science* 185:435–440.

10. J. D. Robertus, J. E. Ladner, J. T. Finch, D. Rhodes, R. S. Brown, B. F. C. Clark, and A. Klug (1974), "Structure of yeast phenylalanine tRNA at 3 Å resolution," *Nature* 250:546–551.

11. A. Rich and S. H. Kim (1978), "The three-dimensional structure of transfer RNA," *Scientific American* 238:52–62.

12. C. V. Dang, D. L. Johnson, and D. C. H. Yang (1982), "High molecular mass amino acyl-tRNA synthetase complexes in eukaryotes," *FEBS Letters* 142:1–6.

13. M. Mirande, B. Çirakoğlu, and J.-P. Waller (1983), "Seven mammalian aminoacyl-tRNA synthetases associated within the same complex are functionally independent," *Eur. J. Biochem.* 131:163–170.

14. W. K. Roberts and W. H. Coleman (1972), "Particulate forms of phenylalanyl-tRNA synthetase from Ehrlich ascites cells," *Biochem. Biophys. Res. Commun.* 46:206–214.

15. J. D. Irwin and B. Hardesty (1972), "Binding of aminoacyl transfer ribonucleic acid synthetases to ribosomes from rabbit reticulocytes," *Biochemistry* 11:1915–1920.

16. J. S. Tscherne, I. B. Weinstein, K. W. Lanks, N. B. Gersten, and C. R. Cantor (1973), "Phenylalanyl transfer ribonucleic acid synthetase activity associated with rat liver ribosomes and microsomes," *Biochemistry* 12:3859–3865.

17. A. T. Alzhanova, A. N. Fedorov, L. P. Ovchinnikov, and A. S. Spirin (1980), "Eukaryotic aminoacyl-tRNA synthetases are RNA-binding proteins whereas prokaryotic ones are not," *FEBS Letters* 120:225–229.

18. R. Thiebe (1983), "No arginyl adenylate is detectable as an intermediate in the aminoacylation of $tRNA^{Arg}$," *Eur. J. Biochem.* 130:525–528.

19. F. Cramer, H. Faulhammer, F. von der Haar, M. Sprinzl, and H. Sternbach (1975), "Aminoacyl-tRNA synthetases from baker's yeast: Reacting site of enzymatic aminoacylation is not uniform for all tRNA," *FEBS Letters* 56:212–214.

20. T. H. Fraser and A. Rich (1975), "Amino acids are not all initially attached to the same position on transfer RNA molecules," *Proc. Nat. Acad. Sci. U.S.A.* 72:3044–3048.

21. M. Yarus and P. Berg (1969), "Recognition of tRNA by isoleucyl-tRNA synthetase: Effect of substrates on the dynamics of tRNA-enzyme function," *J. Mol. Biol.* 42:171–190.

22. D. G. Knorre, E. G. Malygin, M. G. Slinko, V. I. Timoshenko, V. V. Zinoviev, L. L. Kisselev, L. L. Kochkina and O. O. Favorova (1974), "Beef pancreas tryptophanyl-tRNA synthetase: Order of substrate binding in ATP-(^{32}P)-pyrophosphate exchange reaction," *Biochimie* 56:845–855.

23. E. G. Malygin and L. L. Kisselev (1983), "Kinetics and mechanism of reactions catalyzed by aminoacyl-tRNA synthetases," Soviet scientific reviews, section D: Biology reviews ed. V. P. Skulachev, vol. 3, pp. 229–274 (GmbH, Switzerland: Harwood Academic).

24. A. N. Baldwin and P. Berg (1966), "tRNA induced hydrolysis of valyl adenylate bound to isoleucyl tRNA synthetase," *J. Biol. Chem.* 241:839–845.

25. F. van der Haar and F. Cramer (1975), "Isoleucyl-tRNA synthetase from baker's yeast: The 3′-hydroxyl group of the 3′-terminal ribose is essential for preventing misacylation of $tRNA^{Ile}$-C–C–A with misactivated valine," *FEBS Letters* 56:215–217.

26. A. R. Fersht (1977), "Editing mechanisms in protein synthesis: Rejection of valine by the isoleucyl-tRNA synthetase," *Biochemistry* 16:1025–1030.

27. R. Rigler, U. Pachmann, R. Hirsch, and H. G. Zachau (1976), "On the interaction of seryl-tRNA synthetase with $tRNA^{Ser}$: A contribution to the problem of synthetase-tRNA recognition," *Eur. J. Biochem.* 65:307–315.

28. V. V. Vlassov, D. Kern, P. Romby, R. Giegi, and J.-P. Ebel (1983), "Interaction of $tRNA^{Phe}$ and $tRNA^{Val}$ with aminoacyl-tRNA synthetases: A chemical modification study," *Eur. J. Biochem.* 132:537–544.

29. A. Rich and P. R. Schimmel (1977), "Structural organization of complexes of transfer RNAs with aminoacyl transfer RNA synthetases," *Nucleic Acids Res.* 4:1649–1665.

Further reading

Altman, S. (1978). *Transfer RNA.* Cambridge, Mass.: M.I.T. Press.

Clark, B. F. C. (1977). Correlation of biological activities with structural features of transfer RNA. Progress in nucleic acid research and molecular biology (W. E. Cohn, ed.) vol. 20, pp. 1–19. New York: Academic Press.

Kim, S.-H. (1976). Three-dimensional structure of transfer RNA. Progress in nucleic acids research and molecular biology (W. E. Cohn, ed.), vol. 17, pp. 182–216. New York: Academic Press.

Kiselev, L. L. and Favorova, O. O. (1974). Aminoacyl-tRNA synthetases: Some recent results and achievements. Advances in enzymology (A. Meister, ed.), vol. 40, pp. 141–238. New York: Wiley.

Loftfield, R. B. (1972). The mechanism of aminoacylation of transfer RNA. Progress in nucleic acid research and molecular biology (J. N. Davidson and W. E. Cohn, eds.), vol. 12, pp. 87–128. New York: Academic Press.

Nishimura, S. (1972). Minor components in transfer RNA: Their characterization, location, and function. Progress in nucleic acid research and molecular biology (J. N. Davidson and W. E. Cohn, eds.), vol. 12, pp. 49–85. New York: Academic Press.

Ofengand, J. (1977). tRNA and aminoacyl-tRNA synthetases. In *Molecular mechanisms of protein biosynthesis* (H. Weissbach and S. Pestka, eds.), pp. 7–79. New York: Academic Press.

Rich, A. and RajBhandary, U. L. (1976). Transfer RNA: Molecular structure, sequence, and properties. *Ann. Rev. Biochem.* 45:805–860.

Schimmel, P. and Söll, D. (1979). Aminoacyl-tRNA synthetases: General features and recognition transfer RNAs. *Ann. Rev. Biochem.* 48:601–648.

Schimmel, P.; Söll, D.; and Abelson, J., eds. (1979). *Transfer RNA: Structure, properties, recognition*. Cold Spring Harbor, N.Y.: Cold Spring Harbor Laboratory.

Singhal, R. P. and Fallis, P. A. M. (1979). Structure, function, and evolution of transfer RNAs [with appendix giving complete sequences of 178 tRNAs]. Progress in nucleic acid research and molecular biology (W. E. Cohn, ed.), vol. 23, pp. 227–290. New York: Academic Press.

Söll, D.; Abelson, J.; and Schimmel, P., eds. (1980). *Transfer RNA: Biological aspects*. Cold Spring Harbor, N.Y.: Cold Spring Harbor Laboratory.

Sprinzl, M. and Cramer, F. (1979). The –C–C–A end of tRNA and its role in protein biosynthesis. Progress in nucleic acid research and molecular biology (W. E. Cohn, ed.), vol. 22, pp. 1–69. New York: Academic Press.

Chapter 4

Ribosomes and Translation

4-1 First observations

By 1940 Albert Claude[1] had succeeded in isolating from animal cells cytoplasmic RNA-containing granules that were smaller than mitochondria and lysosomes. These granules varied from 50 to 200 mμ in diameter and later Claude began calling them *microsomes*. Chemical analyses indicated that Claude's microsomes were "phospholipid-ribonucleoprotein complexes."

On the other hand, cytochemical studies by Caspersson[2,3] and Brachet[4,5] demonstrated the preferentially cytoplasmic localization of the RNA and the existence of a correlation between the amount of RNA in the cytoplasm and the intensity of protein synthesis.

Later, a number of scientists reported on the isolation of RNA-containing particles, which were much smaller than microsomes from the cytoplasm of animal and plant cells as well as from bacteria. Electron microscopy and sedimentation analysis in the ultracentrifuge indicated that these particles were compact; had a more or less spherical shape; were homogeneous in size, with a diameter of 100 to 200 Å; and exhibited sharp sedimentation boundaries corresponding to sedimentation coefficients of from 30–40S to 70–90S (for references see Petermann (1964)). The first unambiguous evidence that such particles from bacteria are ribonucleoproteins was probably obtained by Schachman, Pardee, and Stanier in 1952[6].

Improved techniques of microtomy and electron microscopy of ultrathin sections of animal cells resulted in the detection of uniform dense granules with a diameter of about 150 Å directly in the cell. Palade's electron microscopic studies[7] demonstrated that small dense granules are abundant in animal cell cytoplasm. These granules were seen either attached to the membrane of the endoplasmic reticulum or freely dispersed throughout the cytoplasm. Claude's microsomes were identified as fragments of the endoplasmic reticulum with these granules attached[8]. It became clear that the Palade granules were ribonucleoprotein particles and that they accounted for most of the cytoplasmic RNA involved in protein synthesis.

Purified preparations of ribonucleoprotein particles were isolated and studied in several laboratories between 1956 and 1958; these investigations included isolating 80S particles from yeast, accomplished by Chao and Schachman[9]; from plants, by Ts'o, Bonner, and Vinograd[10]; and from animals, by Petermann and Hamilton[11]; and isolating 70S particles from bacteria (*E. coli*), by Tissières and Watson[12]. In 1958 the first symposium devoted to these particles and their participation in protein biosynthesis took place (First Symposium of the Biophysical Society at the Massachusetts Institute of Technology, Cambridge, Mass., February 5, 6, and 8, 1958); during this symposium Roberts[13] suggested that ribonucleoprotein particles be called *ribosomes*.

Studies of the functional role played by ribosomes proceeded hand-in-hand with their structural description. The experiments of Zamecnik and co-workers provided the first convincing demonstration that ribonucleoprotein particles of microsomes are responsible for the incorporation of amino acids into newly synthesized proteins[14]. This was followed by other experiments conducted at the same laboratory which demonstrated that the free ribosomes unattached to the endoplasmic reticulum membranes also incorporate amino acids and synthesize the protein released into the soluble phase[15]. The functions of bacterial ribosomes were the subject of intense studies conducted by Roberts' group; the 1959 publication of McQuillen, Roberts, and Britten[16] finally established that proteins are synthesized on ribosomes and then distributed throughout the bacterial cell.

4-2 Localization of ribosomes in the cell

Ribosomes are abundant in cells involved in intense protein synthesis. In the bacterial cell they are dispersed throughout the cytoplasm and account for about 30%, sometimes even more, of its dry weight; about 10^4 ribosomes, on average, are present in one bacterial cell.

The relative content (concentration) of ribosomes in eucaryotic cells is lower; here, the number of ribosomes varies considerably depending on the protein-synthesizing activity of the corresponding tissue or individual cell. Most of the ribosomes are found in the cytoplasm (fig. 29). In the cells with an active protein secretion and a developed network of endoplasmic reticulum, a marked proportion of cytoplasmic ribosomes are attached to the endoplasmic reticulum membrane, specifically to the surface facing the cytoplasmic matrix. The ribosomes are distributed nonuniformly on the reticulum: they may be abundant in one part and virtually nonexistent in others. These ribosomes synthesize proteins which are directly transported into the membrane lumen for subsequent secretion. Protein synthesis for "house-keeping" purposes inside the cell takes place primarily upon the free cytoplasmic ribosomes that are not associated with the membrane but are scattered in the cytoplasmic matrix. That is why the cytoplasm of embryonic, nondifferentiated, rapidly growing or proliferating cells contains mainly free ribosomes.

The formation of all ribosomes present in the cytoplasmic matrix, both membrane-bound and free ones, takes place in the nucleolus of the eucaryotic cell, and ribosomes can naturally also be detected in this compartment of the cell nucleus; it is thought, however, that nucleolar ribosomes are not active in protein synthesis.

In addition, the eucaryotic cell contains different populations of ribosomes in such intracellular organelles as mitochondria and, in the case of plant cells, chloroplasts. Ribosomes of these organelles differ from cytoplasmic ribosomes in that they are slightly smaller and have a different chemical composition and different functional characteristics. These ribosomes are formed directly in the organelles (see below).

4-3 Procaryotic and eucaryotic ribosomes

Two main types of ribosomes can be found in nature (fig. 30). All procaryotic organisms, including gram-positive and gram-negative

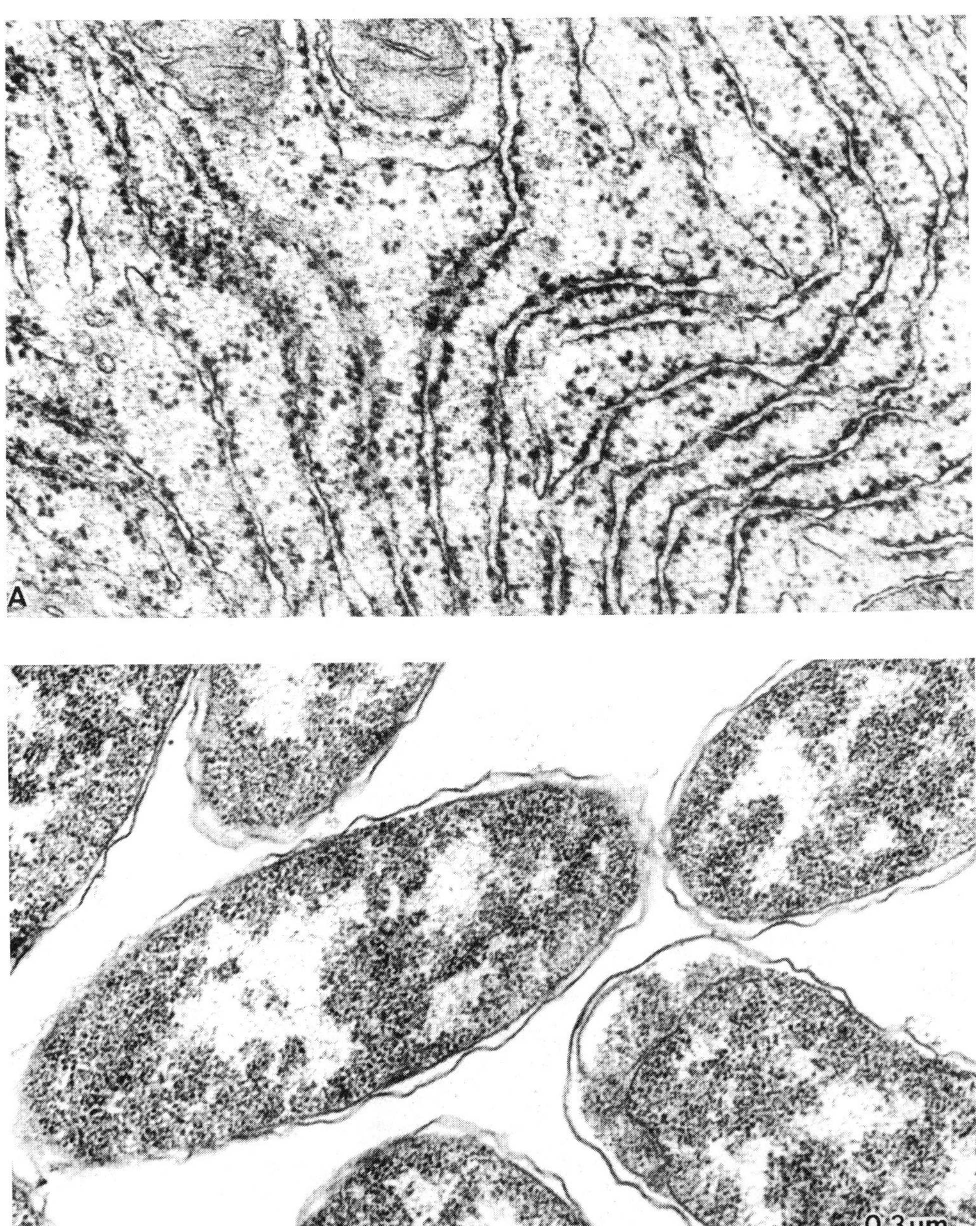

Figure 29 Electron micrographs of ribosomes on ultrathin sections of animal tissue (*top*) and a bacterium (*bottom*). *Top*, Region of rat liver cell cytoplasm. Ribosomes on the membranes of rough endoplasmic reticulum, as well as some clusters of free ribosomes, are seen. Fixed with glutaraldehyde. (Courtesy of Professor Yuri Chentsov, Moscow State University). *Bottom*, Cells of the sea bacterium *Vibrio alginolyticus*. The cytoplasm is full of ribosomes. Fixed with osmium tetraoxide. (Courtesy of Dr. L. Ye. Bakeyeva, Moscow State University)

	PROKARYOTIC TYPE RIBOSOMES:	EUKARYOTIC TYPE RIBOSOMES:
	70S RIBOSOMES OF EUBACTERIA, BLUE-GREEN ALGAE AND CHLOROPLASTS	CYTOPLASMIC 80S RIBOSOMES OF ANIMALS, FUNGI AND PLANTS
MOL. MASS, DALTONS	2.5×10^6	4×10^6
SIZE	200 - 250 Å	250 - 300 Å
RNA : PROTEIN, W / W	2 : 1	1 : 1
	70S RIBOSOMES OF ARCHAEBACTERIA (METABACTERIA)	
	MITOCHONDRIAL 75S RIBOSOMES OF FUNGI	
	MITOCHONDRIAL 55S RIBOSOMES ("MINIRIBOSOMES") OF MAMMALS	

Figure 30 Procaryotic and eucaryotic types of ribosomes.

eubacteria, actinomycetes, and blue-green algae, as well as archebacteria (metabacteria), contain 70S ribosomes. These ribosomes exhibit a sedimentation coefficient of about 70S; their molecular mass is approximately 2.5×10^6 daltons, and their linear dimensions (mean diameter) in a lyophilized state about 200 to 250 Å; in chemical composition they are pure ribonucleoproteins, i.e. they consist of only RNA and protein. The RNA-to-protein weight ratio in them is about 2:1; correspondingly, the partial specific volume of 70S ribosomes is about 0.60 cm^3/g, and buoyant density in CsCl is 1.64 g/cm^3. RNA is present in the ribosomes mainly as a Mg^{2+}, and perhaps partially as a Ca^{2+}, salt; magnesium may account for up to 2% of the ribosomes' dry weight. Furthermore, ribosomes may contain various amounts (up to 2.5% of the dry weight) of such organic cations as spermine, spermidine, cadaverine, and putrescine. The amount of water bound in 70S ribosomes is not high, being about 1 g/g; in other words, ribosomes are rather compact unswollen particles in an aqueous medium.

The morphology of 70S ribosomes of procaryotic organisms is almost universal, and only ribosomes of archebacteria (metabacteria) have been shown to possess some differences from their eubacterial counterparts (see chapter 6).

The cytoplasm of all eucaryotic organisms including animals, fungi, plants, and protozoans contains the somewhat larger 80S ribosomes. The molecular mass of these ribosomes is about 4×10^6 daltons, and the linear dimensions (mean diameter) are about 250 to 300 Å. Like procaryotic ribosomes they contain only two types of biopolymers—RNA and protein—but the protein content is markedly greater; the RNA-to-protein ratio in 80S ribosomes is about 1:1 by weight, the partial specific volume is about 0.65 cm^3/g, and the buoyant density in CsCl is about 1.55 to 1.59 g/cm^3. It is important to point out that the absolute content of both RNA and protein per particle in 80S ribosomes is markedly greater than in 70S ribosomes. The ribosomal RNA of 80S ribosomes is also bound with divalent cations, Mg^{2+} and Ca^{2+}, as well as with small amounts of polyamines and diamines, e.g. spermine, spermidine, and putrescine.

Again, it should be mentioned that the morphological characteristics of all 80S ribosomes regardless of whether they have been obtained from animals, plants, or lower eucaryotes are universal.

The chloroplasts and mitochondria of eucaryotic cells, however, contain ribosomes that differ from the 80S type. The chloroplast ribosomes of higher plants belong to the true 70S type and are difficult to distinguish from the ribosomes of eubacteria and blue-green algae by the above characteristics or by more subtle molecular features. Mitochondrial ribosomes are more diverse; their properties depend on the taxonomic position of the organism from which they originate. Mitochondrial ribosomes of fungi and mammals have been studied in some detail. Mitochondrial ribosomes from fungi (*Saccharomyces* or *Neurospora*) resemble procaryotic 70S ribosomes but are slightly larger (about 75S) and contain relatively more protein; the absolute content of ribosomal RNA seems almost identical to that found for typical 70S ribosomes. Mitochondrial ribosomes of mammals, however, are, significantly lighter than typical 70S ribosomes. The absolute content of ribosomal RNA per particle is also significantly lower. That is why they are sometimes called "miniribosomes." The sedimentation coefficient of miniribosomes from mammalian mitochondria is only about 55S, and the total mass of ribosomal RNA per particle is about two-thirds of that in typical 70S ribosomes. At the same time, mammalian mitochondrial ribosomes contain a high proportion of protein; so their total size does not seem to differ greatly from that of procaryotic ribosomes. On the whole, despite some unusual features mammalian mitochondrial ribosomes are similar to procaryotic 70S ribosomes in some of their characteristics, including functional ones.

4-4 Sequential readout of mRNA by ribosomes: Polyribosomes

Throughout the course of protein synthesis, the ribosome is associated with a limited section of the template polyribonucleotide. Since the ribosome-bound sections of the template are protected from nuclease action, they may be isolated after nuclease treatment of the ribosome-template complexes. Such sections have been found to have a length of 30 to 60 nucleotide residues. It must be noted again that the length of the mRNA coding sequence usually exceeds 300 nucleotides. Therefore, it has been clear for some time that in order to read the entire mRNA coding sequence, the ribosome should *sequentially run over* the template (or thread through itself) from the 5′-terminal part of the coding sequence to the 3′-terminal part. In other words, the ribosome should work as a tape-driving mechanism.

At what rate, then, would the ribosome move along the mRNA? In a bacterial cell (e.g. *E. coli*) a polypeptide with a length of 300 amino acids is synthesized for about 20 seconds at 37°C, i.e. one ribosome runs over about 40 to 50 nucleotides per second. The rate of mRNA readout in eucaryotes can approach 30 nucleotides per second, but regulatory effects may reduce it to 5 to 10 nucleotides per second (see chapter 14).

While moving along the template polynucleotide from the 5′-end to the 3′-end, the ribosome, after some time, moves away from the 5′-terminal section of the template. As a result, this section becomes exposed and is capable of binding with another free ribosome. The second ribosome will start the readout, and moving away from the 5′-terminus will give the third ribosome an opportunity to bind and start reading, etc. In this way, moving along the template one after another, a number of ribosomes simultaneously perform a readout of the same information and, hence, synthesize identical polypeptide chains (of course, at any given moment the chains on different ribosomes are at different stages of completion). This process is schematically presented in fig. 31, where the ribosomes at the 3′-end of the template contain an almost completed polypeptide, the ribosomes located in the middle of the mRNA carry the polypeptide only

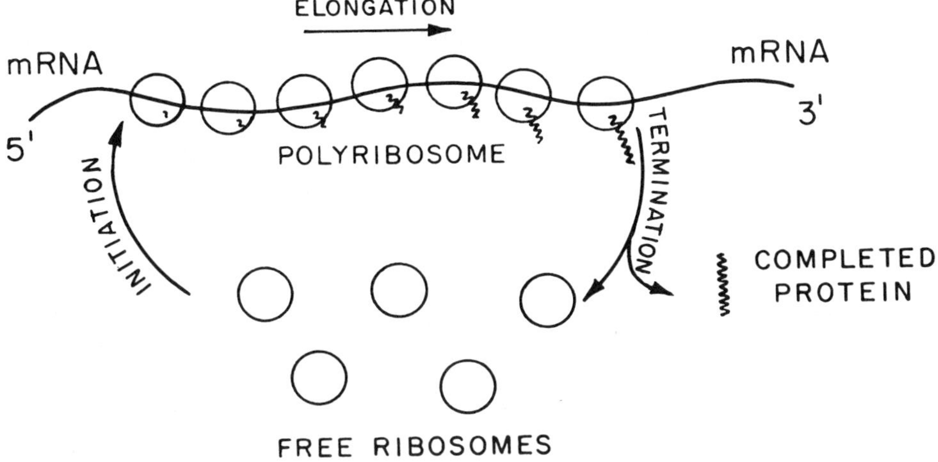

Figure 31 Schematic representation of a polyribosome.

half the length of the complete one, and the ribosomes near the 5′-end contain only short peptides which have just started elongation. A structure in which the template polynucleotide is associated with many translating ribosomes is called the *polyribosome*.

The early electron microscopic observations in the mid-1950s[7,13] demonstrated that ribonucleoprotein granules (ribosomes) are not dispersed uniformly in animal cell cytoplasm or in the preparations of microsome-derived particles but are clustered in groups. Evidence that such aggregates of ribosomes consist of particles that are connected by the mRNA chain and are engaged in translation was provided simultaneously by several groups in 1963[17–21].

Under conditions of intensive protein synthesis, the distance between ribosomes along the mRNA chain within the polyribosome may be extremely short, and so the ribosomes may be packed together very tightly. This means that there may be roughly 50 nucleotide residues of the template per ribosome in the polyribosome. The implication here is that every 1 to 3 seconds, a ribosome finishes synthesizing the protein molecule near the 3′-end of the mRNA coding section and then jumps off the template; correspondingly, one new ribosome will become associated with the template at its 5′-end and will start moving toward the 3′-terminus.

The existence of polyribosomes as a form of translating ribosomes in the cell explains the observation that ribosomes are abundant in the cell while the amount of mRNA is low. Indeed, ribosomal RNA accounts for about 80% of the total cellular RNA, whereas the mRNA content does not, as a rule, exceed 5%. This is easily understandable if one takes into account that the translation machinery of the cell is organized on the basis of polyribosomes: one mRNA is translated by many ribosomes, and one part of the translatable mRNA corresponds to 100 to 200 parts of ribosomal RNA by weight.

4-5 Stages of translation: initiation, elongation, and termination

A ribosome begins to read mRNA from a strictly definite point of its sequence, i.e. from the beginning of its coding region. It has already been noted that this point generally does not coincide with the 5′-terminal mRNA nucleotide and as a rule is located at a certain, sometimes significant, distance from the 5′-end of the polynucleotide chain. The ribosome should in some way identify the readout origin, bind to it, and then begin translation. The series of events that provide for the beginning of translation is called *initiation*. Initiation requires a special initiation codon, initiator tRNA, and proteins, which are referred to as initiation factors.

After initiation the ribosome consecutively reads mRNA codons in the direction of its 3′-end. The mRNA readout implies concomitant synthesis of the polypeptide chain coded by the mRNA. Synthesis takes place on the ribosome by the sequential addition of amino acid residues to the nascent polypeptide chain; it is in this way that the peptide elongation is accomplished. Each new amino acid residue is added to the carboxyl terminus (C-terminus) of the peptide; in other words, the C-terminus of the peptide is the end that grows. The addition of one amino acid residue corresponds to the readout of one nucleotide triplet. This whole process involving the actual translation of mRNA coding region is termed *elongation*.

When a ribosome reaches the mRNA termination codon, synthesis of the polypeptide stops. In the presence of the termination codon the ribosome does not bind any aminoacyl-tRNA; instead, specialized proteins called termination factors come into play. These factors induce the release of the synthesized polypeptide from the ribosome. This stage is designated as *termination*. After termination the ribosome may either jump off the mRNA or continue to slip along it without, however, translating.

When a ribosome comes across a new initiation codon either on a new mRNA chain or on the same chain downstream from the termination codon, a new initiation takes place. Thus each ribosome passes through the whole translation cycle including initiation, elongation, and termination; such an epicycle results in the readout of the whole mRNA coding sequence and synthesis of a complete polypeptide. Thereafter, a given ribosome may repeat the cycle with the same mRNA chain, another mRNA chain, or another coding sequence in the same chain (fig. 31).

4-6 Cell-free translation systems

Protein synthesis on ribosomes may be performed under artificial conditions *in vitro*. The incorporation of amino acids into proteins in cell homogenates, in extracts, and in cell-free systems containing microsomal preparations was demonstrated long time ago; perhaps the first examples were the cell-free systems from animal tissues, specifically from rat liver, described by Siekevitz and Zamecnik in 1951[22] and by Zamecnik in 1953[23] (see also the 1954 work of Zamecnik and Keller[24]). It was shown soon thereafter that the incorporation of amino acids corresponding to protein synthesis in a cell-free system proceeds on ribonucleoprotein particles or ribosomes[14]. Cell-free protein-synthesizing systems with bacterial (*E. coli*) ribosomes were developed almost simultaneously by Schachtschabel and Zillig[25], Lamborg and Zamecnik[26], and Tissières, Schlessinger, and Gros[27] during 1959 and 1960. All of these systems were programmed by the endogenous template; in these systems ribosomes simply continued to synthesize polypeptides upon mRNA molecules to which they were attached at the time of cell disruption. In 1961 Matthaei and Nirenberg improved the system, separated ribosomes from endogenous templates, and introduced the exogenous template[28,29]. One of their main achievements was the use of synthetic polynucleotide templates prepared by polynucleotide phosphorylase, including the simple templates of the poly(U) and poly(A) types.

Today, cell-free protein-synthesizing systems may be reconstituted from well-characterized, highly purified components, including ribosomes, template polynucleotides, and a set of aminoacyl-tRNAs or a system of tRNA acylation, i.e. tRNA, amino acids, ATP, and aminoacyl-tRNA synthetases. In addition, the system should be supplied with a set of special proteins called elongation factors, as well as with GTP. The simplest cell-free ribosomal system of polypeptide synthesis, which can be used to study the fundamental mechanisms of translation, includes only six high-molecular-mass components plus GTP; for example, the poly(U)-directed system may be reconstituted from the following ingredients:

E. coli 70S ribosomes,
Poly(U),
Phe-tRNA,
EF-T_u (protein with a molecular mass of 47,000 daltons),
EF-T_s (protein with a molecular mass of 34,000 daltons),
EF-G (protein with a molecular mass of 83,000 daltons),
GTP.

For translating natural cellular mRNA and viral RNA, the procaryotic cell-free system should be supplemented by a complete set of aminoacyl-tRNA, three proteins necessary for initiating translation (IF-1, IF-2, and IF-3), and three proteins necessary for terminating translation (RF-1, RF-2, and RF-3).

When eucaryotic 80S ribosomes are used for translation, all corresponding protein factors should be of eucaryotic origin. These include the elongation factors, namely EF-1 which consists of three subunits and is equivalent to bacterial EF-T_u plus EF-T_s, and EF-2 equivalent to bacterial EF-G, numerous initiation factors (eIF-1, eIF-2, eIF-3, eIF-4A, eIF-4B, eIF-4C, eIF-5, etc.), and one high-molecular-mass termination factor (eRF). In addition, initiation in eucaryotic systems requires ATP.

The ionic strength and specifically the Mg^{2+} concentration are important factors for the cell-free system. The usual range of Mg^{2+} concentrations, within which ribosomes are active in the cell-free system, extends from 3 to 20 mM; the optimum is somewhere between these values and depends on the ribosome origin and monovalent cation (K^+ or NH_4^+) concentration as well as on the concentration of di- and polyamines; it also depends on the incubation temperature.

The most important information regarding translation and its molecular mechanisms has been, and continues to be, obtained with the aid of cell-free systems.

References

1. A. Claude (1940), "Particulate components of normal and tumor cells," *Science* 91:77–78.

2. T. Caspersson and J. Schultz (1939), "Pentose nucleotides in the cytoplasm of growing tissues," *Nature* 143:602–603.

3. T. Caspersson (1941), "Studien über den Eiweissumsatz der Zelle," *Naturwissenschaften* 29:33–43.

4. J. Brachet (1941), "La détection histochimique et le microdosage des acides pentosenucléiques," *Enzymologia* 10:87–96.

5. J. Brachet (1942), "La localisation des acides pentosenucléiques dans les tissues animaux et les oeufs d'amphibiens en voie de dévelopement," *Arch. Biol.* 53:207–257.

6. H. K. Schachman, A. B. Pardee, and R. Y. Stanier (1952), "Studies on the macromolecular organization of microbial cells," *Arch. Biochem. Biophys.* 38:245–260.

7. G. E. Palade (1955), "A small particulate component of the cytoplasm," *J. Biophys. Biochem. Cytol.* 1:59–68.

8. G. E. Palade and P. Siekevitz (1956), "Liver microsomes," *J. Biophys. Biochem. Cytol.* 2:171–200.

9. F.-C. Chao and H. K. Schachman (1956), "The isolation and characterization of a macromolecular ribonucleoprotein from yeast," *Arch. Biochem. Biophys.* 61:220–230.

10. P. O. P. Ts'o, J. Bonner, and J. Vinograd (1956), "Microsomal nucleoprotein particles from pea seedlings," *J. Biophys. Biochem. Cytol.* 2:451–465.

11. M. L. Petermann and M. G. Hamilton (1957), "The purification and properties of cytoplasmic ribonucleoprotein from rat liver," *J. Biol. Chem.* 224:725–736.

12. A. Tissières and J. D. Watson (1958), "Ribonucleoprotein particles from *E. coli*," *Nature* 182:778–780.

13. R. B. Roberts, ed. (1958), *Microsomal particles and protein synthesis*, First Symposium of the Biophysical Society, at The Massachusetts Institute of Technology, Cambridge, Mass., February 5, 6, and 8, 1958 (London, New York, Paris, and Los Angeles: Pergamon Press).

14. J. W. Littlefield, E. B. Keller, J. Gross, and P. C. Zamecnik (1955), "Studies on cytoplasmic ribonucleoprotein particles from the liver of the rat," *J. Biol. Chem.* 217:111–123.

15. J. W. Littlefield and E. B. Keller (1957), "Incorporation of C^{14} amino acids into ribonucleoprotein particles from the Ehrlich mouse ascites tumor," *J. Biol. Chem.* 224:13–30.

16. K. McQuillen, R. B. Roberts, and R. J. Britten (1959), "Synthesis of nascent protein by ribosomes in *E. coli*," *Proc. Nat. Acad. Sci. U.S.A.* 45:1437–1447.

17. A. Gierer (1963), "Function of aggregated reticulocyte ribosomes in protein synthesis," *J. Mol. Biol.* 6:148–157.

18. J. R. Warner, P. M. Knopf, and A. Rich (1963), "A multiple ribosomal structure in protein synthesis," *Proc. Nat. Acad. Sci. U.S.A.* 49:122–129.

19. F. O. Wettstein, T. Staehelin, and H. Noll (1963), "Ribosomal aggregate

engaged in protein synthesis: Characterization of the ergosome," *Nature* 197:430–435.

20. S. Penman, K. Scherrer, Y. Becker, and J. Darnell (1963), "Polyribosomes in normal and poliovirus-infected HeLa cells and their relationship to messenger-RNA," *Proc. Nat. Acad. Sci. U.S.A.* 49:654–662.
21. J. D. Watson (1963), "The involvement of RNA in the synthesis of proteins," *Science* 140:17–26.
22. P. Siekevitz and P. C. Zamecnik (1951), "In vitro incorporation of 1-C^{14}-DL-alanine into proteins of rat-liver granular fractions," *Fed. Proc.* 10:246.
23. P. C. Zamecnik (1953), "Incorporation of radioactivity from D,L-leucine-1-^{14}C into proteins of rat liver homogenates," *Fed. Proc.* 12:295.
24. P. C. Zamecnik and E. B. Keller (1954), "Relation between phosphate energy donors and incorporation of labeled amino acids into proteins," *J. Biol. Chem.* 209:337–354.
25. D. Schachtschabel and W. Zillig (1959), "Untersuchungen zur Biosynthese der Proteine. I. Über den Einbau ^{14}C-markierter Amonosäuren ins Protein zellfreier Nucleoproteid-Enzym-Systeme aus *E. coli* B," *Hoppe-Seyler's Z. Physiol. Chem.* 314:262–275.
26. M. Lamborg and P. C. Zamecnik (1960), "Amino acid incorporation by extracts of *E. coli*," *Biochim. Biophys. Acta* 42:206–211.
27. A. Tissières, D. Schlessinger, and F. Gros (1960), "Amino acid incorporation into proteins by *E. coli* ribosomes," *Proc. Nat. Acad. Sci. U.S.A.* 46:1450–1463.
28. H. Matthaei and M. W. Nirenberg (1961), "The dependence of cell-free protein synthesis in *E. coli* upon RNA prepared from ribosomes," *Biochem. Biophys. Res. Commun.* 4:184–189.
29. M. W. Nirenberg and J. H. Matthaei (1961), "The dependence of cell-free protein synthesis in *E. coli* upon naturally occurring or synthetic polynucleotides," *Proc. Nat. Acad. Sci. U.S.A.* 47:1588–1602.

Further reading

Bielka, H., ed. (1982). *The eukaryotic ribosome*. Berlin: Akademie-Verlag.

Lengyel, P. (1974). The process of translation: A bird's-eye view. In *Ribosomes* (M. Nomura, A. Tissières, and P. Lengyel, eds.), pp. 13–52. Cold Spring Harbor, N.Y.: Cold Spring Harbor Laboratory.

Petermann, M. L. (1964). *The physical and chemical properties of ribosomes*. Amsterdam, London, and New York: Elsevier.

Spirin, A. S., and Gavrilova, L. P. (1969). *The ribosome*. Berlin, Heidelberg, and New York: Springer-Verlag.

Tissières, A. (1974). Ribosome research: Historical background. In *Ribosomes* (M. Nomura, A. Tissières, and P. Lengyel, eds.), pp. 3–12. Cold Spring Harbor, N.Y.: Cold Spring Harbor Laboratory.

Chapter 5

Chemical Reactions and the Overall Energy Balance of Protein Biosynthesis

As has been demonstrated, proteins are synthesized from amino acids in *three consecutive chemical reactions*. The first two reactions are catalyzed by aminoacyl-tRNA synthetases, and the third and final one is accomplished by the ribosome:

(1) $\mathrm{Aa} + \mathrm{ATP} \xrightarrow{\mathrm{ARSase}} \mathrm{Aa\text{-}AMP} + \mathrm{PP_i}$;

(2) $\mathrm{Aa\text{-}AMP} + \mathrm{tRNA} \xrightarrow{\mathrm{ARSase}} \mathrm{Aa\text{-}tRNA} + \mathrm{AMP}$;

(3) $\mathrm{Aa\text{-}tRNA} + \mathrm{X\text{-}Aa'\text{-}tRNA'} \xrightarrow{\mathrm{RS}} \mathrm{X\text{-}Aa'\text{-}Aa\text{-}tRNA} + \mathrm{tRNA'}$.

In the first reaction the amino acid carboxyl group reacts with the polyphosphate group of ATP, resulting in the replacement of a pyrophosphate residue by the aminoacyl residue; a mixed *anhydride*, the aminoacyl adenylate, is formed. In the second reaction the adenylate residue is exchanged for tRNA, an *ester* bond being formed between the carboxyl group of the aminoacyl residue and the ribose hydroxyl of the tRNA terminal nucleoside. The third reaction catalyzed by the ribosome is the substitution of tRNA residue (tRNA′) by the aminoacyl-tRNA; this results in the formation of an *amide* (*peptide*) bond between the amino group of the aminoacyl-tRNA and the carboxyl group of the other aminoacyl residue (Aa′). If a complete protein molecule consists of n aminoacyl residues, the overall balance of the reactions may be written as follows:

$$n\ \text{Aa} + n\ \text{ATP} \xrightarrow{\text{ARSases, tRNAs, RS}} \text{protein} + n\ \text{AMP} + n\ \text{PP}_i.$$

Furthermore, pyrophosphate is hydrolyzed in the cell by pyrophosphatase to orthophosphate:

$$n\ \text{PP}_i + n\ \text{H}_2\text{O} \rightarrow 2n\ \text{P}_i.$$

The free energy of the hydrolysis of ATP pyrophosphate bonds under standard conditions ($\Delta G^{0\prime}$) is about −7 to −8 kcal/mole. The anhydride bond of the aminoacyl adenylate and the ester bond of the aminoacyl-tRNA possess similar values of free energy of the hydrolysis under standard conditions. The free energy of the hydrolysis of the peptide bond in an infinitely long polypeptide (protein) under standard conditions is equal to only −0.5 kcal/mole. It can therefore be seen that the whole process of protein synthesis involves releasing a considerable amount of free energy; in other words, protein synthesis is a thermodynamically spontaneous and energetically ensured process:

$$n\ \text{Aa} + n\ \text{ATP} \rightarrow \text{protein} + n\ \text{AMP} + n\ \text{PP}_i - n \times (7 \pm 0.5)\ \text{kcal}.$$

If pyrophosphate hydrolysis is added, the overall energy balance will be $-n \times 15$ kcal per mole of protein. Thus, the free-energy gain under standard conditions for a protein of about 200 aminoacyl residues will be roughly 3000 kcal/mole.

An analysis of the energy balance of each of the three reactions shows that the first two reactions do not by themselves achieve any gain in free energy (under standard conditions), and therefore the preribosomal stages should not be shifted markedly toward the synthetic side; the shift, however, will be generated provided pyrophosphate is hydrolyzed in a parallel reaction. The main difference in the free-energy levels between substrates and products is found in the third reaction. This implies that the shift of the overall reaction toward synthesis is provided mainly by the ribosomal stage.

It is surprising that despite the full energy support of protein biosynthesis at the expense of ATP (or the ester bond energy of aminoacyl-tRNA), the ribosomal stage still requires two GTP molecules per amino acid residue:

$$2n\ \text{GTP} + 2n\ \text{H}_2\text{O} \rightarrow 2n\ \text{GDP} + 2n\ \text{P}_i.$$

This gives an additional free-energy gain of about 15 kcal per mole of amino acid (under standard conditions).

Thus, the sum total of all the chemical reactions in protein synthesis may be written as follows:

$$n\ \mathrm{Aa} + n\ \mathrm{ATP} + 2n\ \mathrm{GTP} + 3n\ \mathrm{H_2O} \xrightarrow{\text{ARSases, PPase, tRNAs, RS}} \text{protein} + n\ \mathrm{AMP} + 2n\ \mathrm{GDP} + 4n\ \mathrm{P_i}.$$

The total energy balance of the overall reaction $\Delta G^{0\prime}$ is equal to about -30 kcal per mole of amino acid or -6000 kcal per mole of protein with a length of 200 amino acid residues.

Here, only the chemical aspect of the process has been taken into account. It is important to analyze to what extent this estimate may be changed if we take into account entropy loss due to the ordered arrangement of the amino acid residues along the chain of synthesized protein, and due to the fixed three-dimensional protein structure. It happens that the entropy loss due to amino acid ordering in the polypeptide chain may introduce only a small correction, at most 2.5 kcal per mole of amino acid (see the appendix to this chapter, by Dr. Shakhnovich). As regards the three-dimensional ordering of the chain in the protein molecule, the entropy loss (decrease) is significant here, but it is compensated by the enthalpy gain resulting from noncovalent interactions of amino acid residues. Thus, in any case, the protein synthesis is accompanied by dissipation of a large amount of free energy.

The meaning of the expenditure of such a tremendous excess of energy is an enigma and an extremely interesting problem in molecular biology. Energy excess which is dissipated into heat and not used for any accumulated useful work (in the form of chemical bonds or nonrandom arrangement of residues) should play an important part in the functioning of the protein-synthesizing system. It is likely that this energy excess is necessary to support the high rates and high fidelity of protein synthesis.

Appendix to chapter 5*

Let us consider the ribosome as a polymerizing system that forms a polypeptide chain with a definite sequence of amino acids. If the

* This portion of chapter 5 was written by Dr. E. Shakhnovich.

length of the chain to be synthesized is N, it follows that 20^N sequences can be assembled from 20 species of amino acids. In reality, only one sequence determined by mRNA is synthesized. Therefore, the fraction $w = 20^{-N}$ is selected from all possible variants; the entropy of such amino acid ordering into a chain, calculated per mole of chains, is equal to:

$$S_0 = R \ln w = -RN \ln 20,$$

where R is gas constant. The corresponding free energy per mole of residues at temperature T is as follows:

$$f_0 = -\frac{TS_0}{N} = RT \ln 20 \approx 2.5 \text{ kcal/mole}.$$

Let us now assume that translation proceeds with errors and consider to what extent this lowers the free energy of amino acid ordering into the chain. Let P possible amino acids be incorporated into K arbitrary sites with equal probability. Now, $P^K C_N^K$ types of sequences will be synthesized instead of the unique sequence. Here, the first term reflects the possibility of incorporating each of the P amino acids at each of the K "erroneous" sites. The second term, $C_N^K = N!/(N—K)!K!$, corresponds to the number of ways of selecting K positions from the total number of N positions. Therefore, the number of synthesized sequences expressed as a fraction of all possible sequences will be given by the following expression:

$$w' = \frac{P^K C_N^K}{20^N}.$$

The corresponding entropy per mole of chains will be

$$S' = R \ln w' = RK \ln P - R \ln C_N^K - RN \ln 20.$$

Using Stirling's formula $\ln N! = N \ln N/e$, we may write

$$S' = RK \ln N + RK \ln P - RN \ln 20.$$

The free energy calculated per mole of residues will be

$$f' = -\frac{RTS'}{N} = RT \ln 20 - \frac{RKT}{N} \ln N - \frac{RKT \ln P}{N};$$

$$f' - f_0 = -RKT \frac{\ln N}{N} - RKT \frac{\ln P}{N}.$$

If one assumes that $N \sim 10^3$, $K \sim 10$, and $P \sim 10$, then $f' - f_0 \simeq -4 \times 10^{-2}$ kcal/mole $\simeq -40$ cal/mole. Thus, miscoding will contribute little to the free energy change during translation.

Further reading

Spirin, A. S. (1978). Energetics of the ribosome. Progress in nucleic acid research and molecular biology (W. E. Cohn, ed.), vol. 21, pp. 39–62. New York: Academic Press.

Chetverin, A. B., and Spirin, A. S. (1982). Bioenergetics and protein synthesis. *Biochim. Biophys. Acta* 683: 153–179.

PART II

STRUCTURE OF THE RIBOSOME

Chapter 6

Morphology of the Ribosome

6-1 Size, appearance, and subdivision

When examined by electron microscopy the isolated ribosomes look like compact rounded particles with a linear size of about 250 Å (fig. 32), somewhat larger (up to 300 Å) in the case of eucaryotic ribosomes[1–5]. Ribosomes from different organisms and cells, whether procaryotic or eucaryotic ones, have a strikingly similar appearance[6,7].

A characteristic feature of one of the visible ribosomal projections is a groove dividing the ribosome into two unequal parts (fig. 33). This subdivision reflects the fact that the ribosome consists of two separable subparticles, or *ribosomal subunits*. Under certain conditions, e.g. if the concentration of magnesium ions in the medium is sufficiently low, the ribosome dissociates into two subunits with a mass ratio of about 2:1 (see chapter 10, section 10.1). The procaryotic 70S ribosome disociates into subunits with the sedimentation coefficients 50S (molecular mass 1.65×10^6 daltons) and 30S (molecular mass 0.85×10^6 daltons). The eucaryotic 80S ribosome dissociates into 60S and 40S subunits. Ribosomal subunits may be separated easily in the ultracentrifuge, and then studied individually.

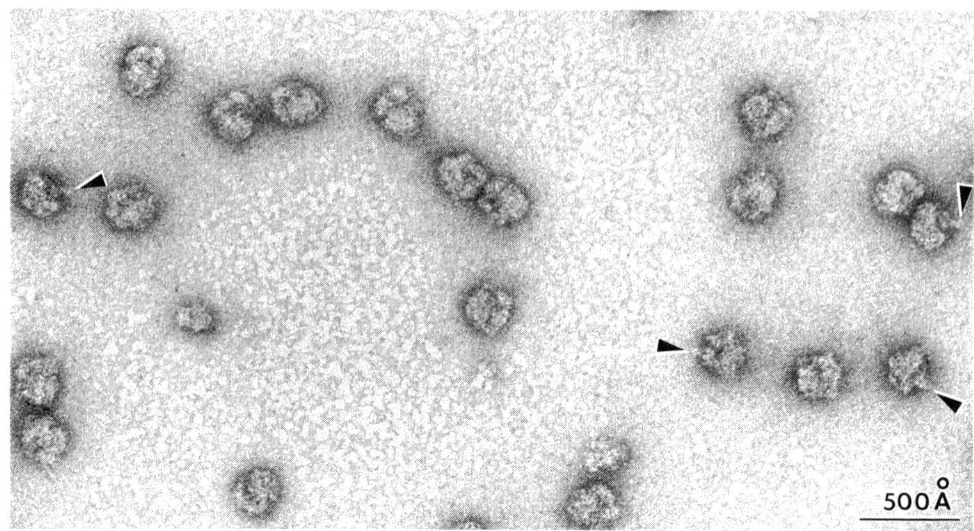

Figure 32 Electron micrograph of the 70S ribosomes isolated from *Escherichia coli*. To achieve the contrast necessary for the particles to be seen in the electron microscope, the isolated 70S ribosomes are applied on an ultrathin carbon film; the film with attached particles is treated by uranyl acetate solution and dried in air. The particles become embedded in uranyl acetate, which fills cavities and grooves. The ribosomal particles having lower electron density than uranyl acetate appear *negatively stained* against the background of uranyl acetate. The arrows indicate the L7/L12 stalk described in the text. (Courtesy of Dr. V. D. Vasiliev)

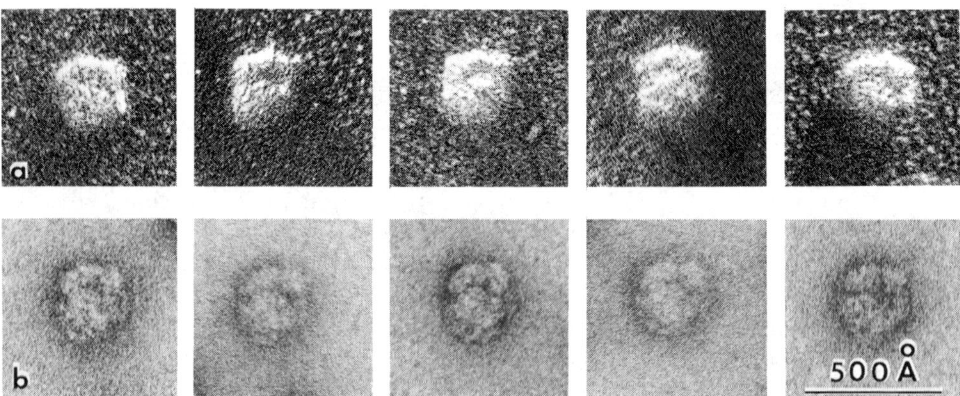

Figure 33 Electron micrographs of individual *Escherichia coli* 70S ribosomes illustrating their subdivision into two unequal subunits; the images are oriented such that the small subunit is at the top and the large subunit is at the bottom. (a) Ribosomes contrasted by metal shadowing[3]. In this case, to achieve the necessary contrast the suspension of isolated 70S ribosomes is applied to a carbon film surface and freeze-dried; the particles are shadowed by metal (tungsten or tungsten-rhenium alloy) using vacuum evaporation at an angle of about 75° to film surface; this yields shadow-cast particles. (b) Negatively stained ribosomes, prepared as described in the legend to fig. 32. (Courtesy of Dr. V. D. Vasiliev)

6-2 Detailed shape of ribosomal subunits

Small subunit

Different electron microscopic projections of the bacterial (*Escherichia coli*) ribosomal 30S subunit are shown in fig. 34, and a photograph of the corresponding morphological model is given in fig. 35. The 30S subunit is somewhat elongated, its length being about 230 Å, and its width about 120 Å. The subunit may be subdivided into lobes which are referred to as the "head," "body," and "side bulge" (or

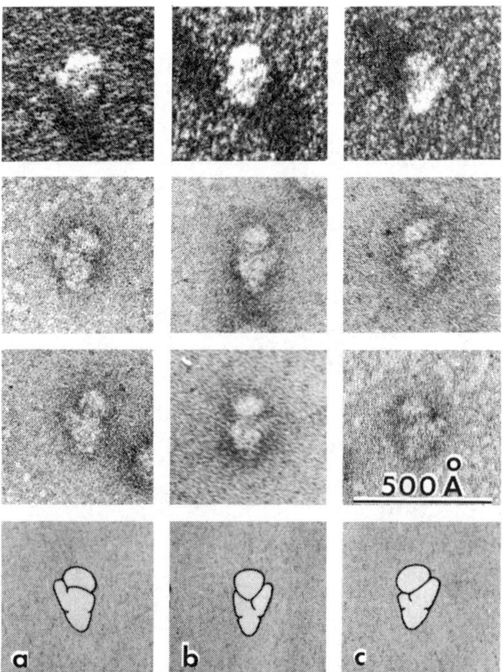

Figure 34 Electron micrographs of individual 30S ribosomal subunits of *E. coli*[8]. The upper row shows metal-shadowed particles, prepared as described in the legend to fig. 33(a). The second and third rows show uranyl acetate-stained particles, prepared as described in the legend to fig. 32. (a) Column of images of the 30S subunits and its contour in the projection when it is viewed from the side opposite to that facing the 50S subunit in the complete ribosome. (b) Column of images of the 30S subunits and its contour in the narrow side projection. (c) Column of images of the 30S subunits and its contour in the projection when it is viewed from the side facing the 50S subunit in the ribosome. (Courtesy of Dr. V. D. Vasiliev)

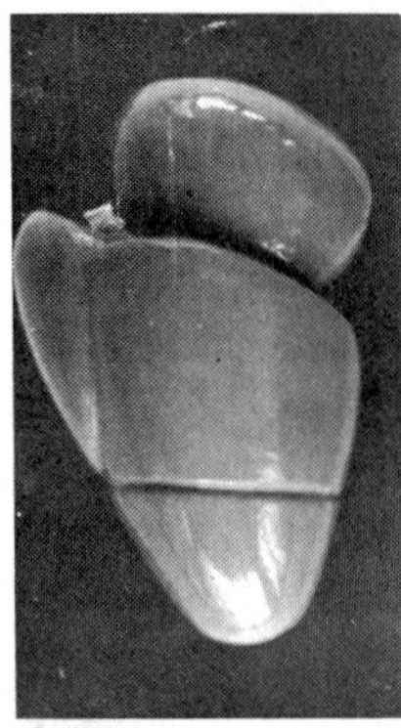
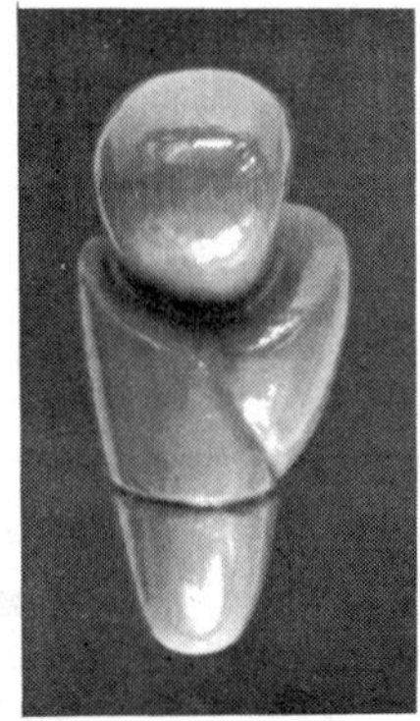
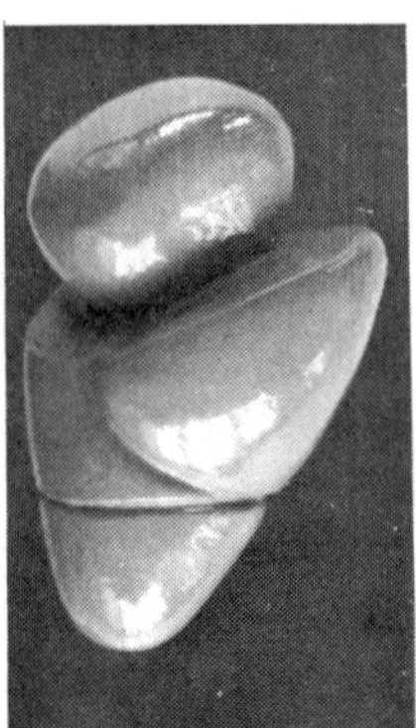

Figure 35 Model of the 30S ribosomal subunit: three different projections[8].

"platform")[8,9]. The groove separating the head from the body is quite distinct.

The eucaryotic 40S subunit has a similar morphology[4–7], although two additional details of structure may be mentioned. The first is a protuberance, or "bill," located on the head on the side opposite from the bulge. Second, the end of the body distal to the head appears to be bifurcated due to the presence of some additional mass; this bifurcation is referred to as the "eucaryotic lobes" (fig. 36).

More reliable information about ribosomal subunits can be derived from electron micrographs if averaged images rather than individual pictures are examined[10,11]. Averaging allows the statistical noise on photographs to be eliminated. This contributes toward a better visualization of the common features in the images of a given particle type. For such an averaging, a number (the more the better) of particle images in the same projection are taken and regularly arranged in strictly identical orientation; the averaging is performed by filtration in optical diffractometer. Alternatively, digital averaging can be done which gives more possibilities; quantitative estimation of similarity of the selected images, their precise alignment, reduction of high-frequency noise and the averaging are performed using computer technique. As a result, all nonreproducible details of images are removed (or "filtered off"), leaving the common elements remain on the averaged image. An example of such an averaging for negatively stained 40S subunits from rat liver is given in fig. 36(b). All three projections show that the head is separated from the remainder of the subunit by a distinct deep groove. The bill present in the head can be seen clearly in two of the projections. In addition, these projections allow the side bulge partially superimposed upon the subunit body to be distinctly seen. The bifurcated tail, or eucaryotic lobes, of the 40S subunit are also prominent.

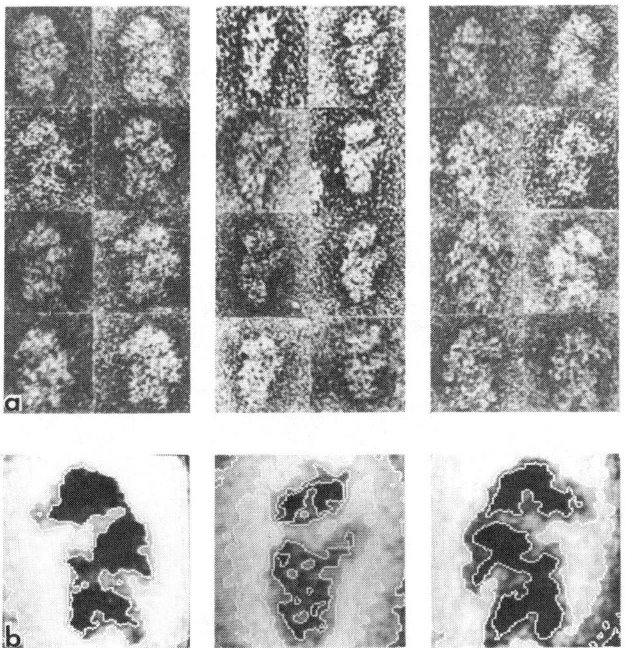

Figure 36 Electron micrographs of the individual 40S subunits of rat liver ribosomes[11]. (a) Uranyl acetate-stained subunits prepared as described in the legend to fig. 32. (b) Averaged images of the subunit stained by uranyl acetate (see text): three different projections. (Courtesy of Dr. N. A. Kiselev, Institute of Crystallography, Moscow)

It should be noted that the small ribosomal 30S subunit of archebacteria (metabacteria) has a morphology which is intermediate between that of the eubacterial 30S subunit and the eucaryotic 40S subunit: the archebacterial subunit has a characteristic bill on the head but does not possess any eucaryotic lobes at the end of the body[7].

Large subunit

The shape of the large subunit is virtually identical in procaryotic and eucaryotic ribosomes[6,9]. Different projections of the 50S subunit are given in fig. 37, and a model is shown in fig. 38[12]. This subunit is more isometric than the smaller one, the linear size being equal to 200 to 230 Å in all directions. Three peripheral protuberances can be distinguished: the central one may be termed the head; the lateral fingerlike protuberance is called the L7/L12 stalk; and still another lateral protuberance, located on the other side of the central proturberance, is referred to as the side lobe or L1 ridge (in the case of *E. coli*

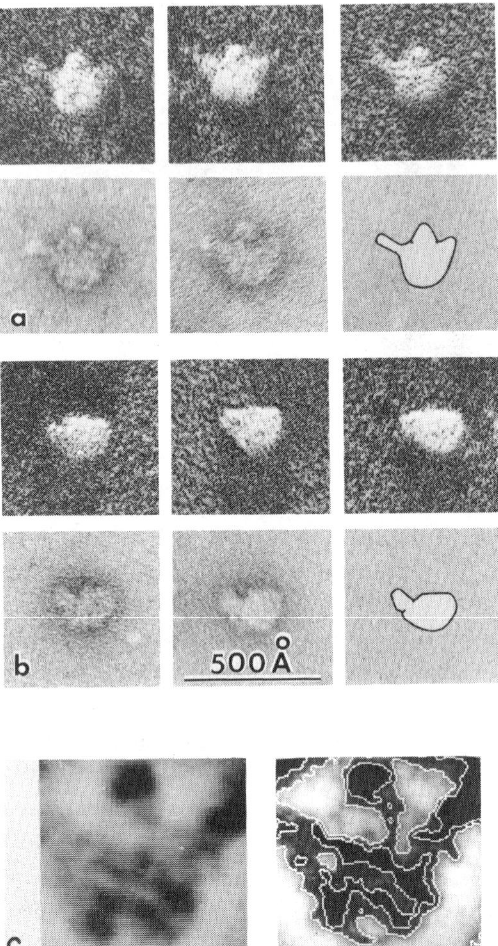

Figure 37 Electron micrographs of the individual 50S subunits of *E. coli* ribosomes[12]. The first and third rows show metal-shadowed subunits prepared as described in the legend to fig. 33(a). The second and fourth rows show subunits negatively stained by uranyl acetate prepared as described in the legend to fig. 32. The fifth row shows averaged images of uranyl acetate-stained particles (see text). (a) 50S subunit and its contour in the "back" projection viewed from the convex side turned away from the 30S subunit in the ribosome. (b) 50S subunit and its contour in the lateral projection. (c) Averaged images of the 50S subunit[13] in the frontal projection viewed from the side facing the 30S subunit in the ribosome. ((a) and (b) courtesy of Dr. V. D. Vasiliev; (c) courtesy of Dr. N. A. Kiselev, Institute of Crystallography, Moscow)

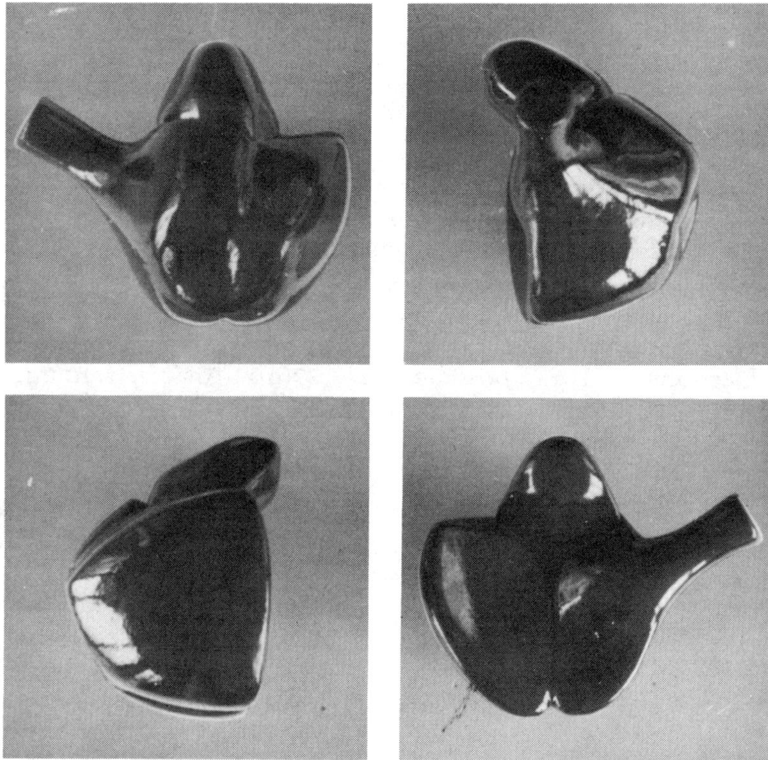

Figure 38 Model of the 50S ribosomal subunit: four different projections[12].

50S ribosomal subunit, the two lateral protuberances contain ribosomal proteins L7/L12 and L1, respectively; see chapter 9 for more details).

Averaged images of the large ribosomal subunit of *E. coli*[13] (fig. 37(c)) allow all details traceable in individual particle images to be clearly revealed. Again, the deep groove is seen to separate the head from the remainder of the subunit, the groove being deeper on the side of the L1 ridge than on the side of the L7/L12 stalk. One can see clearly that the body of the subunit is bifurcated on the side opposite the head. In addition, the averaged images demonstrate that many of the smaller details are not random features but are reproducible in numerous images; this is particularly true of the strandlike structures about 50 Å in diameter that are at the resolution threshold both in the smaller (30S or 40S) and in the larger (50S or 60S) subunits; it is thought that these structures may be the important structural motif of the ribosome[11,13].

6-3 Association of subunits into the complete ribosome

In an intact ribosome the two ribosomal subunits are joined in a very specific manner[4–6,9,14,15] The flattened side of the 50S subunit is involved in the contact between subunits; if the subunit is viewed from this surface, the head of the subunit up, the stalk is on the right (figs. 37 and 38). The subunits are associated in the "head-to-head and the side lobe-to-side lobe" manner[9,15]. The head-to-head association may be seen clearly on an electron micrograph of one of the projections of 70S ribosomes (fig. 39(b)). An electron microscopic

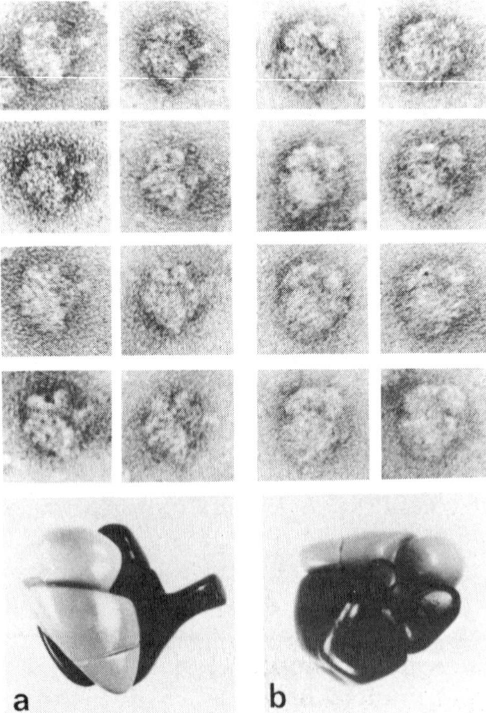

Figure 39 Electron micrographs of the individual 70S ribosomes from *E. coli* and a model of them in two different projections[15]. (a) Overlap projection, when the 30S subunit is above the 50S subunit. (b) Lateral projection viewed from the side of the L7/L12 stalk. The preparation was stained with uranyl acetate as described in the legend to fig. 32. (Courtesy of Dr. V. D. Vasiliev)

image of the "overlap" projection demonstrates that the 30S subunit covers only part of the flattened side of the 50S subunit (fig. 39(a)). The region at the base of the stalk remains exposed. This region appears to accommodate functionally important ribosomal sites (see chapter 11). A photograph of the 70S ribosome model with the coupled 30S and 50S subunits in head-to-head and side lobe-to-side lobe association is presented at the bottom of fig. 39.

References

1. C. E. Hall and H. S. Slayter (1959), "Electron microscopy of ribonucleoprotein particles from *E. coli*," *J. Mol. Biol.* 1:329–332.
2. H. E. Huxley and G. Zubay (1960), "Electron microscope observations on the structure of microsomal particles from *Escherichia coli*," *J. Mol. Biol.* 2:10–18.
3. V. D. Vasiliev (1971), "Electron microscopy study of 70S ribosomes of *Escherichia coli*," *FEBS Letters* 14:203–205.
4. Y. Nonomura, G. Blobel, and D. Sabatini (1971), "Structure of liver ribosomes studied by negative staining," *J. Mol. Biol.* 60:303–324.
5. G. Lutsch, H. Bielka, K. Wahn, and J. Stahl (1972), "Studies on the structure of animal ribosomes. III. Electron microscopic investigations of isolated rat liver ribosomes and their subunits," *Acta Biol. Med. Germ.* 29:851–876.
6. M. Boublik and W. Hellmann (1978), "Comparison of *Artemia salina* and *Escherichia coli* ribosome structure by electron microscopy," *Proc. Nat. Acad. Sci. U.S.A.* 75:2829–2833.
7. J. A. Lake, E. Henderson, M. W. Clark, and A. T. Matheson (1982), "Mapping evolution with ribosome structure: Interlineage constancy and interlineage variation," *Proc. Nat. Acad. Sci. U.S.A.* 79:5948–5952.
8. V. D. Vasiliev (1974), "Morphology of the ribosomal 30S subparticle according to electron microscopy data," *Acta Biol. Med. Germ.* 33:779–793.
9. J. A. Lake (1976), "Ribosome structure determined by electron microscopy of *Escherichia coli* small subunits, large subunits, and monomeric ribosomes," *J. Mol. Biol.* 105:131–159.
10. J. Frank, A. Verschoor, and M. Boublik (1981), "Computer averaging of electron micrographs of 30S ribosomal subunits," *Science* 214:1353–1355.
11. N. A. Kiselev, V. Ya. Stel'mashchuk, E. V. Orlova, M. Platzer, F. Noll, and H. Bielka (1982), "On the fine structure of rat liver ribosome small subunits," *Mol. Biol. Rep.* 8:185–189.

12. V. D. Vasiliev, O. M. Selivanova, and S. N. Ryazantsev (1983), "Structure of the *Escherichia coli* 50S ribosomal subunit," *J. Mol. Biol.* 171:561–569.

13. N. A. Kiselev, E. V. Orlova, V. Ya. Stel'mashchuk, V. D. Vasiliev, O. M. Selivanova, V. P. Kosykh, A. I. Pustovskikh, and V. S. Kirichuk (1983), "Computer averaging of 50S ribosomal subunit electron micrographs," *J. Mol. Biol.* 169:345–350.

14. B. Kastner, M. Stöffler–Meilicke, and G. Stöffler (1981), "Arrangement of the subunits in the ribosome of *Escherichia coli*: Demonstration by immunoelectron microscopy," *Proc. Nat. Acad. Sci. U.S.A.* 78:6652–6656.

15. V. D. Vasiliev, O. M. Selivanova, V. I. Baranov, and A. S. Spirin (1983), "Structural study of translating 70S ribosomes from *Escherichia coli*," *FEBS Letters* 155:167–172.

Chapter 7

Ribosomal RNA

7-1 Significance of ribosomal RNA

The statement that the ribosome is first and foremost its RNA is not much of an exaggeration. A primitive evolutionary ancestor of the ribosome might well have consisted only of RNA, becoming modified gradually by proteins in the course of evolution[1–3]. Two-thirds of the procaryotic ribosome consists of RNA but only a third is comprised of protein. Proteins make up 50% of the evolutionarily more advanced eucaryotic ribosome. It is the ribosomal RNA that seems to determine many structural and functional characteristics of the ribosome. The covalently continuous chains of ribosomal RNAs support the integrity of ribosomal subunits. The specific three-dimensional structure of ribosomal RNA is vital to the shape and morphological characteristics of ribosomal subunits. The association of the subunits into the complete ribosome seems to depend on the specific affinity of two high-molecular-mass ribosomal RNA species toward each other. The arrangement of all ribosomal proteins in the particle is determined by the ribosomal RNA. Finally, the ribosomal RNA plays a crucial part in organizing a number of ribosomal functional sites.

7-2 Ribosomal RNA species

Procaryotic as well as eucaryotic ribosomes contain two different high-molecular-mass ribosomal RNA species[4–6], one in each subunit, and one relatively low-molecular-mass RNA, the so-called 5S RNA[7,8].

In addition, eucaryotic ribosomes contain another relatively low-molecular-mass RNA, the so-called 5.8S RNA[8,9], which is homologous to the 5′-terminal part (about 160 nucleotide residues in length) of the high-molecular-mass RNA present in the large subunit of the procaryotic ribosome[10]. Ribosomes of higher plant chloroplasts contain the so-called 4.5S RNA[11], which is a homolog of the 3′-terminal region (about 100 nucleotides in length) of the high-molecular-mass RNA present in the large ribosomal subunit of bacteria[12]. Thus the 5.8S RNA of eucaryotic ribosomes and the 4.5S RNA of chloroplast ribosomes result from the cleavage or the "processing" of a precursor of the high-molecular-mass RNA of the large ribosomal subunit in the course of ribosome biogenesis[13–16]. These small RNA species participate in the specific folding of the high-molecular-mass RNA in the large ribosomal subunit just as their homologous sections in bacteria do (see below); therefore they should not be regarded as independent ribosomal RNA species. In subsequent presentation they will be considered together with the high-molecular-mass RNA of the large ribosomal subunit. It may well be that in other organisms other cleavage points of the high-molecular-mass RNA covalent chain in a mature ribosome may be present[17].

High-molecular-mass RNA of the small ribosomal subunit

The small (30S) subunit of a bacterial ribosome contains RNA about 1500 to 1600 nucleotide residues long; in *E. coli* the length is 1542 residues[18,19]. This RNA is called 16S ribosomal RNA. Its molecular mass is about 0.5×10^6 daltons. In an isolated state this RNA has a sedimentation coefficient ($S^0_{20,w}$) of 16S at monovalent salt concentrations of about 0.1 M, provided that magnesium ions are absent. Under these conditions, however, the RNA is far less compact than it is within the ribosome. The presence of Mg^{2+} contributes to the extensive folding of the RNA and, as a result, the degree of compactness can approach (but not attain) that of the RNA within the ribosome *in situ*[20]; the sedimentation coefficient ($S^0_{20,w}$) becomes about 22S in a solution containing 20 mM $MgCl_2$. In contrast, when the ionic strength is reduced and the temperature increased, the 16S RNA becomes less compact; its sedimentation coefficient correspondingly drops to about 2 to 5S when, in the absence of salts or at temperatures above 60°C, the molecules undergo unfolding[21].

The 30S ribosomal subunit of higher plant chloroplasts contains 16S RNA of approximately the same size; in *Zea mays* its length is equal to 1490 nucleotide residues[22]. The RNA present in the small ribosomal subunit of higher plant and yeast mitochondria is somewhat larger; in yeasts its length is equal to 1661 nucleotide residues[23]. In contrast, the "miniribosomes" present in mammalian mitochondria contain in their small subunit a relatively short RNA referred to as 12S RNA; its length is 954 and 956 nucleotide residues in man[24] and mouse[25], respectively.

In archaebacteria *Halobacterium* and *Halococcus*, the 16S ribosomal RNA of their 30S subunits has been found to be similar in its size to the eubacterial 16S RNA; its length is 1472–1475 nucleotide residues[26,27].

Eucaryotic 40S ribosomal subunits contain a considerably larger ribosomal RNA; it is called 18S RNA, sometimes 17S RNA. The length of this ribosomal RNA is about 1800 nucleotide residues; in yeasts it is 1789 nucleotides[28], in *Xenopus laevis* 1825 nucleotides[29], and in mammals (rat) 1874 nucleotides[30]; correspondingly, the molecular mass is about 0.6×10^6 daltons. The behavior of the eucaryotic 18S RNA as a function of ionic strength and Mg^{2+} concentration is similar to that of the procaryotic 16S RNA.

High-molecular-mass RNA of the large ribosomal subunit

The large (50S) subunit of the bacterial 70S ribosome contains an RNA molecule with a length of about 3000 nucleotide residues (2904 in *E. coli*)[31]; in other words, it is approximately twice as long as 16S RNA. This RNA is designated as 23S RNA. Its molecular mass is about 10^6 daltons. Its sedimentation coefficients and compactness in an isolated state are determined in a manner similar to that mentioned above for 16S RNA. At an ionic strength of about 0.1 M in the absence of Mg^{2+}, it has a sedimentation coefficient of about 23S and intermediate compactness; the presence of Mg^{2+} as well as polyamines makes it considerably more compact and gives it a sedimentation coefficient of about 31 to 34S; decreased ionic strength and increased temperature have the opposite effect.

A fundamentally similar 23S RNA is present in the large subunit of the 70S ribosomes of higher plant chloroplasts[32]; however, its 100-nucleotide-long 3′-terminal region is cleaved off and exists as a separate 4.5S fragment[11,12].

In mammalian mitochondria the RNA of the large subunit of

miniribosomes is significantly smaller than the bacterial 23S RNA[24,25].

The large (60S) subunit of eucaryotic 80S ribosomes contains a significantly longer RNA molecule compared to the bacterial 23S RNA. This eucaryotic RNA, referred to as the 26S or 28S RNA, has a molecular mass varying from 1.2–1.3 $\times$ 10^6 daltons in fungi and higher plants to about 1.6–1.7 $\times$ 10^6 daltons in birds and mammals[33]. Correspondingly, the 26S RNA of *Saccharomyces* consists of 3392–3393 nucleotide residues[34,35], and the 28S RNA chain of rat comprises 4700–4800 nucleotides[36,37]. The low-molecular-weight 5.8S RNA with a length of 160 nucleotide residues is tightly associated with the high-molecular-mass species[38,39]. It has already been mentioned that this RNA is a homolog of the 5′-terminal sequence of bacterial 23S RNA. The dissociation of 5.8S RNA from 28S RNA is achieved only as a result of unfolding induced by high termperatures or denaturing agents.

5S RNA of the large ribosomal subunit

In addition to 23S (28S) RNA, the large subunit of procaryotic and eucaryotic ribosomes contains a relatively low-molecular-mass RNA species about 120 nucleotide residues in length called the 5S RNA. 5S RNA molecules of gram-negative bacteria, including *E. coli*[40], are exactly 120 nucleotide residues in length, while in the majority of gram-positive bacteria the length of the 5S RNAs is 115 or 116 residues[41]. Eucaryotic 5S RNA as a rule consist of 120 or 121 nucleotide residues[42]. The exception is higher plant 5S RNAs, which are 2 to 4 residues shorter[41]. 5S RNA has not been detected in mitochondrial ribosomes, except in higher plants.

7-3 Primary and secondary structures

16S (18S) RNA

E. coli 16S RNA was the first high-molecular-mass ribosomal RNA to be sequenced. Sequencing was achieved by a direct chemical-enzymatic analysis of RNA nucleotide sequence in Ebel's group[18] and by DNA sequencing of the corresponding cloned gene in Noller's group[19]. The complete nucleotide sequence of *E. coli* 16S RNA is shown in fig. 40.

PAAAUUGAAGAGUUUGAUCAUGGCUCAGAUUGAACGCUGGCGGCAGGCCUAACACAUGCAAGUCGAACGGUAACAGGAAGAAGCUUGCUGCUUUGCUGACG
50 100

AGUGGCGGACGGGUGAGUAAUGUCUGGGAAACUGCCUGAUGGAGGGGGAUAACUACUGGAAACGGUAGCUAAUACCGCAUAACGUCGCAAGACCAAAGAG
150 200

GGGGACCUUCGGGCCUCUUGCCAUCGGAUGUGCCCAGAUGGGAUUAGCUAGUAGGUGGGGUAACGGCUCACCUAGGCGACGAUCCCUAGCUGGUCUGAGA
250 300

GGAUGACCAGCCACACUGGAACUGAGACACGGUCCAGACUCCUACGGGAGGCAGCAGUGGGGAAUAUUGCACAAUGGGCGCAAGCCUGAUGCAGCCAUGC
350 400

CGCGUGUAUGAAGAAGGCCUUCGGGUUGUAAAGUACUUUCAGCGGGGAGGAAGGGAGUAAAGUUAAUACCUUUGCUCAUUGACGUUACCCGCAGAAGAAG
450 500

CACCGGCUAACUCCGUGCCAGCAGCCm^{7}GCGGUAAUACGGAGGGUGCAAGCGUUAAUCGGAAUUACUGGGCGUAAAGCGCACGCAGGCGGUUUGUUAAGUCA
550 600

GAUGUGAAAUCCCCGGGCUCAACCUGGGAACUGCAUCUGAUACUGGCAAGCUUGAGUCUCGUAGAGGGGGGUAGAAUUCCAGGUGUAGCGGUGAAAUGCG
650 700

UAGAGAUCUGGAGGAAUACCGGUGGCGAAGGCGGCCCCCUGGACGAAGACUGACGCUCAGGUGCGAAAGCGUGGGGAGCAAACAGGAUUAGAUACCCUGG
750 800

UAGUCCACGCCGUAAACGAUGUCGACUUGGAGGUUGUGCCCUUGAGGCGUGGCUUCCGGAGCUAACGCGUUAAGUCGACCGCCUGGGGAGUACGGCCGCA
850 900

AGGUUAAAACUCAAAUGAAUUGACGGGGGCCCGCACAAGCGGUGGAGCAUGUGGUUUAAUUCGAUm^{2}Gm^{5}CAACGCGAAGAACCUUACCUGGUCUUGACAUCCA
950 1000

CGGAAGUUUUCAGAGAUGAGAAUGUGCCUUCGGGAACCGUGAGACAGGUGCUGCAUGGCUGUCGUCAGCUCGUGUUGUGAAAUGUUGGGUUAAGUCCCGm^{5}C
1050 1100

AACGAGCGCAACCCUUAUCCUUUGUUGCCAGCGGUCCGGCCGGGAACUCAAAGGAGACUGCCAGUGAUAAACUGGAGGAAGGUGGGGAUGACGUCAAGUC
1150 1200

AUCAUGm^{2}GCCCUUACGACCAGGGCUACACACGUGCUACAAUGGCGCAUACAAAGAGAAGCGACCUCGCGAGAGCAAGCGGACCUCAUAAAGUGCGUCGUAG
1250 1300

UCCGGAUUGGAGUCUGCAACUCGACUCCAUGAAGUCGGAAUCGCUAGUAAUCGUGGAUCAGAAUGCCACGGUGAAUACGUUCCCGGGCCUUGUACACACC
1350 1400

Gm^{4}CmCCGUm^{5}CACACCAUGGGAGUGGGUUGCAAAAGAAGUAGGUAGCUUAACCUUCGGGAGGGCGCUUACCACUUUGUGAUUCAUGACUGGGGUGAAGUCGmUAA
1450 1500

CAAGGUAACCGUAGGGGm$^6_2$Am$^6_2$ACCUGCGGUUGGAUCACCUCCUUA$_{OH}$

Figure 40 Nucleotide sequence of ribosomal 16S RNA of *E. coli*[18,19].

Of course, the most interesting problem was to learn how this 1542-nucleotide-long chain is folded into a specific three-dimensional structure. Unfortunately, even now this problem is far from being completely solved. Nevertheless, attempts to identify elements of the three-dimensional structure, specifically the double-stranded regions with complementary antiparallel strands (Watson–Crick-type helices), have been made, with some positive results.

First of all, the nucleotide residues or oligonucleotide regions which do not participate in complementary pairing and most probably comprise single-stranded sections can be localized in the primary RNA structure. These regions are particularly sensitive to ribonucleases such as pancreatic pyrimidyl RNAase A, fungal RNAase T_1, and bacterial RNAase S_1; they are also quite reactive to base-modifying chemical agents such as glyoxal and kethoxal which modify the unpaired G, bisulphite which modifies the unpaired C, or *meta*-chloroperbenzoic acid which modifies the unpaired A[3,43,44]. Identification of the single-stranded regions enables a search to be made of the adjacent mutually complementary regions between the sensitive sites along the RNA chain.

Furthermore, direct localization of paired chain regions is possible. One of the more fruitful approaches is as follows: the RNA is partially digested by RNAase hydrolyzing single-stranded regions, and the double-stranded fragments obtained are then separated by electrophoresis, initially under nondenaturing conditions (first dimension), and then under conditions when the double-stranded complexes undergo dissociation (second direction); thus, each band after the first direction comprises a double helix, and during electrophoresis in the second dimension it yields two spots corresponding to complementary strands which are identified and then mapped on the primary RNA structure[45]. In this way it is possible not only to detect adjacent complementary regions but also to identify complementary interactions between distant regions of the RNA chain.

Another particularly effective approach to detecting distant interactions involves photoactivated crosslinking of bases present in paired strands within the RNA structure, with subsequent identification of the crosslinked oligonucleotides[46,47].

Finally, the presence of double-stranded regions predicted on the basis of the complementarity of the adjacent or distant chain section may be confirmed or rejected by a comparative analysis of the homologous RNAs. The critical concept underlying this approach is that RNA structures are extremely conservative in evolution. Studies of the 16S RNA from a number of bacterial species as well as from chloroplasts have demonstrated high homology between their primary structures; at the same time, numerous nucleotide replacements occur.

If, despite these replacements, another evolutionary distant 16S RNA shows complementarity between similar sections and correspondingly the predicted helical regions occupy similar positions, this argues strongly in favor of the correct localization of these helices[43,44,48–51]. Identical positioning of helical regions in several different 16S RNA makes such a localization quite convincing.

Combining the experimental and theoretical approaches mentioned above has allowed a model of the *E. coli* 16S RNA secondary structure to be proposed; this structure is shown in fig. 41. Almost identical models have been proposed for the 16S RNA of other eubacteria[43,50,51], higher plant chloroplasts[50,51], and archaebacteria[26,27].

Although it is larger and has a considerably lower sequence homology with bacterial 16S RNA, the 18S RNA present in the cytoplasmic 80S ribosomes of eucaryotes may be accommodated into a secondary structure pattern that is fundamentally similar to that of the 16S RNA; the 18S RNA, however, contains additional helices and their groups[48–51] (fig. 42). Ribosomal RNA molecules of a smaller size, specifically the 12S RNA of mammalian mitochondria, also show homology of the secondary structure with bacterial 16S RNA and eucaryotic 18S RNA; however, some helices or larger pieces are deleted[49–51]. In fig. 42 the conservative core of the secondary structure of the small ribosomal subunit RNA is shown inside the dashed lines; it is more or less common for eucaryotes, bacteria, and mammalian mitochondria.

The following main features of the secondary structure of the ribosomal high-molecular-mass RNA are revealed in the *E. coli* 16S RNA model and appear to hold true for other cases.

1. At least half of all nucleotide residues are involved in the complementary interactions that form the Watson–Crick type of double-stranded structures.
2. The intrachain double-stranded regions are relatively short. The length of a continuous helix rarely exceeds one complete turn, i.e. 10 to 12 nucleotide pairs, while the mean length of the helix is about 7 or 8 nucleotide pairs. The total number of helices in the *E. coli* 16S RNA, according to the scheme of the secondary structure given in fig. 41, is about 60; in other words, there is, on average, one helix per 25 to 30 nucleotide residues (the helices are marked by numbers in boxes).
3. In addition to the standard Watson–Crick G·C and A·U base pairs the helices of ribosomal RNA contain G·U pairs. The helices of ribosomal RNA (just as those of tRNA) may also contain, albeit rarely, G·A pairs, as well as juxtapositions A·C and U·U. Noncanonical pairs occur more frequently in longer RNA helices

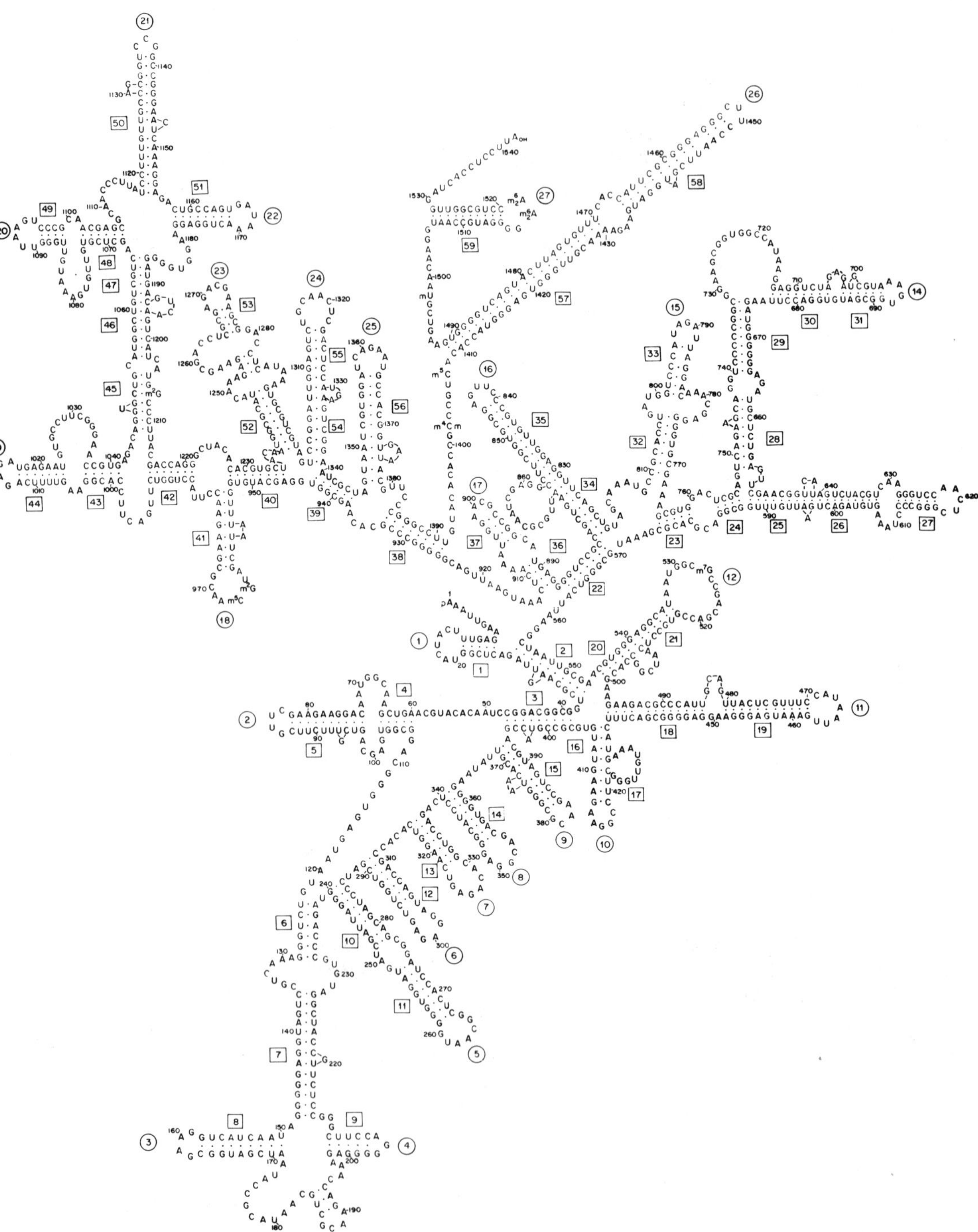

Figure 41 Probable folding of the 16S ribosomal RNA nucleotide sequence of *E. coli* into the secondary structure (arrangement of double-helical regions)[3,43–45] The numbers of helical regions are boxed; the numbers of hairpins are circled.

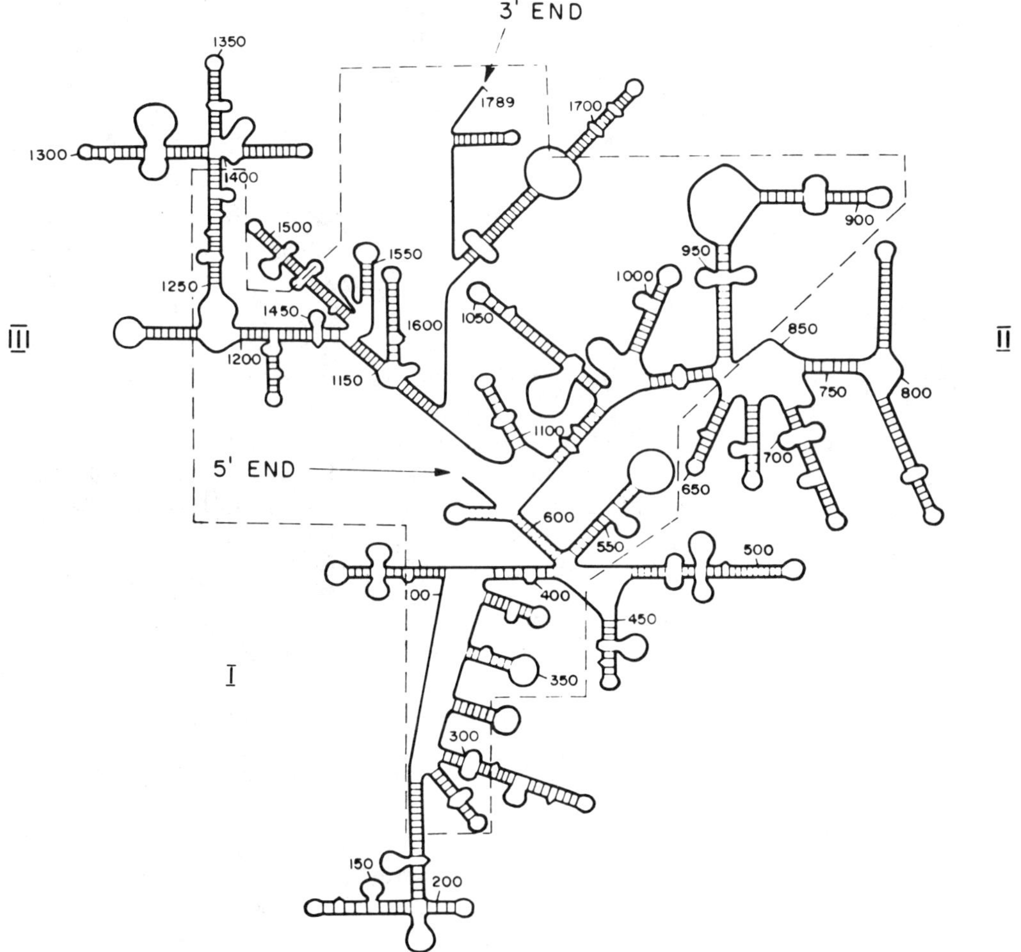

Figure 42 Schematic representation of the probable secondary structure of the yeast 18S ribosomal RNA[48–51]. The 18S ribosomal RNA from animal sources exhibits a similar folding. Conservative part distinguishable in all high-molecular-mass RNAs of the small ribosomal subunits, including the most reduced mitochondrial 12S RNA of mammals, is in the broken-line box.

(e.g. in the longest helix of the 3′-terminal region of *E. coli* 16S RNA, which consists of 22 base pairs, there are 7 G·U pairs and 3 G·A pairs!).

4. RNA helices may contain side loops consisting of one or two nucleotide residues, usually purine nucleotides, e.g. G^{31}, A^{746}, A^{1441}, or $AA^{958-959}$ in *E. coli* 16S RNA (incidentally, the plane(s) of

a base or bases in such a side loop can be oriented perpendicularly to the planes of bases in a helix, thus forming a hydrophobic contact for tertiary interactions). Longer side loops consisting of 3 to 10–14 nucleotides are found as well, but they may be involved in complex structures and so it would be better to consider them as interruptions in helices.

5. The predominant type of RNA helix is the hairpin, i.e. a helix formed by the complementary interaction of two adjacent chain sections. According to the model in fig. 41, the *E. coli* 16S RNA has 27 such hairpins (marked by numbers in circles). The hairpins frequently neighbor the helical regions which are separated from a hairpin by a short noncomplementary segment; the hairpin and the adjacent helix together may form a "compound hairpin." Such compound hairpins in the *E. coli* 16S RNA model are formed by helices 4 and 5; 10 and 11; 16 and 17; 18 and 19; 20 and 21; 25, 26, and 27; 43 and 44; 48 and 49; 52 and 53; 54 and 55; and 57 and 58; thus, compound hairpins account for much of the total secondary structure.

6. There are several distant complementary interactions pairing remote sections of the polynucleotide chain. The most important interactions for the overall folding of the *E. coli* 16S RNA and for the segregation of its domains (see below) are three such distant complementary contacts, i.e. those between sequence positions 27–37 and 547–556 (helix 2), 564–570 and 880–886 (helix 22), and 926–933 and 1384–1391 (helix 38). These double helices fix the beginning and the end of the polynucleotide chain sections which form the main domains of the RNA structure (see below). Less distant complementary interactions should also be mentioned here, such as the pairing between sections 39–47 and 394–403 (helix 3), sections 576–580 and 761–765 (helix 23), and sections 946–953 and 1228–1235 (helix 40); these double helices are also important for stabilizing the domain folding of RNA.

The 16S RNA contains several posttranscriptionally modified (minor) nucleotide residues (see chapter 3, fig. 14). Nearly all of these are found in unpaired chain sections, primarily in the 3′-proximal third of the molecule. The 16S *E. coli* has two residues of m_2^6A (clustered residues in the end loop of the 3′-terminal hairpin, positions 1518 and 1519), one residue of m^7G (position 527), two residues of m^2G (positions 966 and 1207), two residues of m^5C (positions 967 and 1407), one residue of m^4Cm (position 1402), and one residue of mU (position 1498). Many of these minor residues are evolutionary conservative, which follows from comparisons with other species of bacteria. It is noteworthy that the 3′-terminal hairpin containing the helix 10 nucleotide pairs in length is invariant for the 16S and 18S RNAs examined so far; two clustered m^6_2A residues of the hairpin end loop are always present both in procaryotes and in eucaryotes.

Modified residues, as well as helix loops, are thought to play an important part in organizing the ribosomal RNA regions recognized by proteins and in forming the functional sites of the ribosome.

23S (28S) RNA

The primary structure of *E. coli* 23S RNA has been determined by sequencing the corresponding cloned gene[31] as well as by direct chemical-enzymatic analysis[52] (fig. 43). The high-molecular-mass RNAs of the large ribosomal subunits of some other organisms, as well as of chloroplasts and mitochondria, have also been sequenced and have provided the material for evolutionary comparisons[52–54]. The whole armamentarium of techniques used for 16S RNA was employed for studying the 23S RNA secondary structure, and the general rules and features determined for the 23S RNA were found to be fundamentally similar to those of its 16S counterpart. The scheme of the secondary structure model of *E. coli* 23S RNA is given in fig. 44. As with 16S RNA, at least half of the nucleotides in 23S RNA are present in double helices. There are just over 100 individual helices. The most obvious difference from 16S RNA is the probable complementary pairing between the 5′- and the 3′-ends of the 23S RNA; a rather stable and perfect double helix consisting of eight nucleotide pairs seems to hold both ends together, contributing to the fixation of the general chain folding into the final compact structure (see below). As in 16S RNA, G·U pairs are not rare in 23S RNA helices. Moreover, in some cases the helices of 23S RNA may also contain G·A pairs, as well as other noncanonical pairs. The helices may include side loops consisting of one or two nucleotides; A-containing side loops are the most common.

With respect to posttranscriptionally modified (minor) nucleotides, *E. coli* 23S RNA has three residues of ψ, two T residues, two m^6A residues, as well as m^1G, m^7G, Gm, Cm, and unidentified derivatives of U. Some are concentrated at certain regions, e.g. in sequences 745–747 ($m^1G\psi T$) and 1911–1917 (ψAACU*Aψ), while others are scattered throughout the molecule. All of the minor residues, with the exception of one ψ and one T, seem to be present in single-stranded regions.

Just like the high-molecular-mass RNAs of small ribosomal subunits, the eucaryotic 26S (28S) RNA show homology to bacterial 23S

Figure 43 Nucleotide sequence of 23S ribosomal RNA of *E. coli*[31,52].

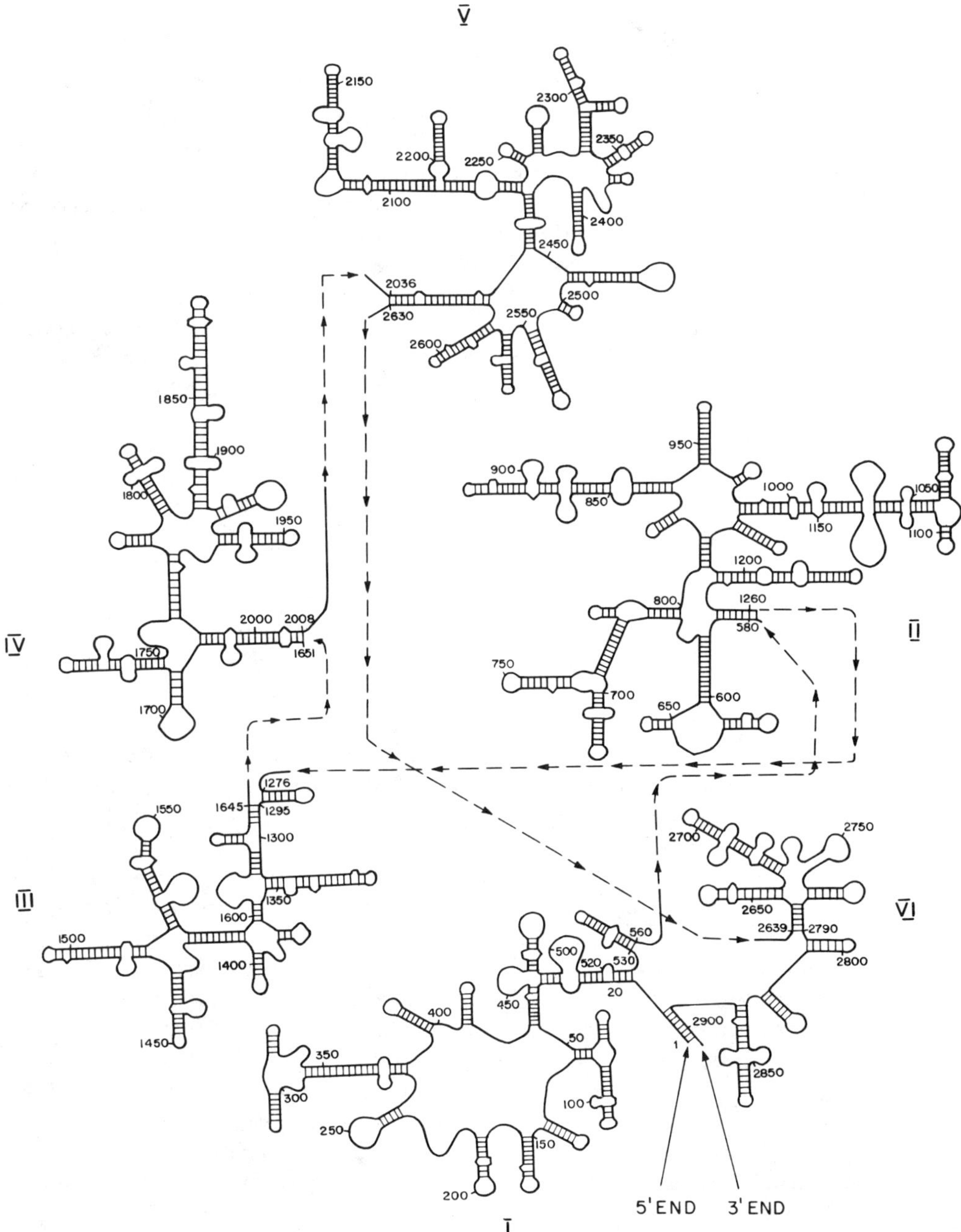

Figure 44 Schematic representation of the probable secondary structure of *E. coli* 23S ribosomal RNA[52–54]. The RNA is divided into six domains, designated by roman numerals.

RNA, particularly in their secondary structure, but contain a number of large inserts[34–37]. The mammalian 28S RNA, which is about 1500 nucleotides longer than yeast 26S RNA, has been shown to contain several distinct G + C-rich segments of different length inserted within the regions of high sequence homology; all G + C-rich segments appear to form strongly base-paired structures, sometimes very long[37]. The most fundamental difference of the eucaryotic ribosomal RNA of the large subunit is the existence of a 5′-terminal, 160-nucleotide-long fragment as a separate 5.8S RNA chain (fig. 45). The eucaryotic 5.8S RNA, however, is so tightly attached to the structure of the 28S RNA, seemingly by stable pairing with it in three regions[55], that it can hardly be discussed as a structurally independent molecule. In fact, the 5.8S RNA, both by itself and together with 28S RNA, forms a secondary structure similar to that for the covalently continuous 5′-terminal part of the bacterial 23S RNA (see section 7.4).

5S RNA

The complete primary structures of 5S RNAs were determined for procaryotes (*E. coli*) and eucaryotes (human) as early as 1967, in Sanger's and Weissman's groups[40,42], respectively, i.e. long before reports were made on the structure of high-molecular-mass ribosomal RNAs. Since then, the primary structure of numerous 5S RNAs from ribosomes of different species has been determined[41]. This has allowed evolutionary comparisons to be made in order to elucidate the phylogenetic relationships of species and to test the predicted secondary structure on the basis of homologous nucleotide replacements.

The 5S RNA generally does not contain posttranscriptionally modified (minor) nucleotides. However, in a number of yeast species, as well as in *Euglena*, pseudouridine is present. In rare cases some other minor nucleotides have been found, e.g. the modified C in the archebacterium *Sulfolobus acidocaldarius*.

A comparison of the nucleotide sequences of 5S RNA from the numerous species illustrates the evolutionary conservatism of ribosomal RNAs. For example, human 5S RNA differs from reptile 5S RNA in only two or three nucleotide residues, whereas the difference from amphibian 5S RNA is seven or eight residues. The 5S RNA of rye differs from its tomato counterpart by two residues while the difference between rye and dwarf bean 5S RNAs amounts to five nucleotide residues. A high similarity has also been observed for the 5S RNA isolated from various species of *Bacillus* genus. The 5S RNA of gram-negative bacteria belonging to the intestinal group and related

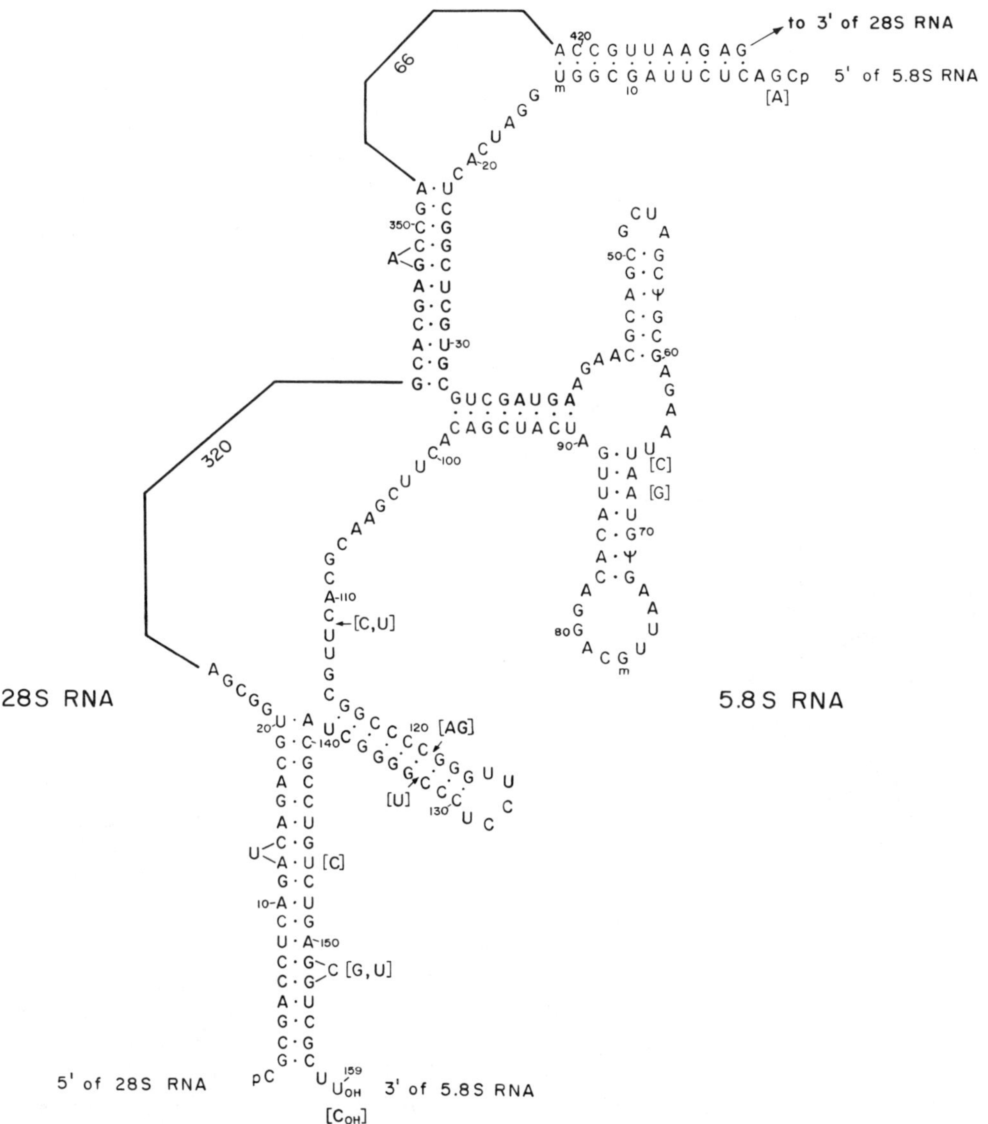

Figure 45 Nucleotide sequence and probable secondary structure of 5.8S ribosomal RNA of eucaryotes (animals). The master sequence refers to 5.8S RNA of mammals[39]; substitutions or insertions of nucleotides observed in 5.8S RNA of other animal taxons (birds, reptiles, amphibia, and fishes)[41] are given in brackets. Possible pairing of 5.8S RNA with the 5′-terminal domain of 28S ribosomal RNA (mouse)[55] is shown.

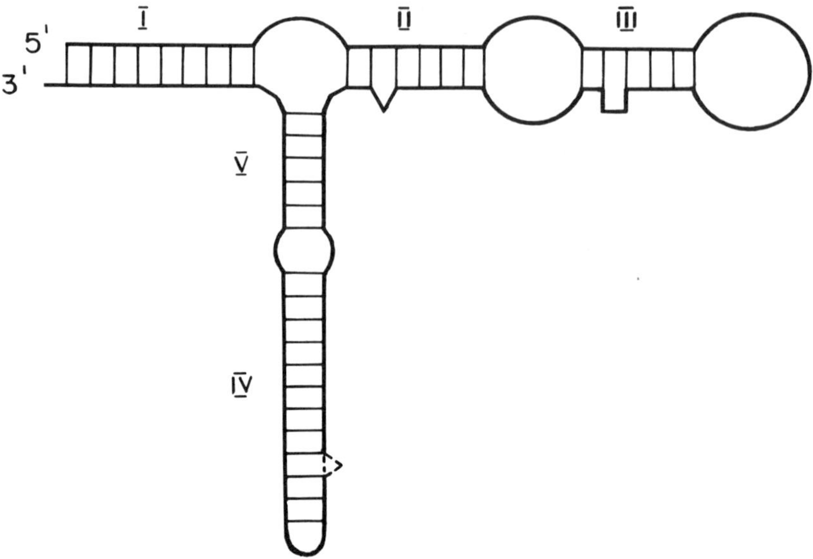

Figure 46 Generalized scheme of the secondary structure of 5S RNA. Roman numerals designate the five helical regions of the 5S RNA molecule.

genera (*Escherichia*, *Salmonella*, *Aerobacter*, *Erwinia*, *Serratia*, *Proteus*, *Photobacter*, etc.) are either identical or differ by just one nucleotide.

While the primary structures of 5S RNAs were being determined, many attempts were made at predicting the secondary structure. The probing of single-stranded regions with nucleases and chemical agents modifying unpaired bases provided the experimental foundation for these attempts. However, the comparative approach based on the folding of many known primary structures of 5S RNA into a similar secondary structure pattern was particularly important. Indeed, all the known primary structures of 5S RNA may be accommodated by the scheme shown in fig. 46 (this scheme is a generalization of a number of fundamentally similar models of the secondary structure of 5S RNA proposed by different authors[56–63]). Figure 47 shows primary and probable (predicted) secondary structures of the 5S RNA of six representatives of the main phyla in living nature: gram-negative bacteria, gram-positive bacteria, archebacteria or metabacteria, plants, fungi, and animals.

Several general features of 5S RNA secondary structure may be traced. The 5′-terminal chain region is complementary to the 3′-terminal region; hence, they can form a stable long double helix consisting of 9 to 11 nucleotide pairs (helix I). The internal nucleotide

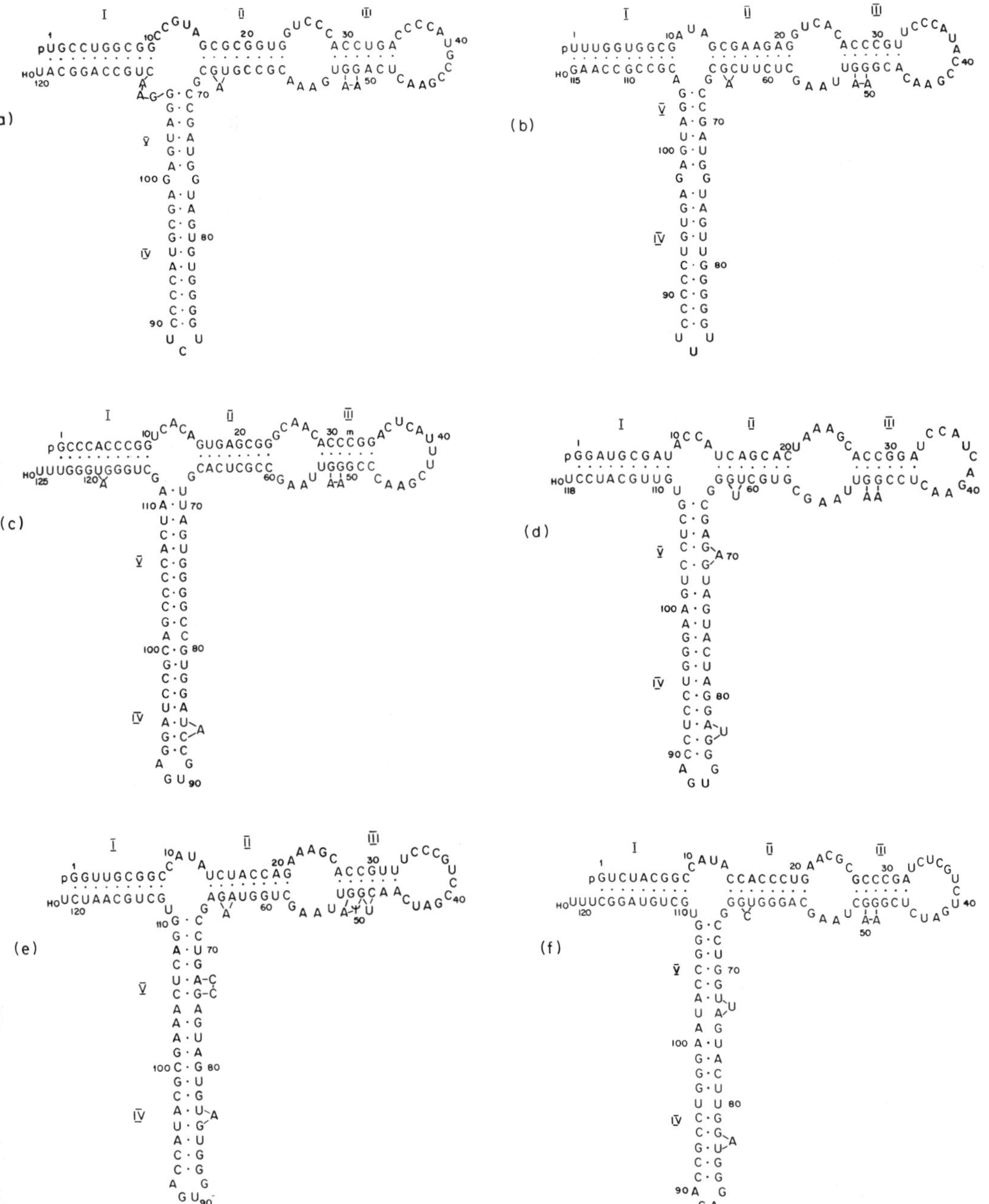

Figure 47 Nucleotide sequences of ribosomal 5S RNAs[41] and their probable folding into secondary structures in representatives of the six main evolutionary groups. (a) Gram-negative bacterium (*Escherichia coli*). (b) Gram-positive bacterium (*Bacillus subtilis*). (c) Archebacterium (*Sulfolobus acidocaldarius*). (d) Plant (rye, *Secale cereale*). (e) Fungus (yeast, *Saccharomyces cerevisiae*). (f) Animal (man, *Homo sapiens*). The double-helical regions are designated by roman numerals.

sequence may be accommodated into two compound hairpins. One compound hairpin divides distinctly into two double-stranded regions: (1) the hairpin proper, consisting of 6 nucleotide pairs with a large unpaired end loop of 11 to 13 residues around position 40 (helix III); and (2) the double-stranded region, consisting of 7 or 8 nucleotide pairs (helix II) connected to the preceding helix by a noncomplementary region. The other compound hairpin, with a small end loop of 2 to 4 nucleotide residues, is an almost continuous double helix. However, it usually contains several imperfections, such as noncanonical pairs, looped-out nucleotide residues, and unpaired nucleotides in its middle part; therefore this hairpin may be subdivided into helix IV and helix V.

The general features of the 5S RNA secondary structure appear to be the same as those mentioned earlier for high-molecular-mass RNAs. The presence of such noncanonical pairs as G·U and G·A should be emphasized. In addition, all internal helices (II through V) frequently contain one- and two-nucleotide side loops; in most cases it is A that loops out, but some other residues may also be found.

The information on sensitivity to nucleases and chemical modification agents shows that the large end loop of helix III and the unpaired region between helices II and III as well as the "poorly paired" residues of the compound helix IV–V are not just unfolded, but appear to be rather tightly structured; it is likely that they take part in organizing the 5S RNA tertiary structure which, however, is still unknown.

It is noteworthy that both the end loops and the side loops often have a more conservative primary structure as compared to the helical regions. The large apex loop of the helical hairpin III contains the semiinvariant sequence

$$\mathrm{C}\overset{\mathrm{U}}{\underset{\mathrm{A}}{\mathrm{C}}}\mathrm{GA}{}^{\mathrm{U}}_{\mathrm{A}}\mathrm{C}$$

directly adjacent to the helical region. (This sequence is represented by CCGAAC in all procaryotes, CAGAAC in plants, CCGAUC in fungi, and CUGAUC in vertebrate animals.) Its tetranucleotide region $\mathrm{GA}{}^{\mathrm{U}}_{\mathrm{A}}\mathrm{C}$, which is adjacent to the helix, is inaccessible to nucleases, suggesting that it is involved in some tertiary structure. Four residues away from it, in the direction of the 3′-end, there is a semiinvariant, two-nucleotide-long side loop; most often this is AA, but in fungi it is comprised of UA, ψA, or CA (CA can also be found in some higher plants). This loop is characteristically flanked by two G, which are paired with two adjacent C in the opposite strand, seemingly providing for the fixation of the double helix at the site of the loop. Another invariant doublet AA is always present three residues further toward the 3′-end in the nonhelical region between helices II and III. The 3′-strand of helix II nearly always contains the invariant, one-nucleotide-long side loop. This side loop is usually A in procaryotes and fungi, U in plants, and C in animals (it may be absent in archebacteria); it is always followed by G paired to C in the opposite chain.

7-4 Structural domains and compact folding of RNA molecules

The firm pairing of the 5′- and 3′-terminal chain sections appears to be a general characteristic of stable RNA structures. The formation of such a *stem helix* completes the formation of the RNA three-dimensional structure and fixes it. In principle, fixation of the ends should inevitably result in the stabilization of internal helices and other types of folding. Locking the ends to each other also contributes to a greater compactness of the whole structure.

The first known example of such an organization was provided by tRNA. As is apparent from the preceding section, ribosomal 5S RNA also possesses a firm stem helix formed by the 5′- and 3′-ends of the chain; this helix appears to lock the general folding of the molecule. The existence of a stable (perfect and sufficiently long) helix formed by distant chain sections in high-molecular-mass ribosomal RNA may be taken as evidence that a relatively independent compact folded structure of the chain section exists between them. Such "self-governed" structural domains, with their 5′- and 3′-ends being paired, are an important conformational motif of both high-molecular-mass ribosomal RNA species.

16S (18S) RNA

Both procaryotic 16S RNA and eucaryotic 18S RNA may be subdivided into three main domains: (1) the 5′-terminal domain, I, which is limited and fixed by a stem helix consisting of 10 or 11 nucleotide pairs (helix 2 formed by sequences 27–37 and 547–556 in the case of *E. coli* 16S RNA; see fig. 41); (2) the middle domain, II, with a stem of 7 nucleotide pairs (helix 22 formed by sequences 564–570 and 880–886 in *E. coli*); and (3) the 3′-proximal domain, III, which is limited by a stem consisting of 7 or 8 nucleotide pairs (helix 38 formed by sequences 826–933 and 1384–1391 in *E. coli*). A subdomain can be defined within domain I which is limited by a helix containing 6 to 9 nucleotide pairs (helix 3 formed by sequences 39–47 and 394–403 in *E. coli*). Similarly, domain II possesses a subdomain limited by its stem (helix 23 formed by sequences 576–580 and 761–765 in *E. coli*). Domain III also contains a subdomain, in this case with a stem of 8 to 10 nucleotide pairs (sequences 946–953 and 1228–1235 in *E. coli*). It is noteworthy that the subdomains existing inside the domains tend to include the 5′-terminal part of corresponding domains. This may well be a general rule facilitating the folding of RNA beginning from its 5′-end: the subdomain may play the part of a compact core which directs the subsequent chain folding. The extreme 3′-terminal sequence of 16S or 18S RNA with a length exceeding 150 nucleotides does not belong to any of the listed domains but may show a marked structural organization; it is sometimes referred to as the additional domain, IV, and its main structural feature is the presence of long and stable double-helical regions, such as a 3′-terminal universal hairpin consisting of 10 nucleotide pairs and a very long compound hairpin comprising about 30 nucleotide pairs.

Electron microscopic examination of isolated 16S RNA under conditions where the molecule is compact, i.e. in the presence of sufficient levels of Mg^{2+}, spermidine, etc., allows the three main structural domains to be visualized. The RNA is seen as a Y-shaped structure with arms of unequal length[64] (fig. 48). In other words, three unequal lobes originating from the center may be seen . If the size and general morphology of the Y-shaped RNA are compared with those of the 30S ribosomal subunit, a good correspondence between their contours is seen, so that one can be easily matched with the other (fig. 49). Then, if both ends of 16S RNA are mapped on the 30S ribosomal subunit and the localization of ribosomal proteins on the 16S RNA primary structure and on the surface of the 30S ribosomal subunit is known, the visually detectable arms or lobes of the Y-shaped RNA can be identified as the three structural domains deduced from its primary and secondary structures. Apparently the lower arm (stem) of the Y corresponds to the 5′-terminal domain I, which forms the body of the 30S ribosomal subunit. The upper longer arm (long branch) of the Y-shaped RNA molecule probably corresponds to the 3′-proximal domain (III), which is a component of the head of the 30S ribosomal subunit. The upper shorter arm (short branch) of the Y may be identified as the middle domain (II), which takes part in forming the side bulge (or platform) of the 30S subunit. According to direct

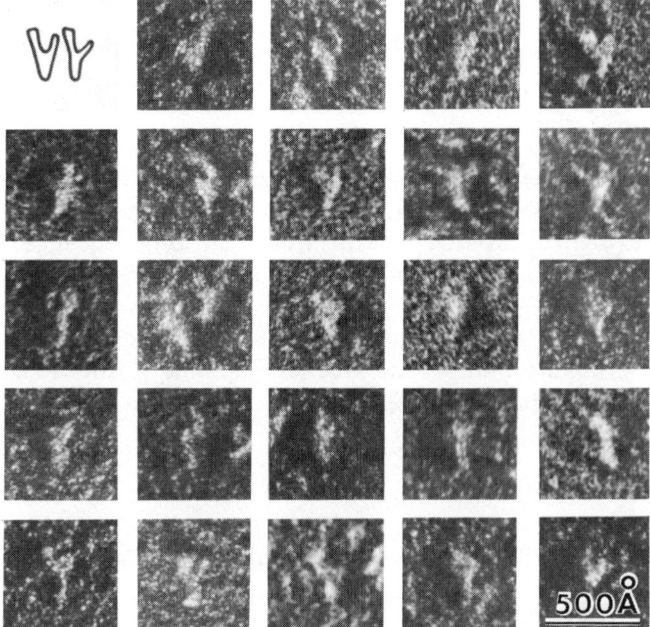

Figure 48 Electron micrographs of the individual molecules of isolated *E. coli* 16S ribosomal RNA in compact form[64], and their Y-shaped contours (*top left*). RNA was freeze-dried and contrasted by metal shadowing (as described in the legend to fig. 33). (Courtesy of Dr. V. D. Vasiliev)

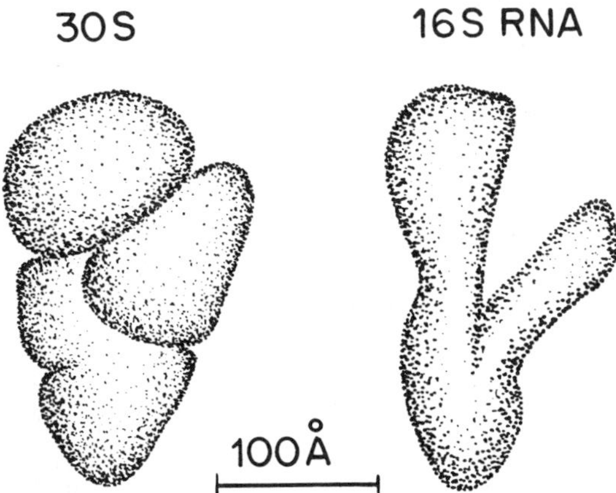

Figure 49 Comparison of the shapes of the 30S ribosomal subunit and the isolated 16S ribosomal RNA in compact form, on the basis of electron microscopy data.

immuno-electron microscopic mapping, the extreme 3′-terminal sequence of 16S RNA is located somewhere between the head and the top of the side bulge, i.e. probably between the ends of the two upper arms of the Y-shaped RNA.

It is quite striking that the overall shape of the isolated RNA in its compact state is unique and resembles the shape of the 30S subunit or at least can be easily inscribed within it. This suggests: (1) that ribosomal RNA is capable of forming not only local secondary structures but also an overall folding pattern, thus determining the morphologically unique conformation of the entire molecule in the absence of proteins; and (2) that it is this self-folding of the ribosomal RNA which critically defines the size, overall shape, and a number of the main morphological features of the ribosomal subunit.

23S (28S) RNA

The chain of 23S (28S) ribosomal RNA folds into six[54] or seven[34,37,52,53] structural domains which are limited, just as in the preceding case, by stem helices formed by pairing between distant chain sections (fig. 44).

In procaryotic 23S RNAs domain I is formed by the 5′-end-adjacent sequence which is about 500 nucleotide residues in length. In eucaryotic 28S RNAs formation of domain I requires interaction with 5.8S RNA which should, as previously indicated, be considered as a homolog of the 160-nucleotide-long 5′-terminal sequence of the procaryotic 23S RNA. The 3′-terminal sequence of 5.8S RNA is firmly paired with the 5′-terminal sequence of 28S RNA over a distance of 15 to 17 nucleotide pairs, and the two RNAs become noncovalently spliced (fig. 45). On the other hand, the 5′-proximal sequence of 5.8S RNA in the region of nucleotide residues 20–30 seems to be paired with an internal region of the 28S RNA chain[55] forming the stem of domain I.

As in the case of 16S (18S) RNA, there is a long 3′-end-adjacent sequence of 23S (28S) RNA which does not belong to any of the main domains and forms only several helical hairpins. In the bacterial 23S RNA, a 110-nucleotide-long 3′-end-adjacent sequence is folded into three hairpins: two simple ones and one compound consisting of two helices separated by an unpaired region (fig. 44). In the chloroplastic ribosomes of higher plants the 23S RNA chain does not contain the 100-nucleotide-long 3′-terminal sequence corresponding to that of bacteria; a separate 4.5S RNA chain is present instead. The 4.5S RNA folds into two hairpins—one simple and one compound consisting of two helices—which are highly homologous to the 3′-terminal hairpins of the bacterial 23S RNA[65].

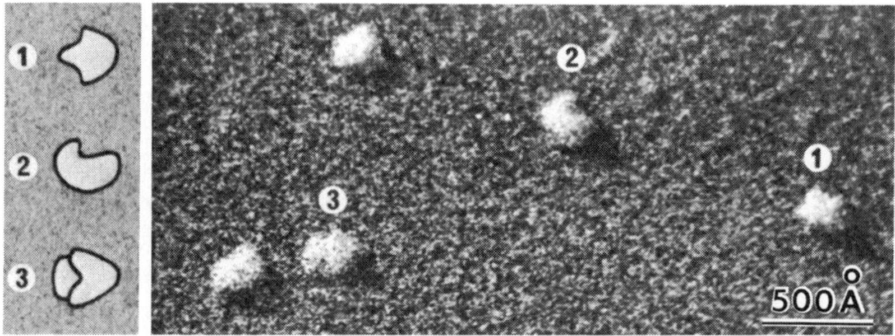

Figure 50 Electron micrograph of isolated *E. coli* 23S ribosomal RNA in compact form[66] and the contours of the molecule (*left*). Three different projections are marked by the numerals 1, 2, and 3. RNA was freeze-dried and contrasted by metal shadowing (as described in the legend to fig. 33). (Courtesy of Dr. V. D. Vasiliev)

As previously mentioned, the 5′-terminus of the bacterial 23S RNA seems to be paired with the 3′-end, yielding a perfect stable helix of eight nucleotide pairs. Correspondingly, in the chloroplastic ribosomes of higher plants the 5′-end of 23S RNA is paired into a similar helix with the 3′-end of 4.5S RNA. In eucaryotic ribosomes the 3′-end of the high-molecular-mass 28S RNA may be paired with the 5′-end of 5.8S RNA.

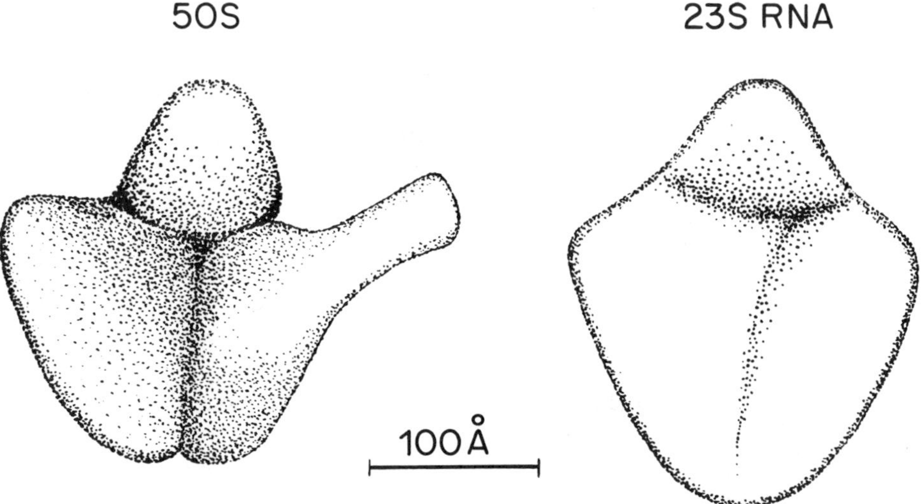

Figure 51 Comparison of the shapes of the 50S ribosomal subunit and the isolated 23S ribosomal RNA in compact form, on the basis of electron microscopy data.

Locking the 3′-end of 23S RNA onto its 5′end appears to be associated with the general compact folding of the entire molecule and should fix the overall folding. An electron microscopic study of isolated *E. coli* 23S RNA under ionic conditions that provide for its compact state has indeed revealed that its shape drastically differs from that of 16S RNA where there is no pairing between the 5′-end and the 3′-end. The 23S RNA molecule appears much more rounded, having only small protuberances[66] (fig. 50), compared with the 16S RNA under similar conditions. The important fact is that the contours of the isolated 23S RNA look like those of the 50S ribosomal subunit, but with the reduced protuberances (fig. 51). Again, the ribosomal RNA chain undergoes a specific self-folding and the general morphology of the ribosomal subunit seems to be determined primarily by its RNA.

References

1. F. H. C. Crick (1968), "The origin of the genetic code," *J. Mol. Biol.* 38:367–379.
2. A. S. Spirin (1976), "Cell-free systems of polypeptide biosynthesis and approaches to the evolution of translation apparatus," *Origins of Life* 7:109–118.
3. H. F. Noller and C. R. Woese (1981), "Secondary structure of 16S ribosomal RNA," *Science* 212:403–411.
4. B. Hall and P. Doty (1959), "The preparation and physical properties of ribonucleic acid from microsomal particles," *J. Mol. Biol.* 1:111–126.
5. C. G. Kurland (1960), "Molecular characterization of ribonucleic acid from *E. coli* ribosomes. I. Isolation and molecular weights," *J. Mol. Biol.* 2:83–91.
6. M. H. Green and B. H. Hall (1961), "A comparison of the native and derived 30S and 50S ribosomes of *Escherichia coli*," *Biophys. J.* 1:517–523.
7. R. Rosset and R. Monier (1963), "A propos de la présence d'acide ribonucléique de faible poids moléculaire dans les ribosomes d'*Escherichia coli*," *Biochim. Biophys. Acta* 68:653–656.
8. B. G. Forget and S. M. Weissman (1967), "Low molecular weight RNA components from KB cells," *Nature* 213:878–882.
9. J. J. Pène, E. Knight, and J. E. Darnell (1968), "Characterization of a low molecular weight RNA in HeLa cell ribosomes," *J. Mol. Biol.* 33:609–623.
10. R. N. Nazar (1980), "A 5.8S rRNA-like sequence in prokaryotic 23S rRNA," *FEBS Letters* 119:212–214.

11. P. R. Whitfeld, C. J. Leaver, W. Bottomley, and B. A. Atchinson (1978), "Low-molecular-weight (4.5S) RNA in higher-plant chloroplast ribosomes," *Biochem. J.* 175:1103–1112.

12. R. M. MacKay (1981), "The origin of plant chloroplast 4.5S ribosomal RNA," *FEBS Letters* 123:17–18.

13. J. Trapman, P. De Jonge, and R. J. Planta (1975), "On the biosynthesis of 5.8S ribosomal RNA in yeast," *FEBS Letters* 57:26–30.

14. M. S. N. Khan and B. E. H. Maden (1976), "Maturation of 5.8S RNA in HeLa cells," *FEBS Letters* 61:10–13.

15. B. R. Jordan, R. Jourdan, and B. Jacq (1976), "Late steps in the maturation of *Drosophila* 26S ribosomal RNA: Generation of 5.8S and 26S RNAs by cleavage occurring in the cytoplasm," *J. Mol. Biol.* 101:85–105.

16. M. R. Hartley (1979), "The synthesis and origin of chloroplast low-molecular weight ribosomal ribonucleic acid in spinach," *Eur. J. Biochem.* 96:311–320.

17. J. Shine and L. Dalgarno (1973), "Occurrence of heat-dissociable ribosomal RNA in insects: The presence of three polynucleotide chains in 26S RNA from cultured *Aedes aegypti* cells," *J. Mol. Biol.* 75:57–72.

18. P. Carbon, C. Ehresmann, B. Ehresmann, and J.-P. Ebel (1978), "The sequence of *E. coli* ribosomal 16S RNA determined by new rapid gel method," *FEBS Letters* 94:152–156.

19. J. Brosius, M. L. Palmer, P. J. Kennedy, and H. F. Noller (1978), "Complete nucleotide sequence of a 16S RNA ribosomal RNA gene from *Escherichia coli*," *Proc. Nat. Acad. Sci. U.S.A.* 75:4801–4805.

20. I. N. Serdyuk, S. Ch. Agalarov, S. E. Sedelnikova, A. S. Spirin, and R. P. May (1983), "On the shape and compactness of the isolated ribosomal 16S RNA and its complexes with ribosomal proteins," *J. Mol. Biol.* 169:409–425.

21. A. S. Spirin (1963), "Some problems concerning the macromolecular structure of ribonucleic acids," Progress in nucleic acid research, ed. J. N. Davidson and W. E. Cohn, vol. 1, pp. 301–345 (New York: Academic Press).

22. Z. Schwarz and H. Kössel (1980), "The primary structure of 16S rDNA from *Zea mays* chloroplast is homologous to *E. coli* 16S rRNA," *Nature* 283:739–742.

23. F. Sor and H. Fukuhara (1980), "Sequence nucleotidique du gène de l'ARN ribosomique 15S mitochondrial de la levure," *C. R. Acad. Sci.* (*Paris*), *Ser. D*, 291:933–936.

24. I. C. Eperon, S. Anderson, and D. P. Nierlich (1980), "Distinctive sequence of human mitochondrial ribosomal RNA genes," *Nature* 286:460–467.

25. R. A. van Etten, M. W. Walberg, and D. A. Clayton (1980), "Precise localization and nucleotide sequence of the two mouse mitochondrial

rRNA genes and three immediately adjacent novel tRNA genes," *Cell* 22:157–170.

26. R. Gupta, J. M. Lauter, and C. R. Woese (1983), "Sequence of the 16S ribosomal RNA from *Halobacterium volcanii*, an archaebacterium", *Science* 221:656–659.
27. H. Leffers, and R. A. Garrett (1984), "The nucleotide sequence of the 16S ribosomal RNA gene of the archaebacterium *Halococcus morrhua*", *EMBO Journal* 3:1613–1619.
28. P. M. Rubtsov, M. M. Musakhanov, V. M. Zakharyev, A. S. Krayev, K. G. Skryabin, and A. A. Bayev (1980), "The structure of the yeast ribosomal-RNA genes. I. The complete nucleotide-sequence of the 18S ribosomal-RNA gene from *Saccharomyces cerevisiae*," *Nucleic Acids Res.* 8:5779–5794.
29. M. Salim and B. E. H. Maden (1981), "Nucleotide sequence of *Xenopus laevis* 18S ribosomal RNA inferred from gene sequence," *Nature* 291:205–208.
30. Y.-L. Chan, R. Gutell, H. F. Noller, and I. G. Wool (1984), "The nucleotide sequence of a rat 18S ribosomal ribonucleic acid gene and a proposal for the secondary structure of 18S ribosomal ribonucleic acid," *J. Biol. Chem.* 259:224–230.
31. J. Brosius, T. J. Dull, and H. F. Noller (1980), "Complete nucleotide sequences of a 23S ribosomal RNA gene from *Escherichia coli*," *Proc. Nat. Acad. Sci. U.S.A.* 77:201–204.
32. K. Edwards and H. Kössel (1981), "The rRNA operon from *Zea mays* chloroplasts: Nucleotide sequence of 23S rDNA and its homology with *E. coli* 23S rDNA," *Nucleic Acids Res.* 9:2853–2869.
33. U. E. Loening (1968), "Molecular weights of ribosomal RNA in relation to evolution," *J. Mol. Biol.* 38:355–365.
34. G. M. Veldman, J. Klootwijk, V. C. H. F. de Regt, R. J. Planta, C. Branlant, A. Krol, and J.-P. Ebel (1981), "The primary and secondary structure of yeast 26S rRNA," *Nucleic Acids Res.* 9:6935–6952.
35. O. I. Georgiev, N. Nikolaev, A. A. Hajiolov, K. G. Skryabin, W. M. Zakharyev, and A. A. Bayev (1981), "The structure of the yeast ribosomal RNA genes. IV. Complete sequence of the 25S rRNA gene from *Saccharomyces cerevisiae*," *Nucleic Acids Res.* 9:6953–6958.
36. Y.-L. Chan, J. Olvera, and I. G. Wool (1983), "The structure of rat 28S ribosomal ribonucleic acid inferred from the sequence of nucleotides in a gene," *Nucleic Acids Res.* 11:7819–7831.
37. A. A. Hadjiolov, O. I. Georgiev, V. V. Nosikov, and L. P. Yavachev (1984), "Primary and secondary structure of rat 28S ribosomal RNA," *Nucleic Acids Res.* 12:3677–3693.
38. G. M. Rubin (1973), "The nucleotide sequence of *Saccharomyces cerevisiae* 5.8S ribosomal ribonucleic acid," *J. Biol. Chem.* 248:3860–3875.

39. R. N. Nazar, T. O. Sitz, and H. Busch (1975), "Structural analyses of mammalian ribosomal ribonucleic acid and its precursors: Nucleotide sequence of ribosomal 5.8S ribonucleic acid," *J. Biol. Chem.* 250:8591–8597.

40. G. G. Brownlee, F. Sanger, and B. G. Barrell (1967), "Nucleotide sequence of 5S-ribosomal RNA from *Escherichia coli*," *Nature* 215:735–736.

41. V. A. Erdmann (1981), "Collection of published 5S and 5.8S RNA sequences and their precursors," *Nucleic Acids Res.* 9:r25–r42.

42. B. G. Forget and S. M. Weissman (1967), "Nucleotide sequence of KB cell 5S RNA," *Science* 158:1695–1699.

43. C. R. Woese, L. J. Magrum, R. Gupta, R. B. Siegel, D. A. Stahl, J. Kop, N. Crawford, J. Brosius, R. Gutell, J. J. Hogan, and H. F. Noller (1980), "Secondary structure model for bacterial 16S ribosomal RNA: Phylogenetic, enzymatic, and chemical evidence," *Nucleic Acids Res.* 8:2275–2293.

44. P. Stiegler, P. Carbon, M. Zuker, J.-P. Ebel, and C. Ehresmann (1981), "Structural organization of the 16S ribosomal RNA from *E. coli*: Topography and secondary structure," *Nucleic Acids Res.* 9:2153–2172.

45. C. Glotz and R. Brimacombe (1980), "An experimentally-derived model for the secondary structure of the 16S ribosomal RNA from *Escherichia coli*," *Nucleic Acids Res.* 8:2377–2395.

46. C. Z. Zwieb and R. Brimacombe (1980), "Localization of a series of intra-RNA cross-links in 16S RNA induced by ultraviolet irradiation of *Escherichia coli* 30S ribosomal subunits," *Nucleic Acids Res.* 8:2397–2411.

47. P. Thammana, C. Cantor, P. L. Wollenzien, and J. E. Hearst (1979), "Cross-linking studies on the organization of the 16S ribosomal RNA within the 30S *Escherichia coli* ribosomal subunit," *J. Mol. Biol.* 135:271–283.

48. A. S. Mankin, A. M. Kopylov, P. M. Rubtzov, and K. G. Skryabin (1981), "Model of the secondary structure of the eucaryotic 18S ribosomal RNA", *Dokl. Adad. Nauk SSSR* 256:1006–1010.

49. A. S. Mankin and A. M. Kopylov (1981), "A secondary structure model for mitochondrial 12S RNA, an example of economy in rRNA structure", *Biochem. Int.* 3:587–593.

50. C. Zwieb, C. Glotz, and R. Brimacombe (1981), "Secondary structure comparisons between small subunit ribosomal RNA molecules from six different species," *Nucleic Acids Res.* 9:3621–3640.

51. P. Stiegler, P. Carbon, J.-P. Ebel, and C. Ehresmann (1981), "A general secondary-structure model for procaryotic and eucaryotic RNAs of the small ribosomal subunits," *Eur. J. Biochem.* 120:487–495.

52. C. Branlant, A. Krol, M. A. Machatt, J. Pouyet, and J.-P. Ebel (1981), "Primary and secondary structures of *Escherichia coli* MRE-600 23S ribosomal RNA: Comparison with models of secondary structure for

maize chloroplast 23S rRNA and for large portions of mouse and human 16S mitochondrial rRNAs," *Nucleic Acids Res.* 9:4303–4324.

53. C. Glotz, C. Zwieb, R. Brimacombe, K. Edwards, and H. Kössel (1981), "Secondary structure of the large subunit ribosomal RNA from *Escherichia coli, Zea mays* chloroplast, and human and mouse mitochondrial ribosomes," *Nucleic Acids Res.* 9:3287–3306.

54. H. F. Noller, J. A. Kop, V. Wheaton, J. Brosius, R. R. Gutell, A. M. Kopylov, F. Dohme, W. Herr, D. A. Stahl, R. Gupta, and C. R. Woese (1981), "Secondary structure model for 23S ribosomal RNA," *Nucleic Acids Res.* 9:6167–6189.

55. T. A. Walker, Y. Endo, W. H. Wheat, I. G. Wool, and N. R. Pace (1983), "Location of 5.8S rRNA contact sites in 28S rRNA and the effect of α-sarcin on the association of 5.8S rNA with 28S rRNA," *J. Biol. Chem.* 258:333–338.

56. K. Nishikawa and S. Takemura (1974), "Nucleotide sequence of 5S RNA from *Torulopsis utilis*," *FEBS Letters* 40:106–109.

57. G. E. Fox and C. R. Woese (1975), "5S RNA secondary structure," *Nature* 256:505–507.

58. H. Hori (1976), "Molecular evolution of 5S RNA," *Molec. Gen. Genetics* 145:119–123.

59. D. A. Stahl, K. R. Luehrsen, C. R. Woese, and N. R. Pace (1981), "An unusual 5S rRNA from *Sulfolobus acidocaldarius* and its implications for a general 5S rRNA structure," *Nucleic Acids Res.* 9:6129–6137.

60. G. M. Studnicka, F. A. Eiserling, and J. A. Lake (1981), "A unique secondary folding pattern for 5S RNA corresponds to the lowest energy homologous secondary structure in 17 different prokaryotes," *Nucleic Acids Res.* 9:1885–1904.

61. S. Douthwaite and R. A. Garrett (1981), "Secondary structure of prokaryotic 5S ribosomal ribonucleic acids: A study with ribonucleases," *Biochemistry* 20:7301–7307.

62. R. A. Garrett and S. O. Olesen (1982), "Structure of eukaryotic 5S ribonucleic acid: A study of *Saccharomyces cerevisiae* 5S ribonucleic acid with ribonucleases," *Biochemistry* 21:4823–4830.

63. T. Pieler and V. A. Erdmann (1982), "Three-dimensional structural model of eubacterial 5S RNA that has functional implications," *Proc. Nat. Acad. Sci. U.S.A.* 79:4599–4603.

64. V. D. Vasiliev, O. M. Selivanova, and V. E. Koteliansky (1978) "Specific self-packing of the ribosomal 16S RNA," *FEBS Letters* 95:273–276.

65. K. Edwards, J. Redbrook, T. Dyer, and H. Kössel (1981), "5S rRNA from *Zea mays* chloroplasts shows structural homology with the 3′-end of prokaryotic 23S rRNA," *Biochem. Int.* 2:533–538.

66. V. D. Vasiliev and O. M. Zalite (1980), "Specific compact self-packing of the ribosomal 23S RNA," *FEBS Letters* 121:101–104.

Further reading

Bogdanov, A. A.; Kopylov, A. M.; and Shatsky, I. N. (1980). The role of ribonucleic acids in the organization and functioning of ribosomes of *E. coli*. Subcellular biochemistry (D. B. Roodyn, ed.), vol. 7, pp. 81–116. New York and London: Plenum Press.

Brimacombe, R.; Maly, P.; and Zwieb, C. (1983). The structure of ribosomal RNA and its organization relative to ribosomal protein, Progress in nucleic acid research and molecular biology (W. E. Cohn, ed.), vol. 28, pp. 1–48. New York: Academic Press.

Brimacombe, R.; Nierhaus, K. H.; Garrett, R. A.; and Wittmann, H. G. (1976). The ribosome of *Escherichia coli*. Progress in nucleic acid research and molecular biology (W. E. Cohn, ed.), vol. 18, pp. 1–44. New York: Academic Press.

Cantor, C. R. (1980). Physical and chemical techniques for the study of RNA structure on the ribosome. In *Ribosomes: Structure, function, and genetics* (G. Chambliss, G. R. Craven, J. Davies, K. Davis, L. Kahan, and M. Nomura, eds.), pp. 23–49. Baltimore: University Park Press.

Erdmann, V. A. (1976). Structure and function of 5S and 5.8S RNA. Progress in nucleic acid research and molecular, biology (W. E. Cohn, ed.), vol. 18, pp. 45–90. New York: Academic Press.

Fellner, P. (1974). Structure of the 16S and 23S ribosomal RNAs. In *Ribosomes* (M. Nomura, A. Tissières, and P. Lengyel, eds.), pp. 169–191. Cold Spring Harbor, N.Y.: Cold Spring Harbor Laboratory.

Maden, B. E. H.; Salim, M.; and Robertson, J. S. (1974). Progress in the structural analysis of mammalian 45S and ribosomal RNA. In *Ribosomes* (M. Nomura, A. Tissières, and P. Lengyel, eds.), pp. 829–839. Cold Spring Harbor, N.Y.: Cold Spring Harbor Laboratory.

Monier, R. (1972). Structure and function of ribosomal RNA. In *The mechanism of protein synthesis and its regulation* (L. Bosch, ed.), pp. 353–394. Amsterdam and London: North-Holland.

——— (1974). 5S RNA. In *Ribosomes* (M. Nomura, A. Tissières, and P. Lengyel, eds.), pp. 141–168. Cold Spring Harbor, N.Y.: Cold Spring Harbor Laboratory.

Noller, H. F., (1980). Structure and topography of ribosomal RNA. In *Ribosomes: Structure, function, and genetics* (G. Chambliss, G. R. Craven, J. Davies, K. Davis, L. Kahan, and M. Nomura, eds.), pp. 3–22. Baltimore: University Park Press.

Osawa, S., and Hori, H. (1980). Molecular evolution of ribosomal components, In *Ribosomes: Structure, function, and genetics* (G. Chambliss, G. R. Craven, J. Davies, K. Davis, L. Kahan, and M. Nomura, eds.), pp. 333–355. Baltimore: University Park Press.

Singhal, R. P., and Shaw, J. K. (1983). Prokaryotic and eukaryotic 5S RNAs: Primary sequences and proposed secondary structures, Progress in nucleic acid research and molecular biology (W. E. Cohn, ed.), vol. 28, pp. 117–209. New York: Academic Press.

Spirin, A. S. (1964). *Macromolecular structure of ribonucleic acids*. New York: Reinhold.

Walker, T. A., and Pace, N. R. (1983). 5.8S ribosomal RNA. *Cell* 33:320–322.

Chapter 8

Ribosomal Proteins

8-1 Diversity: nomenclature

Each of the two ribosomal subunits contains many different proteins most of which are represented by only one copy per ribosome. This is a fundamental difference between the structurally asymmetric ribosomal ribonucleoprotein and symmetric viral nucleoproteins which are formed by the ordered packaging of many identical protein subunits. The discovery in the pioneering studies of Waller[1,2] that the ribosome contains many nonidentical protein molecules established an important principle of the structural organization of ribosomes.

The best technique for analytically separating ribosomal proteins is gel electrophoresis. Even one-dimensional gel electrophoresis under denaturing conditions gives a considerable fractionation of ribosomal proteins by charge and molecular size. Moderately basic polypeptides predominate among ribosomal proteins from most organisms, although several neutral and acidic proteins are always present as well. The molecular masses of ribosomal proteins are usually in the range of 10,000 to 30,000 daltons; just a few proteins have a greater size, up to

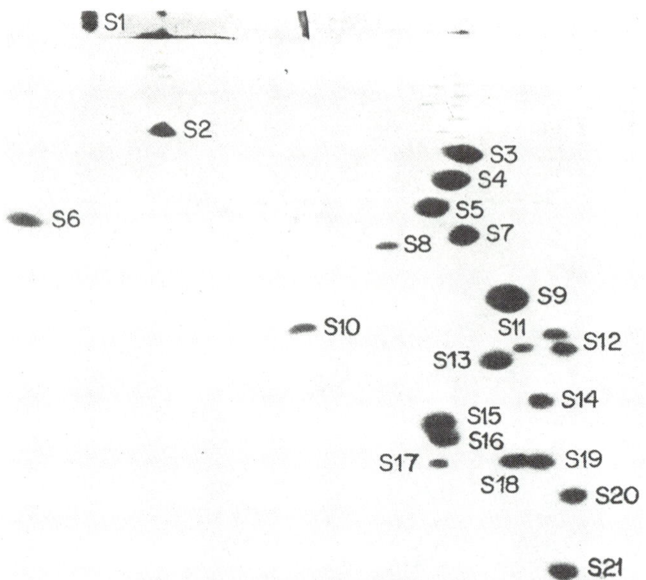

Figure 52 Two-dimensional electrophoretic separation and nomenclature of proteins from the *E. coli* 30S ribosomal subunit[4]. First direction (horizontal): 4% polyacrylamide gel containing 8 M urea, pH 8.6; second direction (vertical, downward): 18% polyacrylamide gel containing 6 M urea, pH 4.6.

about 60,000 or 70,000 daltons. The small subunit (30S) of procaryotic ribosomes contains about 20 proteins, while there are around 30 in the large ribosomal subunit (50S). Eucaryotic ribosomes contain a broader spectrum of proteins: the small (40S) subunit contains about 30 proteins, and the large (60S) about 40. Nearly all of these proteins are present as a single copy per ribosome.

A complete analytical resolution of all ribosomal proteins may be achieved by two-dimensional gel electrophoresis under denaturing conditions. A convenient separation system has been described by Kaltschmidt and Wittmann[3]: they used 8% polyacrylamide gel at pH 8.6 for electrophoresis in the first direction, and 18% gel at pH 4.6 in the second direction. These conditions resulted in the complete separation of all proteins present in the 30S or 50S ribosomal subunits of *Escherichia coli*. An example of such a separation of proteins from 30S and 50S ribosomal subunits in a slightly modified system is given in figs. 52 and 53, respectively. The first electrophoretic separation in a less concentrated gel at neutral or slightly alkaline pH results in the migration of acidic and neutral proteins toward the anode (left), while the basic proteins migrate toward the cathode (right); here the

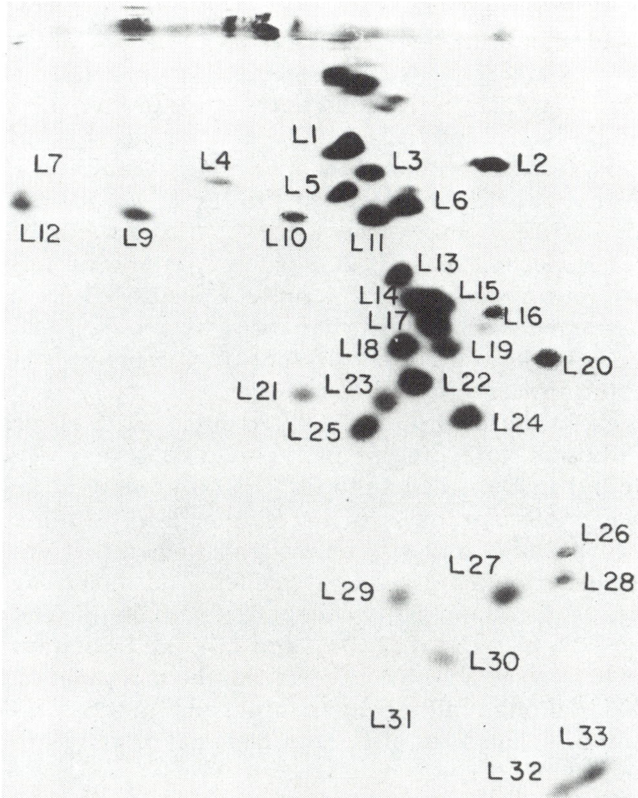

Figure 53 Two-dimensional electrophoretic separation and nomenclature of proteins from the *E. coli* 50S ribosomal subunit[4]. (L36 spot is not seen in this electrophoregram.) Separation conditions are as in fig. 52.

separation is largely on the basis of charge. Electrophoresis in the second direction is conducted in a highly crosslinked gel at acidic pH, and all the proteins migrate toward the cathode (downward); in this case the separation occurs largely on the basis of the molecular size of the components (the smaller the size the greater the mobility).

Separation in the above system provides the basis of the nomenclature of ribosomal proteins. It has been proposed that ribosomal proteins be designated by numbering in a downward direction[4], as seen from the two-dimensional electrophoretic separation patterns (figs. 52 and 53). Proteins of the small ribosomal subunit (30S or 40S) are denoted by the letter *S* (S1, S2, S3, etc.), while proteins of the large subunit (50S or 60S) are designated by the letter *L* (L1, L2, L3, etc.). The small *E. coli* ribosomal subunit contains 21 proteins, from S1 to

S21. The large ribosomal subunit contains 32 different proteins, from L1 to L34; the spot initially referred to as L8 is not an individual protein but a complex between proteins L10 and L12; the spots designated as L7 and L12 correspond to the same protein, L7 being the N-acetylated derivative of L12. Protein S20 of the small ribosomal subunit is identical to protein L26 of the large subunit. Therefore, there are 52 different ribosomal proteins in the *E. coli* 70S ribosome.

The acidic L7/L12 protein present in *E. coli* ribosomes (120 amino acid residues, molecular mass 12,200 daltons[5]) is unique in the sense that there are four molecules of this protein per ribosome; it appears to form a tetramer with a molecular mass of about 50,000 daltons. Analogs of this acidic protein are present in ribosomes of all studied organisms, both procaryotic and eucaryotic.

With the exception of proteins S20/L26 and L7/L12 all proteins in the *E. coli* ribosome are present in a single copy only. The largest protein is found in the small subunit. This is protein S1 (557 amino acid residues, molecular mass 61,160 daltons[6]); it is rather loosely bound in the ribosomes and may be easily lost when they are isolated. Other ribosomal proteins are much smaller. The molecular mass of larger species, e.g. S2 and S3, is about 26,000 daltons. The smallest proteins, e.g. L29, L30, L31, L32, L33, and L34, are basic polypeptides with lengths of 50 to 60 amino acid residues (the molecular mass being 5000 to 7000 daltons); all of them are constituents of the large ribosomal subunit. The sizes of *E. coli* ribosomal proteins are given in table 1[7–10].

Eucaryotic ribosomes contain proteins with molecular masses ranging from 26,000 to 60,000 daltons in addition to smaller proteins (see Bielka (1982)).

8-2 Primary structures

The primary structures of all the *E. coli* ribosomal proteins have been determined. The majority of proteins have been sequenced by Wittmann–Liebold's group in Wittmann's laboratory[7,11,12]. A comparison of the primary structures showed a lack of any sequence similarities: each ribosomal protein is unique by the criterion of its amino acid sequence, and no homologies between different ribosomal proteins have been detected. Thus the *E. coli* ribosome contains 52 different amino acid sequences possessing neither common blocks nor homologous regions.

Table 1 Size of *E. coli* ribosomal proteins

Protein	*Number of amino acid residues*	*Molecular mass (daltons)*	*Radius of gyration (Å)*
S1	557	61,159	58, 63*(?)
S2	240	26,613	
S3	232	25,852	16*
S4	203	23,138	18, 31*
S5	166	17,514	14*
S6	131	15,187	14*
S7	177	19,732	15, 14*
S8	129	13,996	13, 13*
S9	128	14,569	13*
S10	103	11,736	12*
S11	128	13,728	13*
S12	123	13,608	13*
S13	117	12,968	
S14	98	11,191	
S15	87	10,001	13, 12*
S16	82	9,192	12
S17	83	9,573	
S18	74	8,896	
S19	91	10,299	
S20 = L26	86	9,553	
S21	70	8,369	
L1	233	24,599	26*
L2	272	29,730	
L3	209	22,258	22*
L4	201	22,087	20*
L5	178	20,171	
L6	176	18,832	
L7/L12	120	12,206	
L9	148	15,696	
L10	165	17,736	
L11	141	14,872	
L13	142	16,019	
L14	120	13,227	
L15	144	14,981	
L16	136	15,295	
L17	127	14,365	
L18	117	12,770	
L19	114	13,002	
L20	117	13,366	
L21	103	11,564	
L22	110	12,226	

Table 1 *(Continued)*

Protein	*Number of amino acid residues*	*Molecular mass (daltons)*	*Radius of gyration (Å)*
L23	99	11,013	13*
L24	103	11,185	
L25	94	10,694	
L27	84	8,993	
L28	77	8,875	
L29	63	7,272	
L30	58	6,411	
L31	62	6,971	
L32	56	6,315	
L33	54	6,255	
L34	46	5,380	

The values of radii of gyration without asterisks are minimal values obtained for isolated proteins in the compact conformation[8]. The values obtained using neutron scattering for corresponding proteins *in situ*, without protein isolation, are indicated by asterisks[9,10].

No general peculiarities of the primary structures of ribosomal proteins, compared with normal soluble globular proteins, have been found. Most sequences, however, contain a high number of lysine and arginine residues which are sometimes clustered in lysine/arginine-rich blocks; this fact appears to be directly related to the RNA-binding properties of ribosomal proteins and results in the net positive charge of their molecules. Many ribosomal proteins do not contain tryptophan.

The primary structures of ribosomal proteins seem to be extremely conservative in evolution[12–14]. At any rate, there is extensive homology of amino acid sequences between the equivalent ribosomal proteins of two taxonomically distant groups of eubacteria, gram-negative (*Escherichia*) and gram-positive (*Bacillus*); as a rule, at least 50% of the amino acid residues in the polypeptide chains of corresponding proteins are identical. Similarly, considerable homology of amino acid sequences has been found for ribosomal proteins of various evolutionarily distant eucaryotic organisms; animals, higher plants, and fungi reveal homology (40 to 50%) between some of the equivalent ribosomal proteins. It is noteworthy that archebacteria (or metabacteria) also show some homology (up to 20 or 30%) between their ribosomal proteins and the equivalent ribosomal proteins of eucaryotes, but little or no homology with the ribosomal proteins of eubacteria.

8-3 Three-dimensional structures

Generally, ribosomal proteins have compact globular conformations with well-developed secondary and tertiary structures[8,15,16]. Some time ago it was widely believed that ribosomal proteins had an unusual three-dimensional structure and differed markedly from the usual globular proteins; it was reported that they possessed highly extended or very loose conformations, sometimes branched throughout the ribosomal subunit[11]. It soon became clear, however, that the loose conformations of isolated ribosomal proteins are an artifact, caused by the low stability of the native conformation. The conformations of ribosomal proteins in the ribosome are stabilized by interactions with RNA and other ribosomal proteins. Nevertheless, if certain conditions are met, most ribosomal proteins may be isolated in a compact conformation which appears to resemble closely the native one. Furthermore, the compactness of a number of ribosomal proteins has been demonstrated *in situ*, within the ribosomal particle[9,10].

Thus it appears, that ribosomal proteins are basically similar to other soluble globular proteins with regards to compactness and the general degree of their polypeptide chain folding into secondary and tertiary structures. The compactness of several ribosomal proteins compared to some other soluble globular proteins in terms of radii of gyration is demonstrated in fig. 54. The top portion of fig. 55 illustrates the circular dichroism spectrum of an *E. coli* ribosomal protein, e.g. S15; the spectrum demonstrates a high proportion of α-helical secondary structure[15]. The bottom part of fig. 55 shows the nuclear proton magnetic resonance spectrum of the same protein; NMR provides information on a specific environment for the residues of aromatic amino acids, histidine residues, and methyl groups of protein; the conclusion is that the protein has a developed tertiary structure[15].

At the same time, at least some ribosomal proteins may possess noncompact "tails." For example, the *E. coli* protein S6 possesses a strongly acidic C-terminal sequence containing several glutamic acid residues at the end; it is unlikely that this sequence is included in the globular part of this protein. Protein S7 is also a typical compact globular protein with developed secondary and tertiary structures, but in *E. coli* strain K, it has an additional sequence at the C-end, which does not seem to be an indispensable part of the globular structure.

The acidic protein L7/L12 has a number of distinctive features. It has already been pointed out that this protein appears to form a tetramer in the ribosome. In solution it is stable in the dimeric form[17]. The dimer of the L7/L12 protein is a rigid extended rodlike molecule

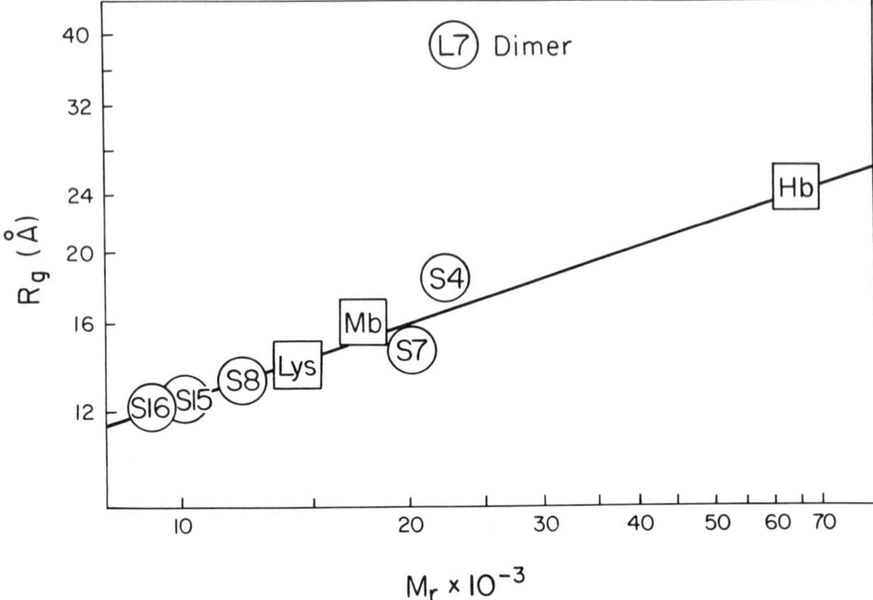

Figure 54 Dependence of the radii of gyration of several *E. coli* ribosomal proteins on their molecular mass[8]. For the sake of comparison the values of radii of gyration versus molecular masses are also given for such typically globular proteins as lysozyme (Lys), myoglobin (Mb), and hemoglobin (Hb). The slope of this dependence is equal to 1/3, which is characteristic of a homologous series of globular compact proteins. (The globular compact conformation does not imply a spherical shape and allows for certain variations in the axial ratio of the equivalent ellipsoid of revolution.)

with a radius of gyration of about 40 Å (the length is about 100 Å, provided the molecular mass is 25,000 daltons). These dimers appear to be packaged side-by-side in the tetramer forming a rodlike stalk in the 50S ribosomal subunit (see chapter 6). The monomeric subunit of the L7/L12 protein consists of two domains: the globular C-terminal one with about 70 to 80 amino acid residues, and the nonglobular extended N-terminal one with approximately 40 amino acid residues; they are connected by an easily cleavable hinge[18,19]. It is likely that the two N-terminal alanine-rich sequences form a rigid bihelical complex consisting of two parallel α-helices; two such pairs may form the main part of the rodlike stalk of the 50S ribosomal subunit in the ribosome. According to one model the helices are antiparallel in the dimer (fig. 56(a)), and the two globular domains in the tetramer form the proximal part of the L7/L12 stalk while two others form the distal part

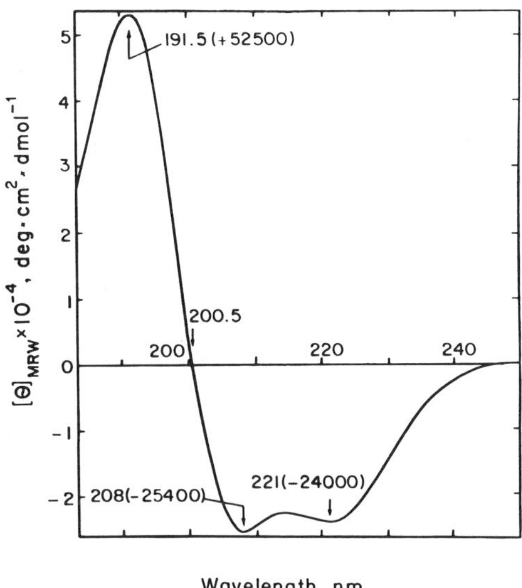

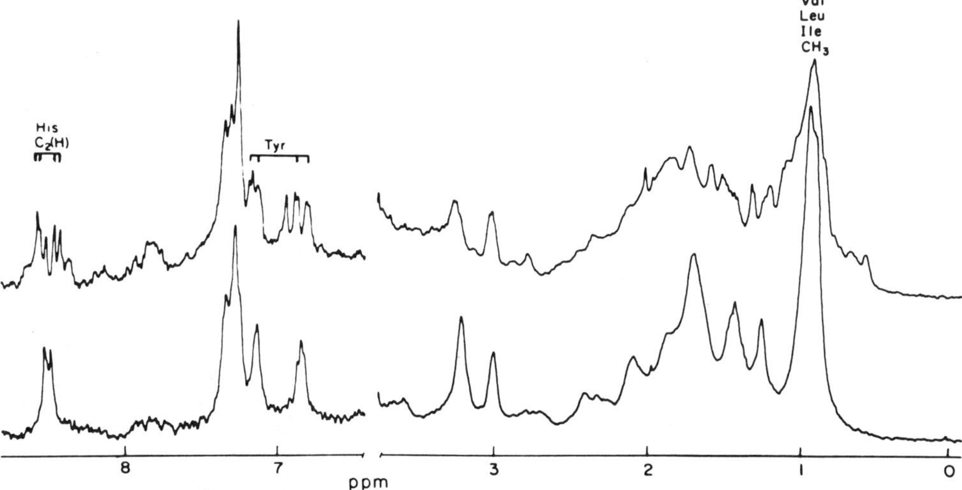

Figure 55 Spectra of circular dichroism and the proton nuclear magnetic resonance of the ribosomal protein S15 demonstrating the existance of a developed secondary and tertiary structure. *Top*, Circular dichroism spectrum. The content of the α-helical conformation calculated from the spectrum is no less than 70%. *Bottom*, Proton NMR spectrum of protein S15 in compact form (*above*) compared to the spectrum of the same protein under denaturing conditions in 5-M urea (*below*). The spectrum of the compact protein shows resonance lines resulting from the interaction of alipahatic amino acid residues with aromatic residues in the extreme high-field spectrum region (0.1 to 0.8 p.p.m.), wide overlapping lines of nonpolar aliphatic residues in the high-field region (0.8 to 3.5 p.p.m.), and shifts of the signals from aromatic residues in the low-field region (6.5 to 8.5 p.p.m.), which are characteristic of protein tertiary folding. (Reproduced from Z. V. Gogia et al. (1974), *FEBS Letters* 105:63–69, with permission.)

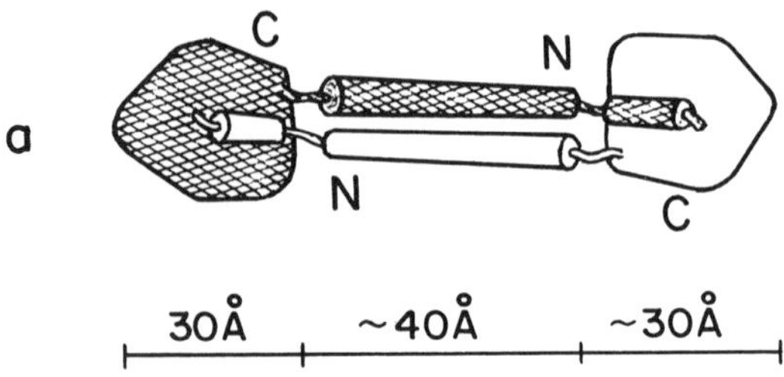

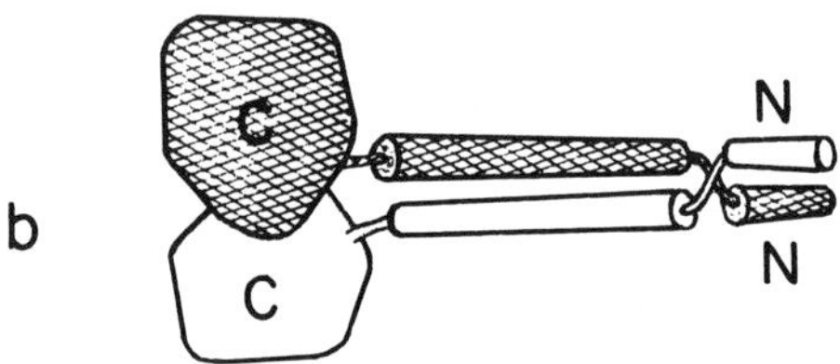

Figure 56 Scheme illustrating two possibilities for dimerization of ribosomal protein L7/L12. (a) Antiparallel arrangement. (b) Parallel arrangement. N and C are the rodlike N-terminal and globular C-terminal domains, respectively.

(dumbbelllike structure)[18]. According to another model all subunits of the L7/L12 protein are positioned parallel to each other (fig. 56(b)), and the four globules form either the base or the tip of the L7/L12 stalk[20,21].

The globular domain of the *E. coli* L7/L12 protein (fragment 47–120) was the first protein element of the ribosome to be crystallized and studied by X-ray analysis. Its three-dimensional structure has been solved at a 1.7-Å level of resolution[21,22]. The secondary structure of the globule includes α-helices and β-sheets; their sequences along the polypeptide chain is as follows: $\beta\alpha\alpha\beta\alpha\beta$. They form two sheets; three α-helices are arranged into one sheet while the antiparallel β-structure consisting of three strands forms the other sheet. The three-dimensional structure of the L7/L12 protein globular domain is shown schematically in fig. 57.

8-4 Protein complexes

A relatively mild technique for dissociating ribosomal proteins from the ribosome includes treatment with monovalent salts such as CsCl,

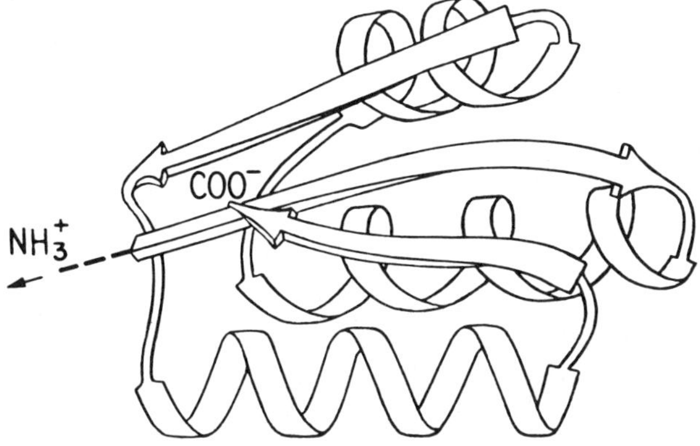

Figure 57 Schematic representation of the three-dimensional structure of the globular C-terminal domain of ribosomal protein L7/L12: ribbon drawing of the course of the polypeptide chain. (Redrawn from A. Liljas (1982), *Prog. Biophys. Molec. Biol.* 40:161–228; and M. Leijonmarck, S. Eriksson, and A. Liljas (1980), *Nature* 286:824–826, with permission.)

LiCl, or NH_4Cl at high concentrations. As a result of such treatment, many proteins dissociate in groups rather than as individual molecules. This reflects a certain cooperativity of protein retention within ribosomal subunits. In a number of cases such groups of proteins may be removed from the particles as stable complexes. A pentamer formed by one molecule of protein L10 and the tetramer of protein L7/L12 is an example of such a stable complex[21–23]. It can be selectively removed from the 50S ribosomal subunit by treatment with 1:1 mixture of 1M NH_4Cl and ethanol. As a result, the 50S subunit loses its stalk. Globular structures of both protein L10 and protein L7/L12 within the complex are markedly more stable than in the individual state. The pentameric complex may be loosely associated with yet another protein, L11.

Another example of the complex is the pair of proteins S6–S18[24]. Individually, these proteins, especially S18, are unstable; it is difficult to prepare S18 in a compact conformation in an isolated state under normal isolation conditions. When these proteins form a complex, however, they stabilize each other. The complex looks like a compact structure of the globular type.

Protein-protein complexes may play an important part in the sturcture and function of the ribosome, participating in the formation of the quaternary structures of ribosomal functional sites.

8-5 Interactions with ribosomal RNA

The quaternary structures of ribosomal functional sites are formed with the participation of ribosomal RNA as well.

The low-molecular-mass 5S ribosomal RNA interacts with a number of proteins in the 50S subunit, forming a complex located in the region of the central protuberance (see chapters 6 and 9). Three proteins—L5, L18, and L25—form a rather stable nucleoprotein complex with the 5S RNA[25]. Other proteins, e.g. L2, L3, L15, L16, L17, L21, L22, L23, and L34, interact with this complex, but more loosely. This large complex even displays some functional characteristics: it can bind tRNA and has a certain affinity to the 30S ribosomal subunit[26].

A similar situation has been observed in the case of eucaryotic ribosomes. The 5S RNA of the rat liver ribosome forms a complex with proteins L5, L6, and L18 of the 60S subunit; proteins L7, L8, and L35 bind to this complex but not so strongly. The eucaryotic 5.8S RNA which is the structural homolog of the procaryotic 23S RNA 5′-terminal sequence may form a complex with virtually the same set of proteins: L5, L6, L7, and L18. As a result, a common complex containing 5S RNA, as well as proteins L5, L6, L18, and others, may be formed[27]. This eucaryotic ribosomal complex possesses some functional activity: it can bind tRNA.

Of particular interest, of course, are the interactions between the ribosomal proteins and the high-molecular-mass ribosomal RNA (16S and 23S procaryotic RNAs or 18S and 28S eucaryotic RNAs), since these RNA species serve as the main covalent backbone and the structural core of the ribosomal subunits. It appears that most ribosomal proteins are in contact with high-molecular-mass ribosomal RNAs. Some special *core proteins* may, however, be distinguished among ribosomal proteins; they bind tightly to the corresponding ribosomal RNA, more or less independently of other proteins. In the case of the *E. coli* 30S ribosomal subunit, proteins S4, S7, S8, S15, S17, and S20 are the core proteins that can independently interact with the 16S RNA[28,29]. Each of these proteins binds only to a specific site on the 16S RNA, recognizing the corresponding nucleotide sequence and three-dimensional structure. The sequence to which a given protein is bound can be identified as follows: an isolated protein is added to the ribosomal RNA, resulting in the specific protein-RNA complex being formed; the complex is digested by ribonuclease, leaving intact only the part of the nucleotide sequence which is protected by the protein; the protected region may then be sequenced and therefore identified. Another technique used for locating proteins on the ribosomal RNA sequence involves the covalent crosslinking (e.g. photo-induced cross-

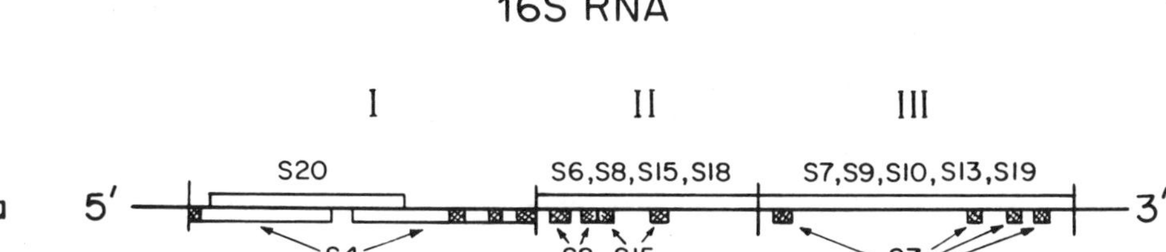

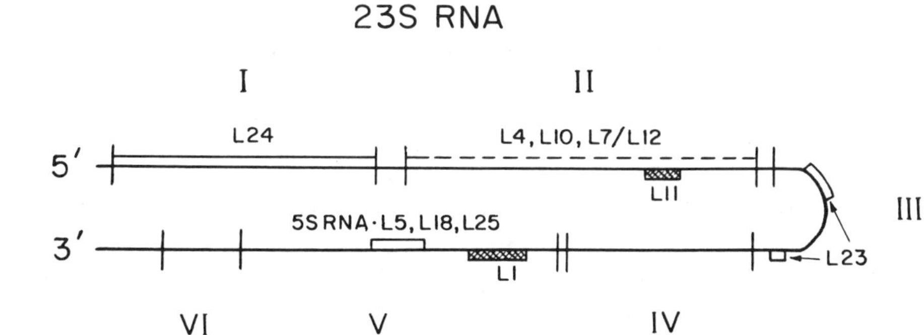

Figure 58 Scheme illustrating the arrangement of the binding sites of core ribosomal proteins along (a) 16S and (b) 23S ribosomal RNA chains[30–32]. Roman numerals designate RNA sections forming the corresponding structural domains.

linking) of protein with RNA directly in the ribosome, followed by the removal of noncrosslinked proteins, the digestion of RNA by RNAase, and the identification of the crosslinked oligonucleotide. The location of the binding sites of the six above-mentioned proteins along the 16S chain is shown schematically in fig. 58(a)[30,31]. It is seen that proteins S4 and S20 (as well as S17 which is not shown in fig.) are complexed in the region of the 5′-terminal third of the 16S RNA (domain I); proteins S8 and S15 interact with the middle part of the 16S RNA (domain II), while protein S7 has its binding site located in the region of the 3′-terminal third of this RNA (domain III).

The regions of the 16S secondary structure recognized by the independently binding proteins may be indicated[32]. Protein S8 recognizes, binds, and protects the long compound helix 25–26 within domain II (see chapter 7, fig. 41). Protein S15 occupies the adjacent long compound helix 28–29–30. Protein S4 recognizes and binds the hairpin including helices 20 and 21, and probably the bases of the two immediately adjacent helices 2 and 18, one of which is the stem of domain I; protein S4 protects from nucleases far more extended chain regions by stabilizing the compact structure of the whole domain I. Protein S7 recognizes and binds the region of domain III, where several helices, 39, 40, 52, and 54–55, are bunched together.

As for *E. coli* 23S ribosomal RNA, proteins L1, L2, L3, L4, L6, L9, L11, L20, L23, L24, and some others specifically bind to various regions, more or less independently of one another. The pentameric complex $(L7/L12)_4 \cdot L10$ should be added to this list. The scheme illustrating the arrangement of the binding sites of some proteins along the 23S RNA chain is given in fig. 58(b)[31,32].

Other ribosomal proteins interacting with ribosomal RNA require the presence of at least one, even several, core RNA-binding proteins for the formation of sufficiently firm complexes. Two patterns are possible: either an intrinsic interaction between a given protein and RNA is insufficient for the stable complex to be formed, and should therefore be supported by protein-protein interaction with the already bound protein; or, alternatively, the protein bound earlier induces (or stabilizes) the local conformation of the RNA required for binding a given protein. For example, the binding of protein S7 to the *E. coli* 16S RNA contributes to a tighter binding of proteins S9, S13, and S19, as well as of S10 and S14, in the 3′-proximal 16S RNA region (domain III); it may well be that S9 and S13–S19 directly interact with S7, while the effect of S7 on the binding of S10 and S14 is less direct[29] (fig. 59). In the case of the *E. coli* 50S subunit, protein L15 has intrinsic affinity to 23S RNA but its tight retention requires proteins L3 and L4; in return, the binding of L15 strengthens the binding of proteins L6, L11, L16, and L18 to the 23S RNA and creates conditions for the binding of proteins L25 and L27 (fig. 60)[33].

It is possible that some ribosomal proteins do not possess an intrinsic affinity to ribosomal RNA and are kept in the complex only by protein-protein interactions.

Thus, both in the small 30S and in the large 50S ribosomal subunits, certain cooperative groups of proteins assigned to definite sites of the three-dimensional ribosomal RNA structure can be revealed. In the case of the *E. coli* 30S ribosomal subunit, one such group is formed by proteins of the 3′-proximal domain of 16S RNA (domain III): S7, S9, S13, S19, S10, and S14 (fig. 59), as well as S3 and S2. It will be demonstrated later that they are all located on the head of the 30S subunit and contribute to the formation of its tRNA-binding site. Another cooperative group of 30S subunit proteins is associated with

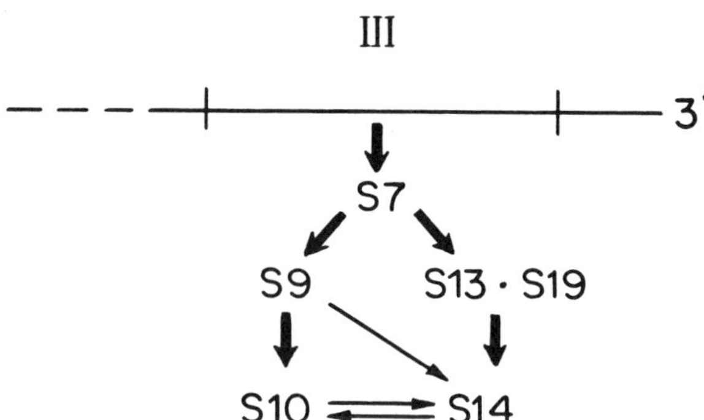

Figure 59 Group of interdependent ribosomal proteins bound in the region of the 3′-proximal domain (III) of 16S ribosomal RNA[29]. The arrows indicate the direction in which one component (RNA or protein) exerts a stimulatory effect on the binding of another component (protein). Major effects are indicated by thick arrows, and lesser effects by thinner ones.

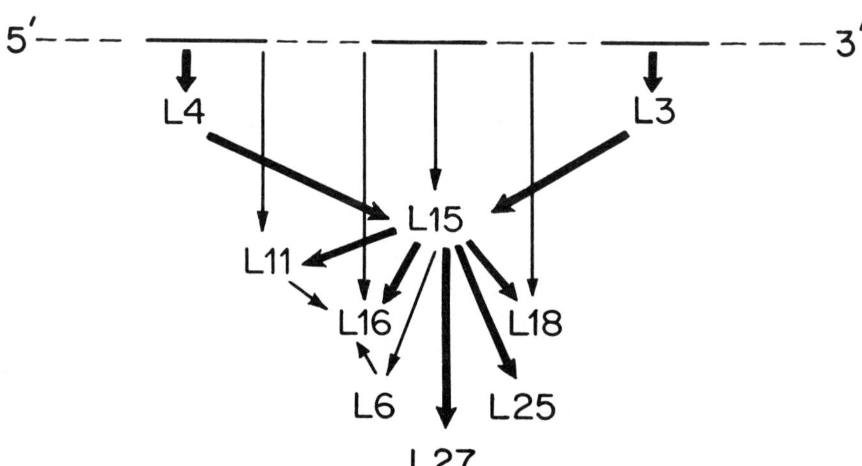

Figure 60 Interdependence of the retention of several ribosomal proteins on 23S ribosomal RNA[33]. Arrows as in fig. 59.

the middle domain of the 16S RNA (domain II) and includes the RNA-binding proteins S8 and S15, the S6–S18 pair, as well as S11 and S21; these proteins are found mainly on the side bulge or platform of the 30S ribosomal subunit. This group, through proteins S5 and S12, is connected with the cooperative group of the 16S RNA 5′-terminal domain (domain I); the latter group includes RNA-binding proteins S4, S17, and S20, as well as proteins S12 and S16, which are constituents of the central body of the 30S ribosomal subunit.

References

1. J.-P. Waller and J. I. Harris (1961), "Studies on the composition of the proteins from *Escherichia coli* ribosomes," *Proc. Nat. Acad. Sci. U.S.A.* 47:18–23.

2. J.-P. Waller (1964), "Fractionation of the ribosomal protein from *Escherichia coli*," *J. Mol. Biol.* 10:319–336.

3. E. Kaltschmidt and H. G. Wittmann (1970), "Ribosomal proteins. VII. Two-dimensional polyacrylamide gel electrophoresis for fingerprinting of ribosomal proteins," *Anal. Biochem.* 36:401–412.

4. E. Kaltschmidt and H. G. Wittmann (1970), "Ribosomal proteins. XII. Number of proteins in small and large ribosomal subunits of *E. coli* as determined by two-dimensional gel electrophoresis," *Proc. Nat. Acad. Sci. U.S.A.* 67:1276–1282.

5. C. P. Terhorst, W. Möller, R. Laursen, and B. Wittmann–Liebold (1973), "The primary structure of an acidic protein from 50S ribosomes of *Escherichia coli* which is involved in GTP hydrolysis dependent on elongation factors G and T," *Eur. J. Biochem.* 34:138–152.

6. M. Kimura, K. Foulaki, A.-R. Subramanian, and B. Wittmann–Liebold (1982), "Primary structure of *Escherichia coli* ribosomal protein S1 and features of its functional domains," *Eur. J. Biochem.* 123:37–53.

7. H. G. Wittmann (1982), "Components of bacterial ribosomes," *Ann. Rev. Biochem.* 51:155–183.

8. I. N. Serdyuk, G. Zaccai, and A. S. Spirin (1978), "Globular conformation of some ribosomal proteins in solution," *FEBS Letters* 94:349–352.

9. V. R. Ramakrishnan, S. Yabuki, I.-Y. Sillers, D. G. Schindler, D. M. Engelman, and P. B. Moore (1981), "Positions of proteins S6, S11, and S15 in the 30S ribosomal subunit of *Escherichia coli*," *J. Mol. Biol.* 153:739–760.

10. K. H. Nierhaus, R. Lietzke, R. P. May, V. Nowotny, H. Schulze, K. Simpson, P. Wurmbach, and H. B. Stuhrmann (1983), "Shape determinations of ribosomal proteins *in situ*," *Proc. Nat. Acad. Sci. U.S.A.* 80:2889–2893.

11. G. Stöffler and H. G. Wittmann (1977), "Primary structure and three-dimensional arrangement of proteins within the *Escherichia coli* ribosome," in *Molecular mechanisms of protein biosynthesis*, ed. H. Weissbach and S. Pestka, pp. 117–202 (New York: Academic Press).

12. H. G. Wittmann, J. A. Littlechild, and B. Wittmann–Liebold (1980), "Structure of ribosomal proteins," in *Ribosomes: Structure, function, and genetics*, ed. G. Chambliss, G. R. Craven, J. Davies, K. Davis, L. Kahan, and M. Nomura, pp. 51–88 (Baltimore: University Park Press).

13. A. T. Matheson, W. Möller, R. Amons, and M. Yaguchi (1980), "Comparative studies on the structure of ribosomal proteins with emphasis on the alanine-rich, acidic ribosomal, "A" protein," in *Ribosomes: Structure, function, and genetics*, ed. G. Chambliss, G. R. Craven, J. Davies, K. Davis, L. Kahan, and M. Nomura, pp. 297–332 (Baltimore: University Park Press).

14. S. Osawa and H. Hori (1980), "Molecular evolution of ribosomal components," in *Ribosomes: Structure, function, and genetics*, ed. G. Chambliss, G. R. Craven, J. Davies, K. Davis, L. Kahan, and M. Nomura, pp. 333–355 (Baltimore: University Park Press).

15. Z. V. Gogia, S. Yu. Venyaminov, V. N. Bushuev, I. N. Serdyuk, V. I. Lim, and A. S. Spirin (1979), "Compact globular structure of protein S15 from *Escherichia coli* ribosomes," *FEBS Letters* 105:63–69.

16. I. N. Serdyuk, Z. V. Gogia, S. Yu. Venyaminov, N. N. Khechinashvili, V. N. Bushuev, and A. S. Spirin (1980), "Compact globular conformation of protein S4 from *Escherichia coli* ribosomes," *J. Mol. Biol.* 137:93–107.

17. W. Möller, A. Groene, C. Terhorst, and R. Amons (1972), "50S ribosomal proteins: Purification and partial characterization of two acidic proteins, A_1 and A_2, isolated from 50S ribosomes of *Escherichia coli*," *Eur. J. Biochem.* 25:5–12.

18. A. T. Gudkov, J. Behlke, N. V. Vtiurin, and V. I. Lim (1977), "Tertiary and quaternary structure for ribosomal protein L7 in solution," *FEBS Letters* 82:125–129.

19. A. Liljas, S. Eriksson, D. Donner, and C. G. Kurland (1978), "Isolation and characterization of stable domains of the protein L7/L12 from *Escherichia coli* ribosomes," *FEBS Letters* 88:300–304.

20. C. A. Luer and K.-P. Wong (1979), "Conformation of *Escherichia coli* ribosomal protein L7/L12 in solution: Hydrodynamic, spectroscopic, and conformation prediction studies," *Biochemistry* 18:2019–2027.

21. A. Liljas (1982), "Structural studies of ribosomes," *Prog. Biophys. Molec. Biol.* 40:161–228.

22. M. Leijonmarck, S. Eriksson, and A. Liljas (1980), "Crystal structure of a ribosomal component at 2.6 Å resolution," *Nature* 286:824–826.

23. A. T. Gudkov, L. G. Tumanova, S. Yu. Venyaminov, and N. N. Khechinashvili (1978), "Stoichiometry and properties of the complex between ribosomal proteins L7 and L10 in solution," *FEBS Letters* 93:215–218.

24. V. Prakash and K. C. Aune (1978), "Molecular interactions between ribosomal proteins: A study of the S6–S18 interaction," *Arch. Biochem. Biophys.* 187:399–405.

25. J. R. Horne and V. A. Erdmann (1972), "Isolation and characterization of 5S RNA-protein complexes from *Bacillus stearothermophilus* and *Escherichia coli* ribosomes," *Molec. Gen. Genetics* 119:337–344.

26. E. Metspalu, M. Ustav, T. Maimets, and R. Villems (1982), "The composition and properties of the *Escherichia coli* 5S RNA-protein complex," *Eur. J. Biochem.* 121:383–389.

27. A. Metspalu, M. Saarma, R. Villems, M. Ustav, and A. Lind (1978), "Interaction of 5S RNA, 5.8S RNA, and tRNA with rat liver ribosomal proteins," *Eur. J. Biochem.* 91:73–81.

28. W. H. Schaup, M. Green, and C. G. Kurland (1970), "Molecular interactions of ribosomal components. I. Identification of RNA binding sites for individual 30S ribosomal proteins," *Molec. Gen. Genetics* 109:193–205.

29. S. Mizushima and M. Nomura (1970), "Assembly mapping of 30S ribosomal proteins from *E. coli*," *Nature* 226:1214–1218.

30. R. A. Zimmermann (1974), "RNA-protein interactions in the ribosome," in *Ribosomes*, ed. M. Nomura, A. Tissières, and P. Lengyel, pp. 225–269 (Cold Spring Harbor, N.Y.: Cold Spring Harbor Laboratory).

31. R. A. Zimmermann (1980), "Interactions among protein and RNA components of the ribosome," in *Ribosomes: Structure, functions, and genetics*, ed. G. Chambliss, G. R. Craven, J. Davies, K. Davis, L. Kahan, and M. Nomura, pp. 135–169 (Baltimore: University Park Press).

32. R. Brimacombe, P. Maly, and C. Zwieb (1983), "The structure of ribosomal RNA and its organization relative to ribosomal protein," Progress in nucleic acids research and molecular biology, ed. W. E. Cohn, vol. 28, pp. 1–48 (New York: Academic Press).

33. K. H. Nierhaus (1980), "Analysis of the assembly and function of the 50S subunit from *Escherichia coli* ribosomes by reconstitution," in *Ribosomes: Structure, function, and genetics*, ed. G. Chambliss, G. R. Craven, J. Davies, K. Davis, L. Kahan, and M. Nomura, pp. 267–294 (Baltimore: University Park Press).

Further reading

Bielka, H., ed. (1982). *The eukaryotic ribosome*. Berlin: Akademie-Verlag.

Brimacombe, R.; Nierhaus, K. H.; Garrett, R. A.; and Wittmann, H. G. (1976). The ribosome of *Escherichia coli*. Progress in nucleic acids research and molecular biology (W. E. Cohn, ed.), vol. 18, pp. 1–44. New York: Academic Press.

Kurland, C. G. (1974). Functional organization of the 30S ribosomal subunit. In *Ribosomes* (M. Nomura, A. Tissières, and P. Lengyel, eds), pp. 309–331. Cold Spring Harbor, N.Y.: Cold Spring Harbor Laboratory.

Nomura, M., and Held, W. A. (1974). Reconstitution of ribosomes: Studies of ribosome structure, function, and assembly. In *Ribosomes* (M. Nomura, A. Tissières, and P. Lengyel, eds.), pp. 193–223. Cold Spring Harbor, N.Y.: Cold Spring Harbor Laboratory.

Subramanian, A.-R. (1983). Structure and functions of ribosomal protein S1. Progress in nucleic acid research and molecular biology (W. E. Cohn, ed.) vol. 28, pp. 101–142. New York: Academic Press.

Wittmann, H. G. (1974). Purification and identification of *Escherichia coli* ribosomal proteins. In *Ribosomes* (M. Nomura, A. Tissières, and P. Lengyel, eds.), pp. 93–114. Cold Spring Harbor, N.Y.: Cold Spring Harbor Laboratory.

Wittmann, H. G., and Wittmann–Liebold, B. (1974). Chemical structure of bacterial ribosomal proteins. In *Ribosomes* (M. Nomura, A. Tissières, and P. Lengyel, eds.), pp. 115–140. Cold Spring Harbor, N.Y.: Cold Spring Harbor Laboratory.

Wool, I. G. (1980). The structure and function of eukaryotic ribosomes. In *Ribosomes: Structure, function, and genetics* (G. Chambliss, G. R. Craven, J. Davies, K. Davis, L. Kahan, and M. Nomura, eds.), pp. 797–824. Baltimore: University Park Press.

Wool, I. G., and Stöffler, G. (1974). Structure and function of eukaryotic ribosomes. In *Ribosomes* (M. Nomura, A. Tissières, and P. Lengyel, eds.), pp. 417–460. Cold Spring Harbor, N.Y.: Cold Spring Harbor Laboratory.

Chapter 9

Mutual Arrangement of Ribosomal RNA and Proteins (Quaternary Structure)

9-1 Peripheral localization of proteins on the RNA core

In contrast to the RNA present in viral nucleoproteins, the RNA of ribosomal particles is not entirely covered by a protein envelope. As was demonstrated many years ago, extended regions of ribosomal RNA in the ribosome are exposed to the environment and are open to the action of various agents, e.g. nucleases. This fundamental difference compared to viral particles is understandable, since the ribosome is a functional structure, where RNA should actively participate in interactions with external factors and is not used for storing genetic information.

At the same time, protein and RNA in the ribosome are not just "scrambled." The high-molecular-mass RNA of each ribosomal subunit is self-folded into a compact structure with a unique shape (see chapter 7), and it appears that proteins do not associate with the "inside" of this structure. Hence, ribosomal proteins are positioned mainly *on* the compactly folded high-molecular-mass RNA. This implies that proteins occupy a preferentially outside position on the RNA core.

This principle of ribosomal organization was first deduced from experiments conducted to measure the radii of gyration (R_g) of ribosomal subunits. The radius of gyration measured by the diffuse small-angle X-ray scattering was found to be markedly lower than expected on the basis of the size of the subunit assuming that it was a uniformly dense body. It followed from this observation that a more electron-dense component of the particle (e.g. RNA) lay nearer the center of gravity of the particle, while a less dense component (e.g. protein) tended to be closer to the periphery[1]. Furthermore, measurements of the radii of gyration of ribosomal subunits using different types of radiation, e.g. X rays, neutrons, and light, demonstrated that the greater the contribution to the total scattering by the protein component compared to RNA (the relative scattering capacity of the protein increases in the series from X rays to neutrons to light), the greater the value of the particle's radius of gyration (fig. 61)[2]. Finally, neutron-scattering experiments in solvents with a different scattering capacity for neutrons, i.e. with different proportions of H_2O and D_2O, allowed for direct measurement of the radii of gyration of either the RNA or the protein components *in situ*[3–5]. The basis is that H_2O and D_2O are known to differ greatly in their scattering capacity for neutrons, while the scattering capacities of biological macromolecules are intermediate between those of H_2O and D_2O. Because of this, a proportion between H_2O and D_2O in the medium can be selected when the scattering values of a given macromolecule, either protein or RNA, and the solvent are equal, i.e. a given type of macromolecule is not "seen" by neutrons or is contrast-matched. Experiments have shown that the neutron scattering of protein is matched by 40 to 42% D_2O, whereas RNA is not "seen" by neutrons in 70% D_2O. Correspondingly, measurements of the radii of gyration of ribosomal subunits in 42% D_2O yield values only for ribosomal RNA *in situ*, while measurements in 70% D_2O give the radius of gyration of the total protein component of the particle. In the case of *E. coli* 50S ribosomal subunits, these values were found to be equal to 65 Å and 100 Å for RNA and protein, respectively (fig. 61). In other words, the RNA is located preferentially at the center as a core, while protein, on average, occupies a more peripheral position. In the case of the 30S ribosomal subunit, this difference is less pronounced—65 Å and 80 Å for RNA and protein, respectively (fig. 61). This smaller difference is understandable since the 16S RNA, despite its lower mass, has a less compact shape or is less isometric than the 23S RNA of the 50S ribosomal subunit. Furthermore, some protein material may be

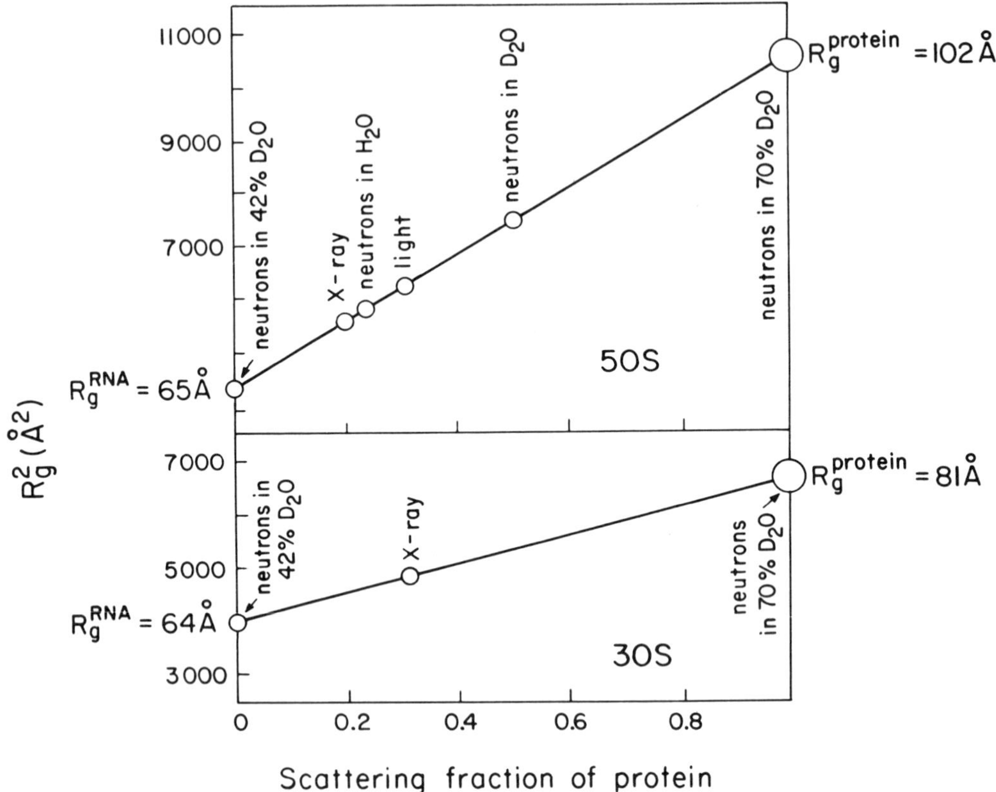

Figure 61 Dependence of the radii of gyration of the ribosomal particles measured by X-ray scattering, neutron scattering, and light scattering on the relative contribution of the protein component into the scattering[2,5]. The 50S subunits are on the upper plot and the 30S subunits are below. (Provided by Dr. I. N. Serdyuk, Institute of Protein Research, Pushchino, U.S.S.R.)

located between the lobes (branches) of the 16S RNA in the 30S ribosomal subunit.

A more central position of RNA and a more peripheral position of proteins in ribosomal subunits can also be demonstrated using a specialized technique of electron microscopy when the particles are embedded into media of a different electron-scattering density[6]. In this case the same principle of contrasting that is employed for diffuse scattering in the solution is used. When the particles are embedded into glucose, the electron-scattering densities of the medium and protein component are approximately equal, and only RNA can be seen. In other media both protein and RNA can be visualized. This technique has confirmed that RNA forms the central core of the ribosomal subunit; in addition, it has again been demonstrated that not all RNA is covered with proteins, and in a number of regions RNA is arranged on the surface of the particle.

The RNA core in the ribosomal subunit seems to be dense, i.e. the extent of RNA folding *in situ* is high. It follows from the value of the radius of gyration and the scattering curve that the volume of RNA in the 50S subunit is equal to only 2×10^6 Å^3. This value is only twice as much as the "dry" volume of RNA. A similar conclusion has been made for 16S RNA in the 30S ribosomal subunit. Therefore, the density of RNA packaging in the ribosomal particle is approximately equal to that found for the crystalline packaging of hydrated RNA helices or tRNA[5].

9-2 Topography of proteins

After the core position of ribosomal RNA is determined, elucidation of protein distribution on the surface of the particle, i.e. of *protein topography*, becomes the next crucial step toward the quaternary structure of the ribosome. A large number of experimental approaches to the study of protein topography have been developed. These approaches will now be discussed using the *E. coli* 30S ribosomal subunit as an example.

Identification of neighboring proteins

Some information regarding protein neighbors can even be taken from the data on protein binding sites upon the primary and secondary structure of ribosomal RNA[7] (see chapter 8, section 8.3). Indeed, if the binding sites of proteins on RNA are located close to each other, it is clear that these proteins are neighbors in the ribosome. For example, the previously discussed proteins S8 and S15 recognize and bind adjacent sections of the chain and adjacent hairpins in the secondary structure of 16S RNA (chapter 8, section 8.3); therefore it may be concluded that proteins S8 and S15 are neighbors in the topographic sense as well. Their neighbors are proteins S6 and S18, which for their binding require the preceding binding of proteins S8 and S15 and have the recognition sites in the same region of the RNA sequence.

Another example of a group of neighboring proteins includes proteins S4, S16, S17, and S20 which are located close to each other on the 16S chain within domain I.

A more universal approach makes use of bifunctional chemical reagents which are capable of crosslinking neighbor proteins with each other[8]. After treatment of ribosomal subunits with such reagents, the identification of proteins in the crosslinked pairs provides the means of establishing that the corresponding proteins are neighbors in the ribosome. Diimidoesters of a different carbon chain length have been very widely used as bifunctional crosslinking agents:

I

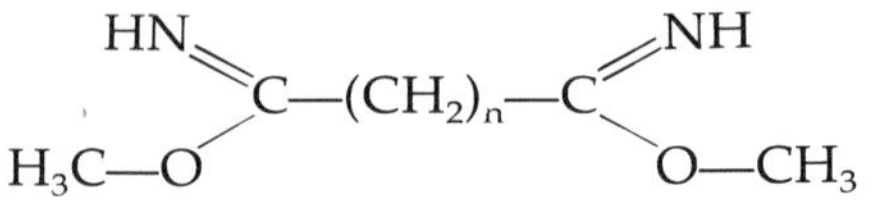

$$\begin{array}{ccc} HN & & NH \\ & \diagdown\!\!\!\!\diagdown \;\; C{-}(CH_2)_n{-}C \;\; /\!\!\!/ & \\ H_3C{-}O & & O{-}CH_3 \end{array}$$

The ester groups of such a reagent are effectively attacked by the ε-amino groups of lysyl residues present in ribosomal proteins, resulting in the formation of amidine bonds instead of ester bonds. Using reagents of a different length, e.g. dimethylsuberimidate ($n = 6$) or dimethyladipimidate ($n = 4$), permits a rough estimation of the distance between neighboring proteins. The identification of proteins in crosslinked pairs may present some problems because the corresponding proteins are not in the individual state. One possible solution involves immunological identification of the partners within the pair without separating them. Another approach makes use of cleavable crosslinks. For example, ribosomal particles may be treated by sulfhydryl derivatives of the imidoester, such as methyl-4-mercaptobutyrimidate:

Ia

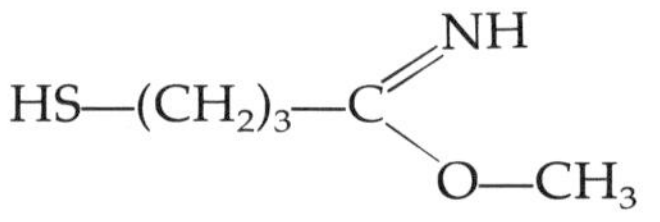

$$HS{-}(CH_2)_3{-}C\begin{array}{l} /\!\!\!/ NH \\ \diagdown O{-}CH_3 \end{array}$$

It reacts with protein amino groups. The subsequent oxidation yields pairs of proteins crosslinked by disulfide bridges.

Ib

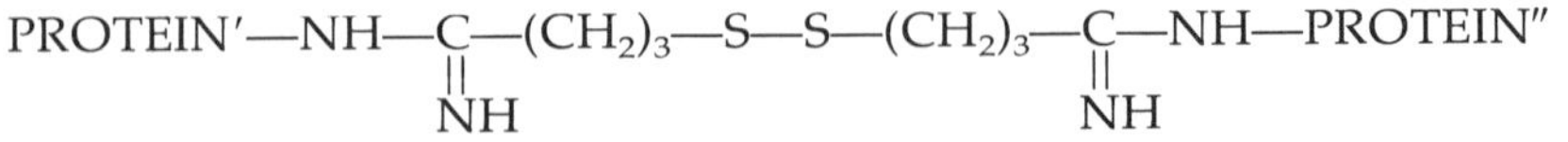

$$\text{PROTEIN}'{-}NH{-}\underset{\underset{NH}{\|}}{C}{-}(CH_2)_3{-}S{-}S{-}(CH_2)_3{-}\underset{\underset{NH}{\|}}{C}{-}NH{-}\text{PROTEIN}''$$

Pairs of proteins crosslinked in this way are isolated, the disulfide bonds reduced, and individual proteins identified electrophoretically. The following pairs of proteins crosslinked in the 30S ribosomal subunit by "short" crosslinking agents have been noted particularly frequently: S2–S3, S3–S10, S4–S5, S4–S17, S5–S8, S8–S15, S6–S18, S18–S21, S7–S9, S7–S13, S13–S19; they may be considered to be immediate neighbors in the ribosome. Several other pairs, which are formed better when crosslinking reagents with the greater molecular length are used, should be noted as well: S3–S5, S3–S9, S4–S8.

The general scheme illustrating protein neighbors in the 30S

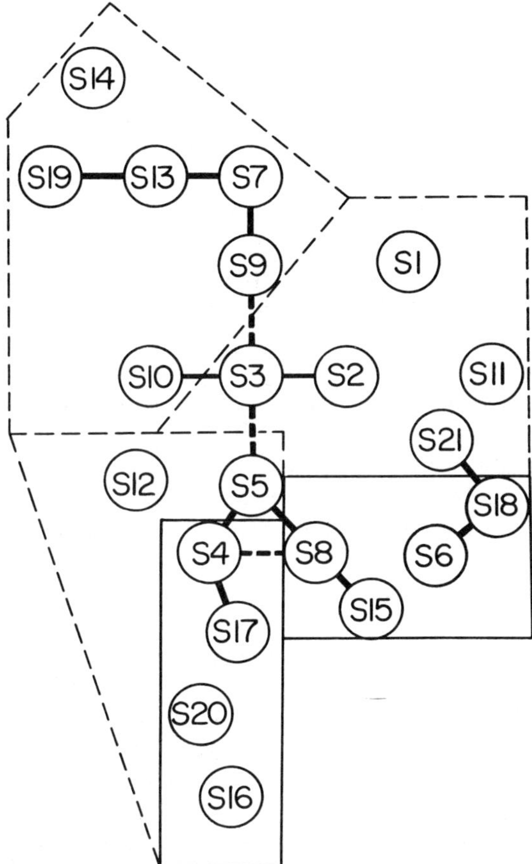

Figure 62 Scheme showing proteins neighboring in the 30S ribosomal subunit. Proteins connected by a solid line are crosslinkable with a short reagent (most direct contact); those connected by a broken line are crosslinkable with longer reagents. The proteins in a solid-line box have adjacent binding sites on the ribosomal RNA sequence. The three groups of proteins within the broken-line boxes correspond to the three RNA domains and three particle lobes.

ribosomal subunit as provided by these approaches is given in fig. 62. Circles connected with lines designate crosslinked proteins; groups of proteins neighboring on RNA are boxed.

Measuring distances between proteins

The problem of the mutual arrangement of proteins in the ribosome may be solved even more comprehensively by measuring the distances between the proteins. This approach is not limited to determining the nearest neighbors. Technically, however, this approach appears far more complex.

One way of measuring the distances between proteins is as follows. Two proteins in the ribosome are selectively labeled by fluorescent groups[9]. Then, depending on the type of labels, the overlapping of their excitation and emission spectra, and the distance between them, some spectral interaction between the fluorescent groups may take place (singlet–singlet energy transfer). By recording changes in the fluorescence spectra as a measure of the spectral interaction one may, in this way, accurately estimate the distance between labeled proteins. Use of this technique has confirmed that proteins S7 and S9, S8 and S15, S6 and S18, and S20 and S16/S17 are located very close to each other, and are probably in direct contact. Such measurements have also demonstrated that proteins S4 and S20, S4 and S16/S17, S13 and S19, and some others are located close to each other. In contrast, labels introduced into proteins S4 and S18, S4 and S19, S13 and S16/S17, S15 and S19, S19 and S20, and others were found to be far from each other within the ribosome.

Another approach to measuring the distances between ribosomal proteins is based on the use of neutron scattering by ribosomal particles containing selectively deuterated pairs of proteins[10]. Since protonated and deuterated proteins exhibit different neutron scattering, comparing the scattering of correspondingly unlabeled and labeled ribosomal particles allows the contribution of the deuterated pair to be distinguished and used for estimating the distance between mass centers of the two proteins, as well as the degree of asymmetry (or compactness) of each of the proteins *in situ*. In selecting the solvent composition (proportion of H_2O and D_2O) in order to match the scattering of protonated proteins, one can further increase the apparent relative contribution of the deuterated pair. Using measured distances between mass centers of proteins in numerous deuterated pairs, Moore, Engelman, and co-workers employed triangulation in constructing a model of the three-dimensional arrangement of ribosomal proteins in the *E. coli* 30S subunit (fig. 63)[11,12]. Results demonstrated that proteins S7, S9, and S10 form a tight group, whereas proteins S4, S5, S8, and S12 are grouped in another cluster, with protein S3 located in between (compare with fig. 62). These results provide one of the most accurate and fundamental contributions to our knowledge of the arrangement of proteins in the ribosomal particle.

Immuno-electron microscopy

The above approaches provide evidence of the arrangement of proteins with respect to each other but without reference to the

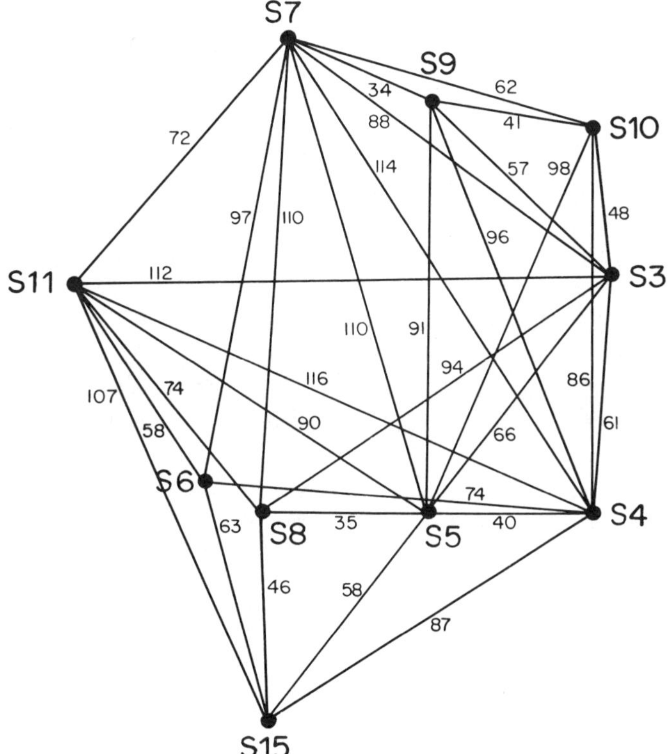

Figure 63 Triangulation of proteins present in the 30S ribosomal subunit, on the basis of neutron-scattering data[11,12]. Numbers indicate the distances between the protein mass centers, measured in angstroms.

morphology of the ribosomal particle. The use of electron microscopy for visualizing proteins on the ribosome allows the location of a protein on a morphologically visible contour of the ribosomal particle to be determined; combined with the above data, this provides an opportunity for superimposing the entire network of protein topography (figs. 62 and 63) on visible projections of the particle. Electron microscopic visualization of proteins on the ribosome makes use of specific antibodies against individual ribosomal proteins[13,14]. The bivalent antibody bound to a given protein may interact with two identical ribosomal particles, yielding their dimer through the bridge of the antibody molecule. By observing dimers under an electron microscope, one may identify sites on the surface responsible for the joining; these sites correspond to the localization site of a given protein on the surface (figs. 64 and 65). In a number of cases, provided the resolution is sufficiently high, one can directly see the attachment

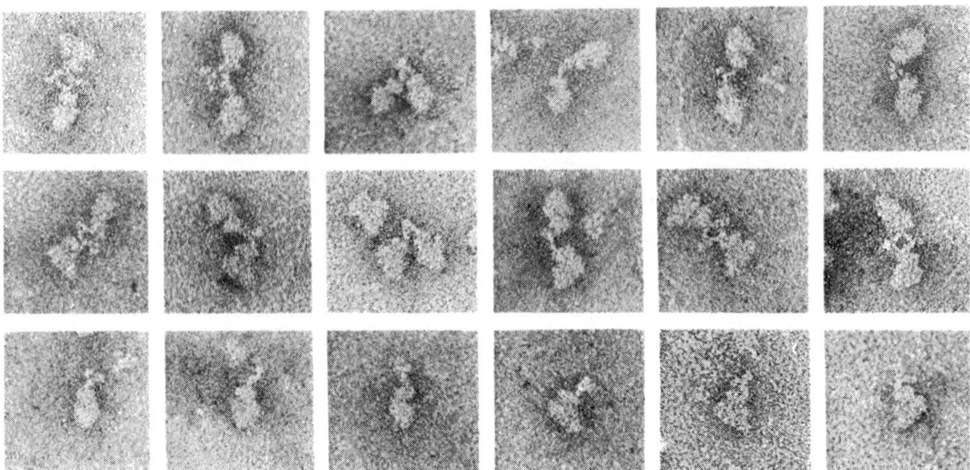

Figure 64 Electron micrographs of 30S ribosomal subunits after reaction with antibodies against protein S14. The two upper rows present the particles dimerized through a bivalent antibody molecule; the lower row shows single particles with an antibody molecule attached. (From J. A. Lake et al. (1974), *Proc. Nat. Acad. Sci. U.S.A.* 71:4688–4692, with permission; original was kindly provided by Dr. James Lake, University of California, Los Angeles.)

of the Y-shaped antibody molecule to a certain region on the ribosomal surface. Using this approach, it has been established that protein L7/L12 forms the lateral rodlike stalk of the 50S ribosomal subunit[15], protein L1 is located in another lateral protuberance (side lobe) of the 50S subunit[16,17], and the 5S RNA-protein complex is detected in the central protuberance, or head, of the 50S subunit[18,19] (fig. 66).

Great efforts have been made to localize all of the proteins of the *E. coli* 30S ribosomal subunit. Despite the feasibility of obtaining specific antibodies against each of the 21 individual proteins, the task was far from simple and the technique yielded many false localizations. It should be pointed out that this method, which appears so direct and illustrative, may result in artifactual information due to the insufficient purity of antibodies, the nonspecific binding of antibodies to certain regions of the ribosomal surface, distortion of the specific position of the antibody molecule on the ribosome caused by the orientation of the ribosomal particle on the substrate, etc. Nevertheless, some reliable results have been obtained. The first reliable achievement was the identification of the position of protein S14 on the head of the 30S subunit[13,14] (fig. 64); the antigenic determinant of the protein was detected on the head surface on the side opposite the side bulge or platform of the 30S subunit (fig. 67). The antigenic determinants of

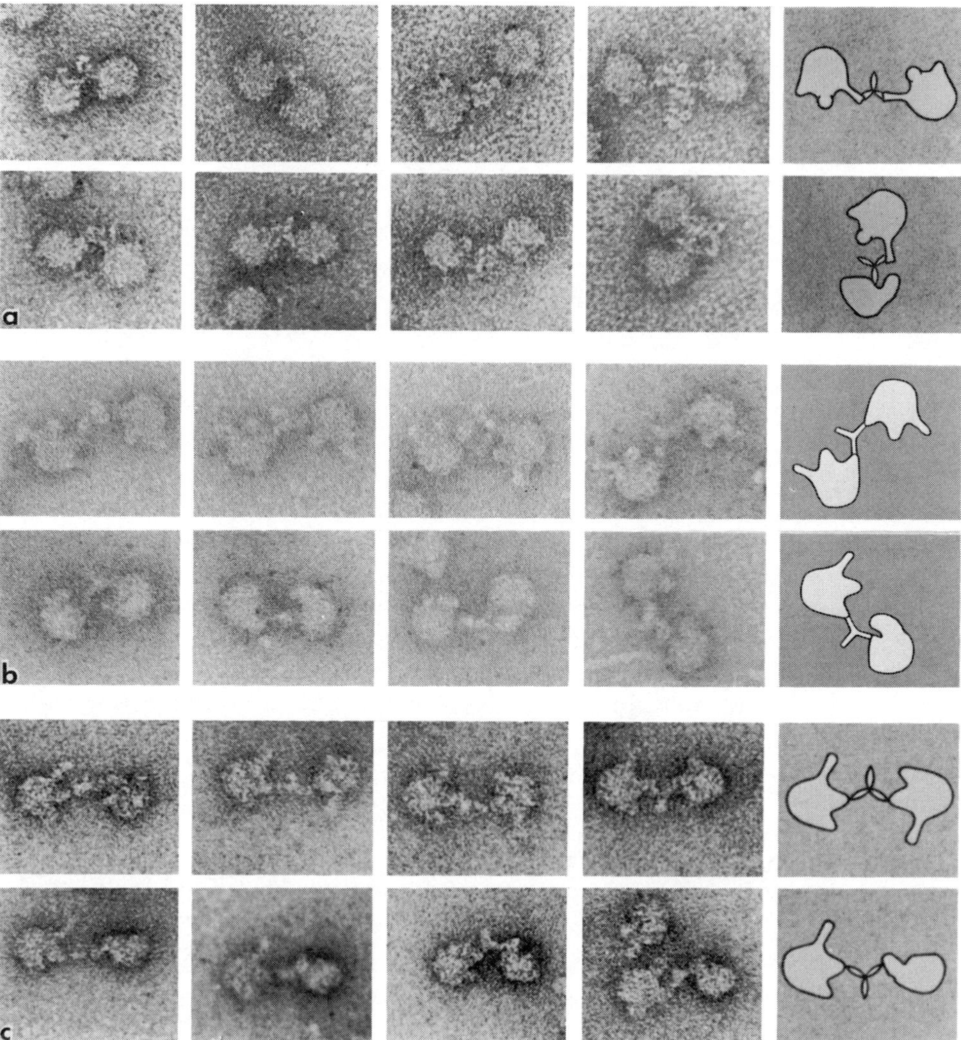

Figure 65 Electron micrographs of 50S ribosomal subunits reacted with antibodies. (a) Antibodies against protein L7/L12. (From J. A. Lake et al. (1974), *Proc. Nat. Acad. Sci. U.S.A.* 71:4688–4692, with permission; original was kindly provided by Dr. James Lake, University of California, Los Angeles.) (b) Antibodies against the protein L1. (Courtesy of Dr. Georg Stöffler, Max-Planck-Institut für Molekulare Genetik, West Berlin.) (c) Antibodies reacted with the 5S RNA-protein complex. The 50S particles are viewed from their convex ("back") side. (Courtesy of Dr. V. D. Vasiliev.)

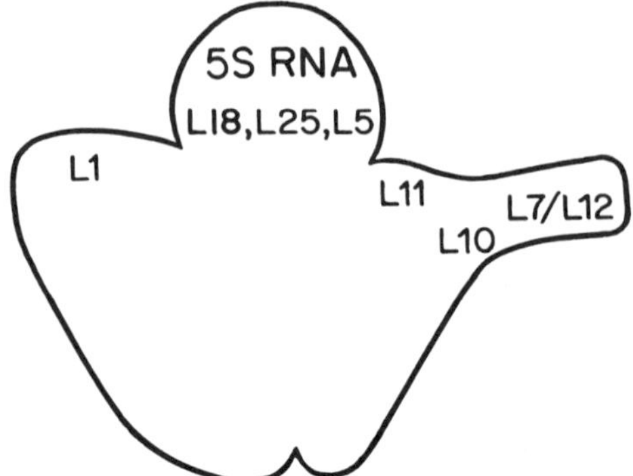

Figure 66 Approximate localizations of several ribosomal proteins on the contour of the 50S ribosomal subunit.

proteins S10 and S19 were localized nearby. Protein S3 was localized below this group of proteins, near the groove separating the head from the body. Protein S5 was localized even lower, also close to the groove but on the body of the subunit. Proteins S6 and S11 were localized on the other side of the 30S subunit, i.e., on its side bulge or platform. Protein S8, according to the data provided by immunoelectron microscopy, is also located near the side bulge, somewhere

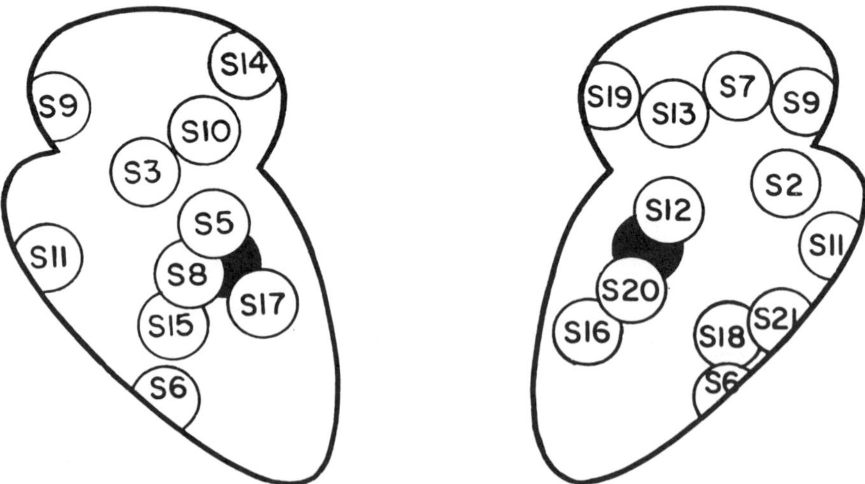

Figure 67 Approximate localization of ribosomal proteins on the contours of the 30S ribosomal subunit. Protein S4 is designated by the filled circle. *Left,* Side of the 30S subunit turned away from the 50S subunit. *Right,* Side of the 30S subunit facing the 50S subunit.

between the bulge and the body, on the external side of the 30S subunit (on the side turned away from the 50S subunit)[19,20].

The localization of protein S4 is particularly interesting. Many artifacts have been obtained with this protein, including numerous false antigenic determinants throughout the particle. Eventually it was found that antigenic determinants of the S4 protein are not easily accessible for antibodies from the particle surface, but become accessible after proteins S5 and S12 are removed[21]. The position of protein S4 is very close to that of protein S5, and it appears that protein S5 as well as S12 partially screens protein S4 from the surface. Generally speaking, protein S4 shows a more or less central position in the 30S ribosomal subunit (fig. 67).

9-3 Topography of RNA

Immuno-electron microscopy

By using either specific haptens, e.g. dinitrophenyl or carbohydrate groups bound to certain sites of the ribosomal RNA, or antibodies against modified (minor) bases of RNA, one may employ immuno-electron microscopy to study the topography of RNA on the surface of the ribosomal particles. First of all, this approach provided the successful localization of ribosomal RNA termini which are easily labeled selectively by hapten groups.

The 3′-end of the 16S RNA on the *E. coli* 30S ribosomal subunit was mapped in the region of the tip of the side bulge (platform), or somewhere between the bulge and the head[22,23] (fig. 68(a)). Dimethyl-adenine (m_2^6A) of the apex loop of the 16S RNA 3′-terminal hairpin was mapped in approximately the same region[24]. The 5′-end of the 16S RNA was localized on the body of the 30S subunit on the side opposite the side bulge[25] (fig. 68(b)).

The 23S RNA 3′-end on the surface of the *E. coli* 50S subunit was mapped in the region of the L7/L12 stalk base, on the external side (the side turned away from the 30S subunit)[26] (fig. 69(b)). (This site probably corresponds to both ends of the 23S RNA, which are locked together.) The 3′-end of the 5S RNA was mapped on the head or central protuberance of the 50S subunit[18] (fig. 69(a)); this seems to define localization of the entire 5S RNA-protein complex including proteins L5, L18, and L25.

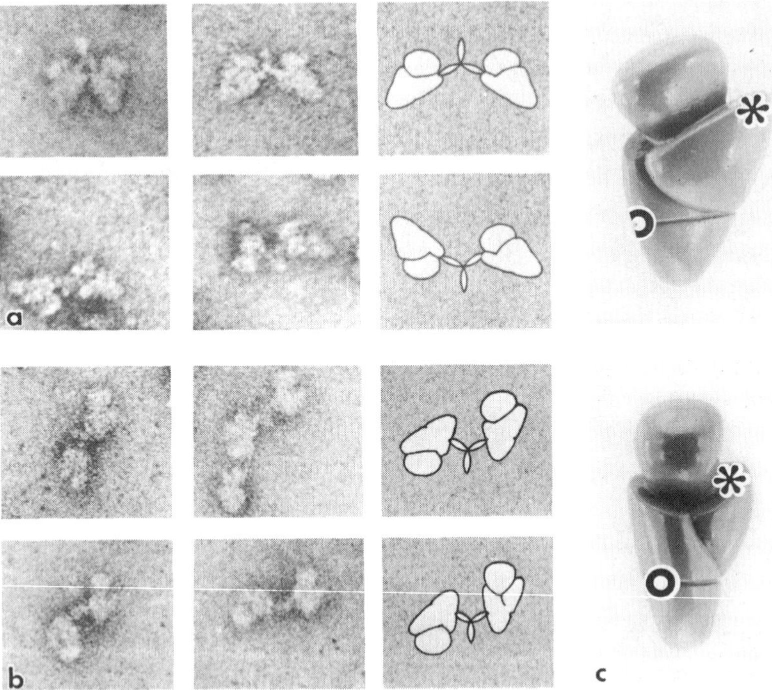

Figure 68 Electron micrographs of 30S ribosomal subunits with antibodies bound to modified 16S RNA (a) 3′-end and (b) 5′-end. (c) Photographs of two projections of the 30S ribosomal subunit model illustrate the localization of 16S RNA 3′-end (asterisk) and its 5′-end (circle). (Courtesy of Dr. V. D. Vasiliev.)

Assignment to protein topography

Data regarding protein topography and protein binding sites on the primary and secondary structure of ribosomal RNA allow the approximate topography of the protein-binding regions of RNA on the ribosomal particle to be deduced. These assignments simply repeat the mapping of corresponding ribosomal proteins. For example, the compound hairpin including helices 20 and 21 of domain I as well as helix 2 comprising the stem of domain I (the paired 5′- and 3′-ends of this domain) (chapter 7, fig. 41) should be located in the region of protein S4 (fig. 67). This region is actually the central point of the three-dimensional structure of 16S RNA: its three main domains branch off from this point. On the morphological picture of the Y-shaped 16S RNA (chapter 7, figs. 48 and 49) this corresponds to the point of bifurcation.

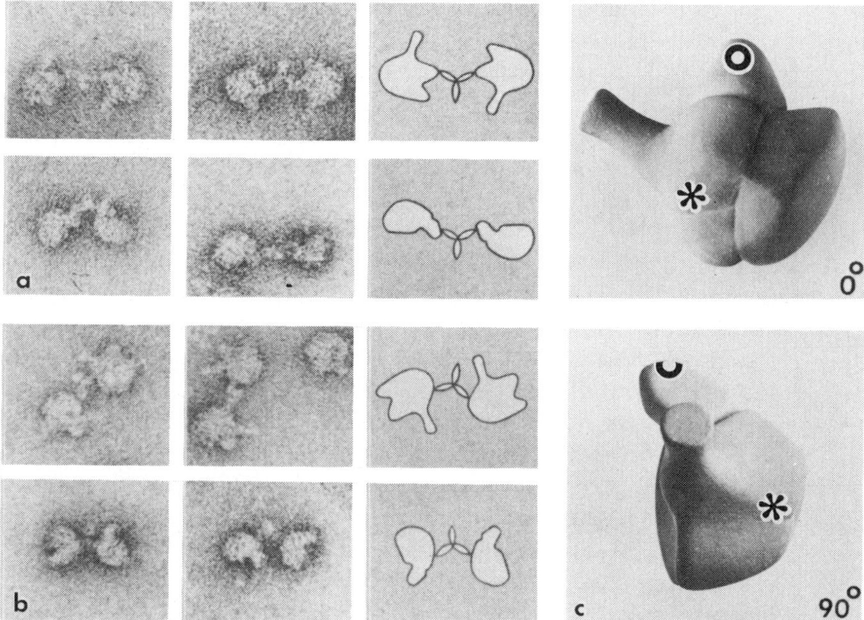

Figure 69 Electron micrographs of the 50S ribosomal subunits with antibodies attached to the modified 3′-ends of (a) 5S RNA and (b) 23S RNA. (c) Photographs of two projections of the 50S subunit model illustrate the localization of the 5S RNA 3′-end (circle) and the 23S RNA 3′-end (asterisk). (Courtesy of Dr. V. D. Vasiliev.)

The compound helix 25–26 of domain II (chapter 7, fig. 41) is located in the region of protein S8 (fig. 67). Since proteins S8 and S4 are located close to each other (see section 9.2, as well as fig. 62), the compound helix 25–26 of domain II should be located somewhere in the neighborhood of the hairpin 20–21 of domain I (as shown schematically in chapter 7, fig. 41).

In contrast, protein S7, its immediate neighbors, i.e. proteins S9 and S13, and their environment comprising proteins S10, S14, and S19, are known to form a cluster remote from the central S4 protein (section 9.2). Correspondingly, the nodular region of 16S RNA domain III (where helices 39, 40, 52, and 55 are grouped), which binds protein S7, should be located distantly from the stem of domain I and from the bifurcation point of the Y-shaped RNA.

Discussing general aspects of topography of the 16S RNA main domains and their correspondence to the main morphological lobes of the 30S ribosomal subunit, i.e. the body, side bulge, and head, one can use available data about mapping proteins on RNA and on the 30S subunit. These data are as follows. (1) Proteins S4, S16, S17, and S20 are bound to the 5′-terminal domain (I) and at the same time are revealed on the body of the 30S subunit. (2) Proteins S8, S15, S6, and S18 interact with the middle domain (II) of 16S RNA while on the morphological image of the 30S subunit they are located either directly on the side bulge (platform) or on the line of contact between the side bulge and the body. (3) A group of proteins including S7, S9, S10, S13, S14, and S19 is attached to 16S RNA in the region of its 3′-proximal domain (III), and all these proteins are found in the head of the 30S subunit. It can be deduced from this evidence that the three main structural domains of 16S RNA generally correspond to the three main morphologically visible lobes of the 30S ribosomal subunit: the 5′-terminal domain (I) forms the core of the subunit body, the middle domain (II) contributes to the formation of the side bulge or platform, and the 3′-proximal domain (III) fills the head of the subunit. It appears that the extreme 3′-terminal region of 16S RNA including helices 57, 58, and 59 protrudes from the head base, or "neck," to the tip of the side bulge or platform, as evidenced by the immunoelectron microscopy data on the mapping of the 3′-end and 3′-terminal hairpin (fig. 68). Correspondence between the 16S RNA domains and the morphological lobes of the 30S subunit is shown schematically in fig. 70.

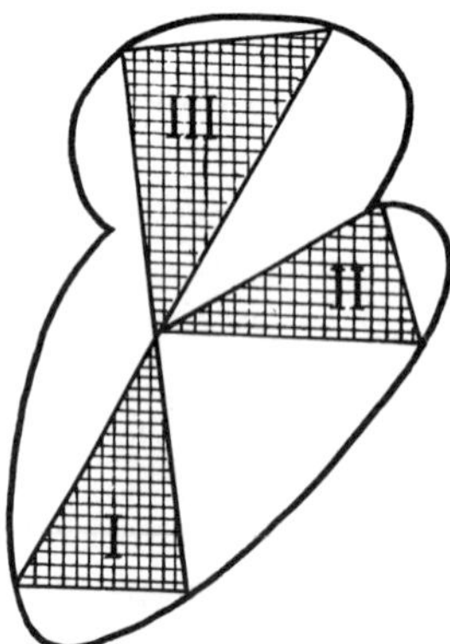

Figure 70 Scheme illustrating the correspondence between the three main morphological lobes of the 30S ribosomal subunit (body, side bulge, and head) and the three main structural domains of 16S RNA (I, II, and III, respectively).

9-4 Quaternary structure

Determination of the precise mutual arrangement of all the structural elements of each ribosomal subunit, including proteins and their groups, the compact domains of RNA, individual RNA helices, etc., is waiting for suitable crystals of the particles to be prepared and studied by X-ray, electron, and neutron diffraction analyses. It is encouraging that ribosomal particles can, in principle, be crystallized.

Based on the data outlined above, plus additional available information, one can, however, attempt to construct tentative models which account for the mutual arrangement of proteins and RNA in the ribosome. These models may, to some extent, reflect the quaternary structure of the ribosomal subunits at low resolution. Model-building studies are particularly appropriate for the 30S ribosomal subunit, since there is far more information on this subunit than on the 50S subunit; also, it should be pointed out that the 30S ribosomal subunit is about half the size of its 50S counterpart. Figure 71 shows one tentative model accounting for the arrangement of the Y-shaped 16S RNA molecule and 21 ribosomal proteins approximated by spheres with diameters corresponding to their molecular masses[27].

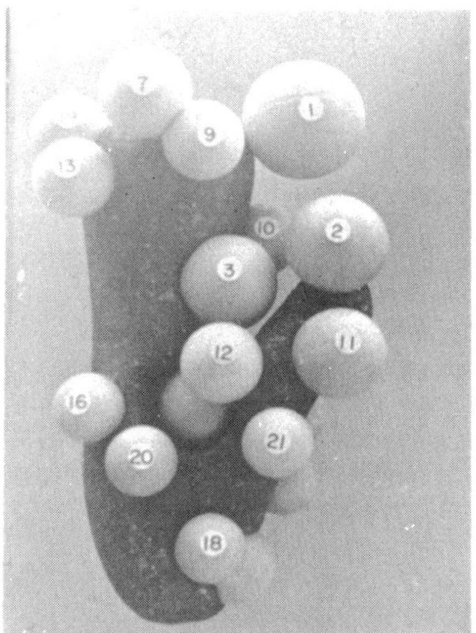

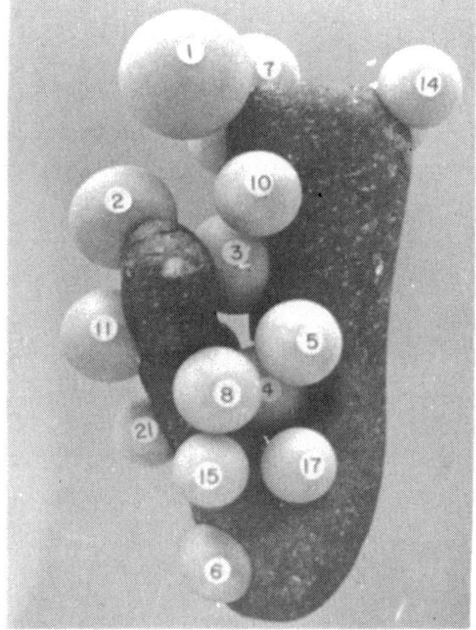

Figure 71 Two projections of the tentative model of the mutual arrangement of the Y-shaped RNA and 21 proteins (quaternary structure) in the 30S ribosomal subunit[27].

Although data on diffuse X rays or neutron scattering by ribosomal particle suspensions cannot be interpreted directly in structural terms, these data may be used to test the models suggested. For example, the scattering curves for the particles corresponding to the model can be calculated theoretically using different scattering contributions of protein and RNA. The size and shape of the RNA in the model can be tested by comparing the theoretically calculated curve at protein zero contribution, with the experimental curve of neutron scattering in 42% D_2O, when the protein is contrast-matched. On the other hand, the pattern of the mutual arrangement of proteins in the model may be tested by comparing the curve calculated for the model at a zero contribution of RNA with the experimental curve of neutron scattering in 70% D_2O, when the RNA is contrast-matched. Finally, by setting different contributions of protein and RNA while calculating the scattering of a given model, one can obtain a set of theoretical curves for comparison with the experimental curves of neutron scattering in H_2O, D_2O, and at their different proportions and with the X-ray scattering curves as well. In this way, mutual position of proteins and RNA in the model is tested. If there is no correspondence between the calculated and the experimental curves the model should be rejected. The model of mutual orientation of RNA and proteins in the 30S subunit shown in fig. 71 does not contradict the experimental curves of neutron and X-ray scattering down to the resolution of about 40 to 50 Å[27].

References

1. I. N. Serdyuk, N. I. Smirnov, O. B. Ptitsyn, and B. A. Fedorov (1970), "On the presence of a dense internal region in the 50S subparticle of *E. coli* ribosomes," *FEBS Letters* 2:324–326.
2. I. N. Serdyuk and A. K. Grenader (1975), "Joint use of light, X-ray, and neutron scattering for investigation of RNA and protein mutual distribution within the 50S particle of *E. coli* ribosomes," *FEBS Letters* 59:133–136.
3. H. B. Stuhrmann, J. Haas, K. Ibel, B. De Wolf, M. H. J. Koch, R. Parfait, and R. R. Crichton (1976), "New low resolution model for 50S subunit of *Escherichia coli* ribosomes," *Proc. Nat. Acad. Sci. U.S.A.* 73:2379–2383.
4. P. Beadry, H. V. Peterson, M. Grunberg–Manago, and B. Jacrot (1976), "A neutron study of the 30S-ribosome subunit and of the 30S-IF3 complex," *Biochem. Biophys. Res. Commun.* 72:391–397.
5. I. N. Serdyuk, A. K. Grenader, and G. Zaccai (1979), "Study of the internal

structure of *Escherichia coli* ribosomes by neutron and X-ray scattering," *J. Mol. Biol.* 135:691–707.

6. W. Kühlbrandt and P. N. T. Unwin (1982), "Distribution of RNA and protein in crystalline eukaryotic ribosomes," *J. Mol. Biol.* 156:431–448.
7. R. A. Zimmermann (1980), "Interactions among protein and RNA components of the ribosome," in *Ribosomes: Structure, function, and genetics*, ed. G. Chambliss, G. R. Craven, J. Davies, K. Davis, L. Kahan, and M. Nomura, pp. 135–169 (Baltimore: University Park Press).
8. R. R. Traut, J. M. Lambert, G. Boileau, and J. W. Kenny (1980), "Protein topography of *Escherichia coli* ribosomal subunits as inferred from protein crosslinking," in *Ribosomes: Structure, functions, and genetics*, ed. G. Chambliss, G. R. Craven, J. Davies, K. Davis, L. Kahan, and M. Nomura, pp. 89–110 (Baltimore: University Park Press).
9. K.-H. Huang, R. H. Fairclough, and C. R. Cantor (1975), "Singlet energy transfer studies of the arrangement of proteins in the 30S *Escherichia coli* ribosome," *J. Mol. Biol.* 97:443–470.
10. D. M. Engelman, P. B. Moore, and B. P. Schoenborn (1975), "Neutron scattering measurements of separation and shape of proteins in 30S ribosomal subunits of *Escherichia coli*: S2–S5, S5–S8, S3–S7," *Proc. Nat. Acad. Sci. U.S.A.* 72:3888–3892.
11. P. B. Moore, J. A. Langer, B. P. Schoenborn, and D. M. Engelman (1977), "Triangulation of proteins in the 30S ribosomal subunit of *Escherichia coli*," *J. Mol. Biol.* 112:199–234.
12. V. Ramakrishnan, S. Yabuki, I.-Y. Sillers, D. G. Schindler, D. M. Engelman, and P. B. Moore (1981), "Positions of proteins S6, S11, and S15 in the 30S ribosomal subunit of *Escherichia coli*," *J. Mol. Biol.* 153:739–760.
13. J. A. Lake, M. Pendergast, L. Kahan, and M. Nomura (1974), "Localization of *Escherichia coli* ribosomal proteins S4 and S14 by electron microscopy of antibody-labeled subunits," *Proc. Nat. Acad. Sci. U.S.A.* 71:4688–4692.
14. G. W. Tischendorf, H. Zeichhardt, and G. Stöffler (1975), "Architecture of the *Escherichia coli* ribosome as determined by immune electron microscopy," *Proc. Nat. Acad. Sci. U.S.A.* 72:4820–4824.
15. W. A. Strycharz, M. Nomura, and J. A. Lake (1978), "Ribosomal proteins L7/L12 localized at a single region of the large subunit by immune electron microscopy," *J. Mol. Biol.* 126:123–140.
16. E. R. Dabbs, R. Ehrlich, R. Hasenbank, B.-H. Schroeter, M. Stöffler–Meilicke, and G. Stöffler (1981), "Mutants of *Escherichia coli* lacking ribosomal protein L1," *J. Mol. Biol.* 149:553–578.
17. J. A. Lake and W. A. Strycharz (1981), "Ribosomal proteins L1, L17, and L27 from *Escherichia coli* localized at single sites on the large subunit by immune electron microscopy," *J. Mol. Biol.* 153:979–992.
18. I. N. Shatsky, A. G. Evstafieva, T. F. Bystrova, A. A. Bogdanov, and V. D. Vasiliev (1980), "Topography of RNA in the ribosome: Location of the

3′-end of 5S RNA on the central protuberance of the 50S subunit," *FEBS Letters* 121:97–100.

19. G. Stöffler and M. Stöffler–Meilicke (1983), "The ultrastructure of ribosomes: An immunological approach," in *Modern methods in protein chemistry*, ed. H. Tschesche, pp. 409–457 (Berlin and New York: Walter de Gruyter Verlag).
20. L. Kahan, D. A. Wilkelmann, and J. A. Lake (1981), "Ribosomal proteins S3, S6, S8, and S10 localized on the external surface of the small subunits by immune electron microscopy," *J. Mol. Biol.* 145:193–214.
21. D. A. Winkelmann, L. Kahan, and J. A. Lake (1982), "Ribosomal protein S4 is an internal protein: Localization by immunoelectron microscopy on protein-deficient subribosomal particles," *Proc. Nat. Acad. Sci. U.S.A.* 79:5184–5188.
22. I. N. Shatsky, L. V. Mochalova, M. S. Kojouharova, A. A. Bogdanov, and V. D. Vasiliev (1979), "Localization of the 3′-end of *Escherichia coli* 16S RNA by electron microscopy of antibody-labelled subunits," *J. Mol. Biol.* 133:501–515.
23. H. M. Olson and D. G. Glitz (1979), "Ribosome structure: Localization of 3′-end of RNA in small subunit by immunoelectron microscopy," *Proc. Nat. Acad. Sci. U.S.A.* 76:3769–3773.
24. S. M. Politz and D. G. Glitz (1977), "Ribosome structure: Localization of N^6,N^6-dimethyladenosine by electron microscopy of a ribosome-antibody complex," *Proc. Nat. Acad. Sci. U.S.A.* 74:1468–1472.
25. L. V. Mochalova, I. N. Shatsky, A. A. Bogdanov, and V. D. Vasiliev (1982), "Topography of RNA in the ribosome: Localization of the 16S RNA 5′-end by immune electron microscopy," *J. Mol. Biol.* 159:637–650.
26. I. N. Shatsky, A. G. Evstafieva, T. F. Bystrova, A. A. Bogdanov, and V. D. Vasiliev (1980), "Topography of RNA in the ribosome: Localization of the 3′-end of the 23S RNA on the surface of the 50S ribosomal subunit by immune electron microscopy," *FEBS Letters* 122:251–255.
27. A. S. Spirin, I. N. Serdyuk, J. L. Shpungin, and V. D. Vasiliev (1979), "Quaternary structure of the ribosomal 30S subunit: Model and its experimental testing," *Proc. Nat. Acad. Sci. U.S.A.* 76:4867–4871.

Further reading

Kurland, C. G. (1974). Functional organization of the 30S ribosomal subunit. In *Ribosomes* (M. Nomura, A. Tissières, and P. Lengyel, eds.), pp. 309–331. Cold Spring Harbor, N.Y.: Cold Spring Harbor Laboratory.

Lake, J. A. (1980). Ribosome structure and functional sites. In *Ribosomes:*

Structure, function, and genetics (G. Chambliss, G. R. Craven, J. Davies, K. Davis, L. Kahan, and M. Nomura, eds.), pp. 207–236. Baltimore: University Park Press.

Liljas, A. (1982). Structural studies of ribosomes. *Prog. Biophys. Molec. Biol.* 40:161–228.

Moore, P. B. (1980). Scattering studies of the three-dimensional organization of the *E. coli* ribosomes. In *Ribosomes: Structure, function and genetics* (G. Chambliss, G. R. Craven, J. Davies, K. Davis, L. Kahan, and M. Nomura, eds.), pp. 111–134.

Stöffler, G.; Bald, R.; Kastner, B.; Lührmann, R.; Stöffler–Meilicke, M.; and Tischendorf, G. (1980). Structural organization of the *Escherichia coli* ribosome and localization of functional domains. In *Ribosomes: Structure, function, and genetics* (G. Chambliss, G. R. Craven, J. Davies, K. Davis, L. Kahan, and M. Nomura, eds.), pp. 171–205. Baltimore: University Park Press.

Traut, R. R.; Heimark, R. L.; Sun, T.-T.; Hershey, J. W. B.; and Bollen, A. (1974). Protein topography of ribosomal subunits from *Escherichia coli*. In *Ribosomes* (M. Nomura, A. Tissières, and P. Lengyel, eds.), pp. 271–308. Cold Spring Harbor, N.Y.: Cold Spring Harbor Laboratory.

Wittmann, H. G. (1983). Architecture of prokaryotic ribosomes. *Ann. Rev. Biochem.* 52:35–65.

Chapter 10

Structural Transformations of Ribosomes (*in vitro*)

Interactions between the ribosomal RNA of two subunits, intramolecular RNA interactions, as well as interactions between different ribosomal proteins and between the ribosomal proteins and RNA within a ribosome, greatly depend on the conditions of the medium, particularly on its ionic composition. Divalent cations, especially magnesium ions, play an important part.

10-1 Dissociation of ribosomes into subunits

The first attempts at isolating ribosomes from various sources have already made it clear that the stability of the particles requires magnesium or calcium ions to be present in the medium in sufficient concentrations. A reduction in the concentration of these divalent cations in the solution results in the ribosomes being dissociated into

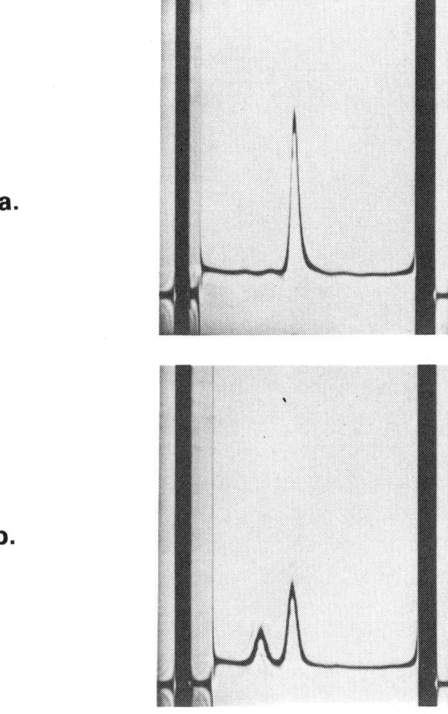

Figure 72 Sedimentation patterns of the original *E. coli* ribosomes and the products of their dissociation achieved by lowering the Mg^{2+} concentration in the medium. (a) 70S ribosomes in 10 mM $MgCl_2$, 100 mM NH_4Cl. (b) 30S and 50S subunits in 1 mM $MgCl_2$, 100 mM NH_4Cl.

component subunits[1,2]:

$$70S \rightarrow 50S + 30S;$$
$$80S \rightarrow 60S + 40S.$$

Dissociation can be induced by a considerable increase in the concentration of monovalent cations in the medium. Dissociation of *Escherichia coli* 70S ribosomes following a decrease of the Mg^{2+} concentration is illustrated by the sedimentation patterns shown in fig. 72.

The discovery of ribosome dissociation was a landmark in the study of ribosomes. The possibility of separating the subunits from each other and isolating them provided the starting point for structural investigations of individual large and small subunits, for the isolation and study of 16S (18S) and 23S (28S) ribosomal RNA, as well as for the isolation and study of ribosomal proteins present in each subunit. It has also become possible to probe into partial functional activities of isolated ribosomal subunits; such activities include mRNA binding, the binding of aminoacyl-tRNA, peptidyl transferase reaction, the binding and hydrolysis of GTP, etc.

The need for sufficient concentrations of Mg^{2+} or Ca^{2+} in the medium (at least 1 mM) in order to keep the ribosomes in an associated state is largely due to the structural and functional characteristics of ribosomal RNA. RNA in the ribosome exists primarily as magnesium salt. The bound magnesium ions are responsible for neutralizing negative charges of RNA phosphate groups; as a consequence, the polyelectrolyte properties of RNA are not manifested and RNA–RNA interactions may be allowed. A reduction of the Mg^{2+} concentration in the medium or an increase in the level of competitive monovalent cations results in Mg^{2+} being partially displaced from the ribosome, and this primarily affects weak RNA–RNA interactions. It is likely that the dissociation of ribosomes into subunits is caused, first of all, by a disruption or weakening of some local contacts between the 16S (18S) RNA of the small subunit and the 23S (28S) RNA of the large subunit. Indeed, isolated 16S RNA and 23S RNA under conditions where they possess a compact conformation, i.e. at a high Mg^{2+} concentration in the medium, seem to be capable of interacting with each other[3].

Experiments on the chemical modifications of ribosomes and their subunits with kethoxal have shown that two main 16S RNA regions take part in subunit association: the first region includes the two adjacent compound hairpins containing helices 29–30–31 and 32–33 of the middle domain (II); the second region is the 3′-terminal sequence[4] (see chapter 7, fig. 41). It has been found that certain unpaired guanylic acid residues present in these regions (e.g. those between helices 29 and 30, 30 and 31, 32 and 33, 32 and 34, as well as in the end loop of helix 33, on one hand; and those between helices 38 and 57, 57 and 59, and in the end loop of helix 59, on the other) react with kethoxal in the isolated 30S subunit but are protected from reacting after the 30S subunit has formed a couple with the 50S subunit. If these residues are modified by kethoxal, the 30S subunit is no longer capable of associating with the 50S subunit. Association with the 50S subunit also results in some rearrangement of the RNA structure in the 30S subunit, leading to the greater accessibility of nucleotide residues in the compound helix 45–46–47 of the 16S RNA domain III. On the basis of the mapping of these RNA regions (see chapter 9, section 9.3), one may suggest that the 30S subunit takes part in the contact with the 50S subunit primarily with its side bulge (platform) and head; the exposed regions of the 16S RNA molecule in these sites seem to be responsible for the interaction with the partner subunit.

Regions participating in the interaction with the 30S subunit may also be pinpointed in the 23S RNA molecule of the 50S subunit. These regions are mainly found in the domain formed by the 23S RNA sequence 2043–2625 (domain V)[5]. It is noteworthy that the same 23S RNA domain appears to interact with the 5S RNA-protein complex located in the head of the 50S subunit and with the protein L1 present in the side lobe. It is highly likely that the head and side lobe of the 50S subunit are involved in the association with the 30S subunit (a "head-to-head" and "side lobe-to-side lobe" association) (see chapter 6).

As was mentioned in preceding chapters, the head of the 50S subunit contains the 5S RNA-protein complex. It has been found that the 5S RNA-protein complex is capable of interacting with the 30S subunit by itself[6]. This may be taken as evidence that the 5S RNA-protein complex contributes to the head-to-head association of ribosomal subunits.

The part played by individual ribosomal proteins in subunit association is not sufficiently understood. It may include both the stabilization of the RNA structure responsible for association and the direct contributions to the intersubunit contacts.

It has long been known that the concentration of Mg^{2+} required to maintain ribosomal subunits in an associated state largely depends on the functional state of the ribosome and on the ligands bound to it. The subunits of the translating ribosome are coupled particularly firmly, and their dissociation requires a marked reduction of the Mg^{2+} concentration in the medium, down to 10^{-4} M or even lower[7]. Translating ribosomes in the posttranslocation state appear to dissociate somewhat easier than pretranslocation state ribosomes. In addition to possible variations in the mode of subunit coupling in different functional states, the bound tRNA may contribute directly to the stability of the associated state of ribosomes[8]. The difference in the stability of pre- and posttranslocation ribosomes may be explained by the fact that pretranslocation particles contain two tRNA residues per ribosome, whereas in posttranslocation ribosomes there is only one tRNA residue. The contribution of tRNA to stabilizing the ribosomal subunit's association is easily understandable: each tRNA-binding site on the ribosome is formed by both subunits (see chapter 11), and therefore the bound tRNA provides an additional "bridge" which stabilizes subunit coupling.

Ribosomal subunits without ligands (tRNA) are coupled more weakly. In any case, free nontranslating *E. coli* 70S ribosomes completely dissociate into subunits in Mg^{2+} concentrations of about 5 mM and lower, whereas at higher Mg^{2+} concentrations they are in a

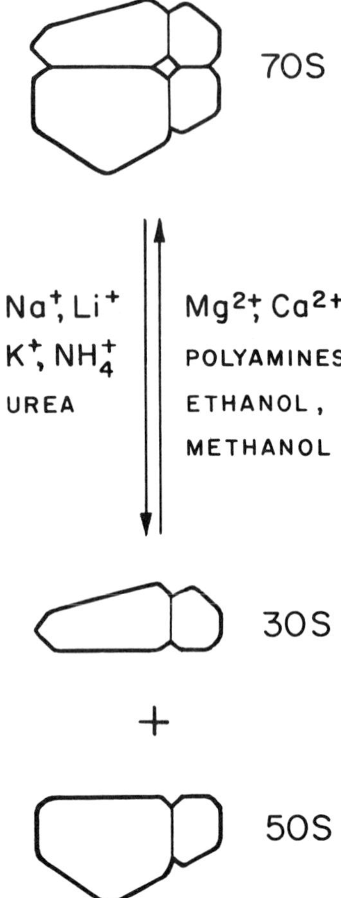

Figure 73 Scheme of ribosome dissociation into subunits. The factors contributing to and counteracting dissociation are indicated.

dynamic equilibrium with the subunits[9]:

$$70S \rightleftarrows 50S + 30S.$$

Eucaryotic 80S ribosomes are far less sensitive to lower Mg^{2+} concentrations in the medium.

Certain environmental factors contributing to and counteracting the association of ribosomal subunits should also be mentioned (fig. 73). As already pointed out, the increased concentration of monovalent cations such as K^+, NH_4^+, etc. always promotes ribosomal dissociation; the higher is the monovalent cation concentration the more Mg^{2+} is needed to keep the ribosomes in an associated state. Direct competition resulting in the displacement of Mg^{2+} from the ribosomes seems to play an important part in this. However, an interesting specificity has been found: sodium and lithium ions promote ribosomal dissociation particularly strongly, even in low concentrations.

To keep ribosomes in an associated state, the Mg^{2+} of the medium can be replaced, at least partially, by Ca^{2+}, as well as by Mn^{2+} or Co^{2+} (it should be mentioned that these cations have similar ionic radii). At the same time, Sr^{2+}, Ba^{2+}, Cd^{2+}, Hg^{2+}, Ni^{2+}, Zn^{2+}, and other di- and polyvalent cations are either ineffective for maintaining the associated state or exhibit dissociating action[10]. Organic polycations such as diamines (putrescine and cadaverine) and polyamines (spermidine and spermine) stabilize the association between subunits, acting as powerful synergists of Mg^{2+}. Di- and polyamines are always present in various amounts in ribosomes and seem to contribute to the stabilization of the ribosomal RNA tertiary structure and of RNA–RNA interactions.

It is noteworthy that numerous water-soluble organic solvents, e.g. methanol, ethanol, and dimethylsulfoxide, may act as Mg^{2+} synergists as well, and are capable of partially replacing the Mg^{2+} in the medium in order to keep ribosomes in the associated state[11]. The anti-dissociating action of these compounds is roughly proportional to their hydrophobicity. Adding these solvents to the medium is also a very effective way of maintaining the isolated ribosomal RNA in a compact state[12].

The dissociation of 70S or 80S ribosomes into constituent subunits *in vivo* takes place only after termination of translation. In this case the coupling between subunits becomes weaker due to the release of ligands, e.g. peptidyl-tRNA and deacylated tRNA. Furthermore, dissociation appears to be additionally catalyzed by special protein factors. The dissociation of ribosomes into subunits after translation has terminated is a necessary step in their cyclic reutilization in the cell (see chapter 17).

10-2 Unfolding of subunits

If more Mg^{2+} are removed from the ribosomal subunits, the compactness of the particles will be destroyed without the dissociation of protein from RNA. This is the so-called *unfolding* of ribosomal particles. To achieve unfolding, two main techniques of removing

Mg^{2+} from ribosomal particles have been proposed. One method makes use of the competitive displacement of bound Mg^{2+} by a high concentration of monovalent cation, e.g. 1 M NH_4^+, with a subsequent decrease in ionic strength[13,14]. An alternative technique makes use of the direct removal of Mg^{2+} by a chelating agent, e.g. ethylenediamine tetraacetate, at a low ionic strength[15].

The reduction of bound Mg^{2+} while the ionic strength is low allows for the manifestation of polyelectrolyte properties of ribosomal RNA. An increase in the electrostatic repulsion of negatively charged phosphate groups concentrated in the relatively small volume of the particle leads to compact RNA packaging being disrupted. It can be demonstrated that in this process long-range interactions are disrupted while most of the RNA secondary structure survives, provided the ionic strength is not too low.

At the same time, under such conditions the proteins are still attached to the loosened high-molecular-mass RNA; in other words the process is that of *ribonucleoprotein* unfolding. The survival of many RNA-protein interactions in the course of unfolding provides evidence that local recognition sites on RNA, including elements of the local three-dimensional structure necessary for retaining proteins in the particles, still exist after the overall pattern of folding has been distorted. Also, it follows that Mg^{2+} does not have a decisive role in RNA-protein interactions. Moreover, a decrease in the ionic strength required for unfolding may even stabilize the ionic interactions between basic groups of proteins and negatively charged groups of RNA; these interactions seem to play an important part in keeping the ribosomal proteins within the ribosomal nucleoprotein (see section 10.3). It is possible, however, that in the course of the unfolding of ribosomal subunits, particularly under conditions of extensive unfolding when the ionic strength is very low and when the RNA helices are beginning to disrupt, certain proteins may undergo redistribution over the RNA chain and now are held by nonspecific electrostatic interactions. It should be added that some weakly bound proteins as well as 5S RNA may be released from the ribonucleoprotein in the course of its extensive unfolding.

A characteristic feature of the unfolding of ribosomal subunits is its cooperativity. Unfolding passes through definite discrete stages, and can best be followed by measuring sedimentation coefficients of the particles. A lowering of sedimentation coefficients during unfolding is accompanied by an increase in intrinsic viscosity, a decrease in the diffusion coefficient, and an increase in the radii of gyration. When the bound Mg^{2+} has been largely displaced from the particles, the gradual decrease in ionic strength results in the following pattern of changes. At first the initial 50S and 30S components are transformed directly into 35S and 26S components, respectively; this transition is quite abrupt, of the all-or-none type, and under intermediate conditions both forms, initial and unfolded, may coexist (fig. 74). Upon a further decrease in ionic strength the 35S and 26S components undergo another jumplike transition into 22S and 15S components, respectively. The latter components behave like typical polyelectrolytes fully resembling the noncompact forms of 23S and 16S RNA; progressive removal of salts from the solution results in their further, now smooth, unfolding which is accompanied by a disruption of the secondary structure and a decrease in sedimentation coefficients, down to values of about 5S in water (fig. 74). This sequence of events may be represented as follows:

$$50S \xrightarrow{\text{sharp}} 35S \xrightarrow{\text{sharp}} 22S \xrightarrow{\text{gradual}} 5S;$$

$$30S \xrightarrow{\text{sharp}} 26S \xrightarrow{\text{sharp}} 15S \xrightarrow{\text{gradual}} 5S.$$

The scheme of the 50S subunit unfolding is given in fig. 75.

Discovery of the unfolding of ribosomal subunits[13] provided the first insight into several fundamental features of the structural organization of the ribosomes. First, it demonstrated that each ribosomal subunit is, in essence, a compactly folded ribonucleoprotein strand, where ribosomal RNA plays the role of a covalently continuous backbone. Furthermore, it became clear that the folding of the ribonucleoprotein strand into a compact ribosomal structure is governed mainly by intramolecular RNA–RNA interactions since the disruption of these interactions results in unfolding. Now, when there is evidence that RNA forms the core of the particle, while proteins are located more to the periphery, an analogy between the folding of the ribonucleoprotein strand into a compact ribosomal particle and the folding of a polypeptide chain into a protein globule can be drawn. In both cases strand folding implies a segregation of two chemically different phases, one of which (corresponding to RNA in the ribosome or hydrophobic groups in the protein) becomes preferentially localized inside while the other (proteins in the ribosome or hydrophilic groups in the protein) is exposed outside.

The all-or-none type of the first unfolding stages may be taken as evidence that the ribonucleoprotein begins to lose its compact state when the critical links responsible for fixing the overall fold (gross structure) are ruptured. The first stage of unfolding may involve a disruption in the interactions between the main lobes of the ribosomal particle or between the domains of high-molecular-mass RNA. Subsequent unfolding stages may include a more or less cooperative "melting" of the intradomain tertiary structure. This may be followed

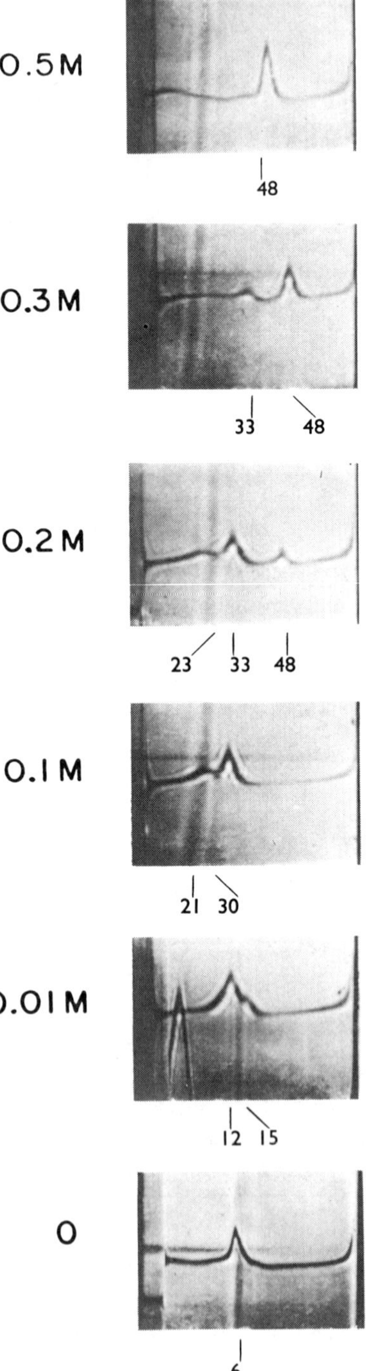

0.5M
48
0.3M
33
48
0.2M
23
33
48
0.1M
21
30
0.01M
12
15
0
6

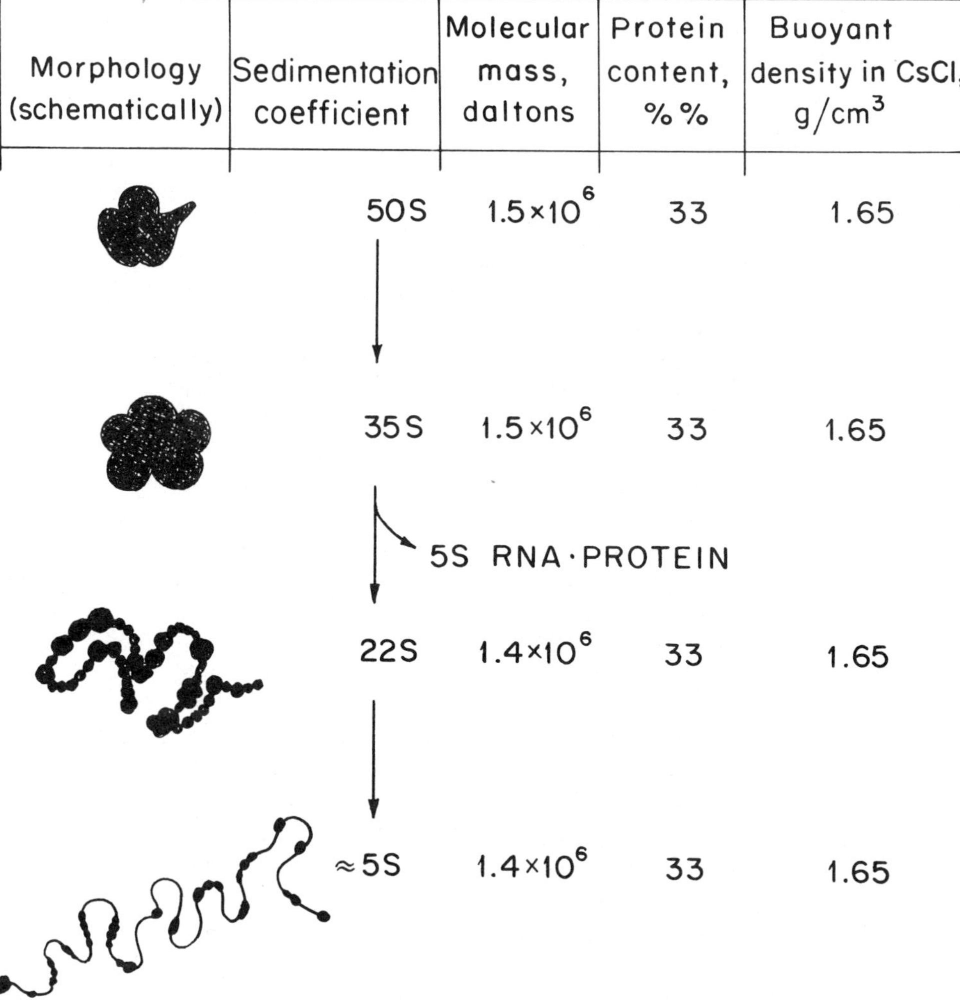

Figure 75 Scheme of the 50S ribosomal subunit unfolding achieved by Mg^{2+} removal or by lowering of the ionic strength.

Figure 74 (*Opposite page*) Sedimentation patterns of 50S ribosomal subunits in the course of their unfolding achieved by the lowering of ionic strength. Initially, the 50S subunits were treated with a high concentration of NH_4Cl (0.5 M) to deplete Mg^{2+}. The unfolding is traced as discrete transitions to the states with progressively decreased sedimentation coefficients, without fragmentation of the particles. The NH_4Cl concentrations are indicated on the left; values of sedimentation coefficients are given below each diagram. (Reproduced from L. P. Gavrilova, D. A. Ivanov, and A. S. Spirin (1966), *J. Mol. Biol.* 16:473–489, with permission.)

by the gradual melting of individual helices at later stages when the salts are being more exhaustively removed.

Generally, the ribosomal ribonucleoprotein behaves rather like high-molecular-mass ribosomal RNA, despite the presence of a large number of proteins. The compact structure of the ribonucleoprotein, however, is always far less sensitive to the removal of Mg^{2+} and to a decrease in the ionic strength, as well as to heating, compared to the structure of free RNA. Moreover, ribosomal RNA cannot attain the state of maximal compact folding characteristic of RNA within the ribosomal particle (or, at least, this state remains unstable) until it binds a certain minimal set of proteins[16,17] (see section 10.3). It appears that the specificity of ribonucleoprotein folding is determined by the RNA, i.e. by the formation of the RNA tertiary structure, while some proteins stabilize more local elements of the structure including the vital RNA helices and interdomain interactions. This contribution from proteins may be responsible for the more cooperative nature of ribosomal particle unfolding compared to the unfolding of isolated RNA. The stabilization of certain important parts of the tertiary and secondary RNA structures by specifically bound proteins may well be a general principle applying not only to the structural organization of ribosomes but to the functioning of many other protein-nucleic acid systems as well (see, for example, chapter 16, section 16.7).

The discovery that ribosomal particles may be unfolded into ribonucleoprotein strands without dissociating ribosomal proteins provided the basis for the important conclusion that it is the covalently continuous RNA which holds on itself numerous ribosomal proteins, thus providing the structural backbone for their arrangement[13,18]. When this principle was formulated, it was indeed new, since previous studies on viral nucleoproteins demonstrated an opposite example, that of the symmetric self-packaging of proteins into a specific quaternary structure where RNA has just adapted to it. In contrast, the formation of the ribosomal particle cannot proceed by the self-assembly of identical protein subunits or subunits of a few types into a symmetric structure. Nearly all of the protein subunits of the ribosome are different and are present in only one copy per particle. In other words, there are many different ribosomal proteins. Therefore, another principle of self-assembly could be expected here: the site-specific positioning of numerous different proteins on a unique covalently continuous backbone or its three-dimensional fold. Experiments on the disassembly and reassembly of ribosomal particles, which logically followed the discovery of unfolding, have been performed (see section 10.3) and have provided full confirmation of the scaffold role of the ribosomal RNA in the arrangement of ribosomal proteins[18].

10-3 Disassembly and reassembly of subunits

Disassembly

If ribosomal subunits are incubated at a high ionic strength with a sufficiently high Mg^{2+} concentration, the compactness of the subunits is retained, but ribosomal proteins partly dissociate from them. It appears that this dissociation is primarily the result of a weaker holding of proteins on the RNA backbone due to their electrostatic interactions being suppressed. Both during incubation at a high salt concentration[19] and upon a stepwise increase in ionic strength[20], groups of proteins sequentially split from the particles, resulting in the formation of a series of protein-deficient derivatives. This is a stepwise disassembly of ribosomal particles.

Since disassembly is accompanied by an increase in the RNA to protein ratio of the particles, it can be conveniently followed by measuring the buoyant density of the particles, using isopycnic equilibrium centrifugation in a CsCl density gradient (the greater the RNA to protein ratio the higher the buoyant density value). Figure 76 shows the appearance of a series of protein-deficient derivatives with increased buoyant densities, as a result of the *E. coli* 50S and 30S ribosomal subunits being treated with 5M CsCl in the presence of 2 mM $MgCl_2$.

The scheme illustrating the stepwise dissociation of proteins from the 30S subunit by increasing the LiCl concentrations at 5 mM $MgCl_2$ is given in fig. 77. Initially, incubation with 1 M salt results in the release of such relatively loosely bound proteins as S1, S2, S3, S14, and S21; the removal of these proteins yields 28S particles with a protein content of 30% and a buoyant density in CsCl equal to 1.67 g/cm^3. Incubation in 2 M LiCl leads to the splitting off of the next portion of proteins, S5, S9, S10, S12, S13, and S20; the resulting 25S particles contain almost half the initial proteins (20%), and their buoyant density in CsCl is equal to 1.74 g/cm^3. In the range from 3 to 3.5 M LiCl, proteins S6, S18, S11, S19, and then S16 and S17, are released; the residual 23S particles contain just four ribosomal RNA-binding proteins: S4, S7, S8, and S15. The removal of the latter group of proteins from RNA requires more drastic treatment, e.g. a combination of high salt and urea.

The dissociation of most of the proteins from the ribosomal particles does not induce an evident disruption of the overall tertiary structure and compactness of the ribosomal RNA. Electron microscopic observations in the course of stripping *E. coli* 30S ribosomal

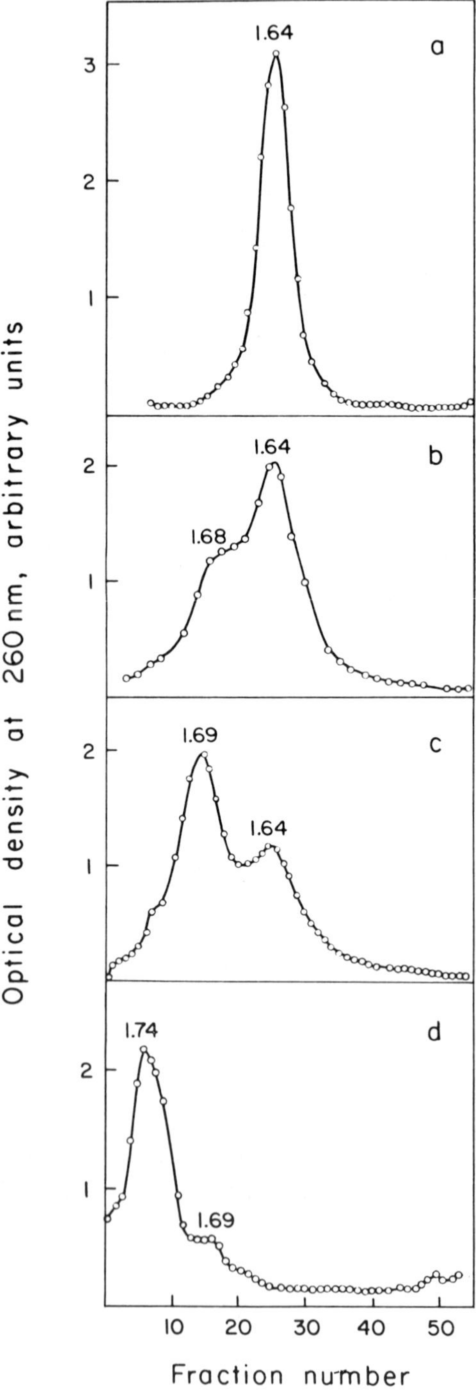

Optical density at 260 nm, arbitrary units
a
1.64
3
2
1
b
1.64
1.68
2
1
c
1.69
1.64
2
1
d
1.74
1.69
2
1
10
20
30
40
50
Fraction number

subunits have demonstrated that removing half of all the proteins does not lead to significant morphological changes in the particles: they retain the same size, axial ratio (2:1), and their characteristic subdivision into a head, body, and side bulge[21]. Moreover, morphologically similar particles can be seen after 15 of the 21 ribosomal proteins have been removed[22]. Measuring the compactness of the ribosomal RNA in particles with different protein content using X-ray and neutron scattering confirms the electron microscopic observations: the 16S RNA retaining only 6 proteins, specifically S4, S7, S8, S15, S16, and S17, maintains the compactness and shape characteristic of RNA within the 30S ribosomal subunit[16].

Removal of those core RNA-binding proteins, however, affects the stability of RNA conformation more drastically: as follows from the measurement of the radius of gyration, the compactness of RNA decreases somewhat, corresponding to an increase in the linear size of about one-quarter. Nevertheless, free 16S RNA at a sufficient Mg^{2+} concentration and ionic strength, like 16S RNA complexed with the "coremost" protein S4 and the 16S RNA carrying the four proteins S4, S7, S8, and S15, is still quite compact and retains its specific overall folding pattern[16]; it can be visualized as a characteristic Y-shaped particle the contours of which can be inscribed in those of the 30S ribosomal subunit[12] (see chapter 7, section 7.4). This implies that the general pattern of 16S RNA folding is governed and maintained by its internal intramolecular interactions, although the stabilization of the eventual completely folded conformation requires the set of six core RNA-binding proteins.

Other ribosomal proteins may, of course, contribute to the folding

---

Figure 76 (*Opposite page*) Density distribution of ribosomal particles in the course of their disassembly (loss of proteins) achieved by incubation in CsCl. (a) Ribosomes fixed with formaldehyde and centrifuged in CsCl to equilibrium; the only component, with density of 1.64 g/cm^3, corresponds to the original ribosomes. (b) Ribosomes incubated in 5 M CsCl for 15 minutes then fixed with formaldehyde and centrifuged in CsCl to equilibrium; the component corresponding to protein-deficient particles with densities of 1.68 to 1.69 g/cm^3 appears. (c) Ribosomes incubated in 5 M CsCl for 1 hour, then fixed with formaldehyde and centrifuged in CsCl to equilibrium; the accumulation of the protein-deficient particles with densities equal to 1.68 to 1.69 g/cm^3, as well as of particles even more deficient in protein, with densities equal to 1.72 to 1.74 g/cm^3, is seen. (d) Unfixed ribosomes centrifuged in CsCl for 36 hours; the main component consists of particles greatly depleted of proteins, with densities of 1.72 to 1.74 g/cm^3. (Redrawn from A. S. Spirin, N. V. Belitsina, and M. I. Lerman (1965), *J. Mol. Biol.* 14:611–615, with permission.)

MORPHOLOGY	SEDIMENTATION COEFFICIENT (IN COMPACTIZATION CONDITIONS)	SPLIT PROTEINS	PROTEIN CONTENT, %%	BUOYANT DENSITY IN CsCl, g/cm^3
	30S SUBUNIT		38	1.62
		S1, S2, S3, S14, S21		
	28S RNP		30	1.67
		S5, S9, S10, S12, S13, S20		
	25S RNP		20	1.74
		S6, S18, S11, S19		
	25S RNP		15	1.76
		S16, S17		
	23S RNP		12	1.78
		S7, S8, S15		
	22S RNP		4	≈ 1.8
		S4		
	22S RNA		0	≈ 1.9

Figure 77 Scheme of the disassembly of the 30S ribosomal subunit achieved by high salt concentrations (e.g. by an increased concentration of LiCl in the presence of 5 mM $MgCl_2$).

and stabilization of ribosomal RNA, but they rather affect its local structures.

Similar trends may be noted during the stripping of the *E. coli* 50S ribosomal subunit. The 23S RNA retains its initial compactness until the stage when just 9 of the 32 proteins, i.e. L2, L3, L4, L13, L17, L20, L21, L22, and L23, remain in the particle[17]. The further removal of proteins leads to a reduction in compactness which, nevertheless, remains reasonably high, and the overall shape of the molecule does not undergo any marked changes.

Thus, the step-wise stripping or disassembly of ribosomal particles clearly demonstrates that high-molecular-mass RNA plays the role of scaffold for the arrangement of ribosomal proteins. The phenomenon of unfolding has shown that the RNA chain serves as a covalently continuous backbone of the particle, carrying all ribosomal proteins. The phenomenon of disassembly, during which the basic compactness and shape of RNA remain unchanged, suggests that the RNA tertiary structure forms a three-dimensional scaffold for the proper spatial arrangement of ribosomal proteins.

Reassembly

Disassembly is reversible, implying that under proper ionic conditions the ribosomal particles can be reassembled[23]; this includes the recovery of their functional activities[24–26]. The reconstitution of ribosomal particles, both 30S and 50S, can be achieved from isolated ribosomal RNA and the complete set of individual ribosomal proteins[27,28].

Conditions for the reassembly of ribosomal particles include (1) a moderate ionic strength (below 0.5), (2) a rather high Mg^{2+} concentration (10 to 30 mM), and (3) an increased temperature. Nomura and associates were the first to achieve the complete reconstitution of biologically active *E. coli* 30S subunits from individual RNA and proteins[27,29]; they used 0.3 to 0.33 M KCl containing 20 mM $MgCl_2$ and incubated the mixture at 40°C for 20 minutes. They found an ionic strength of about 0.4 to be optimal for reconstitution. A higher ionic strength suppresses interactions between the proteins and RNA, while at a lower ionic strength the contribution of competing nonspecific interactions between basic proteins and the negatively charged polynucleotide increases markedly. The relatively high concentration of Mg^{2+} appears to be necessary primarily for the maintenance of the RNA tertiary and secondary structure which provides the scaffold for the arrangement of proteins. In general, the reconstitution buffer of Nomura et al. provides conditions under which ribosomal RNA is sufficiently compact in the isolated state and maintains its unique shape[16]. Elevated temperature is believed to be necessary for facilitating the structural rearrangement of an intermediate ribonucleoprotein complex from a less compact to a more compact conformation[29,30].

The first proteins to bind to ribosomal RNA in the course of self-assembly are the core proteins which are capable of binding to RNA independently of each other. In the case of *E. coli* 30S subunits, these are proteins S4, S7, S8, S15, S17, and S20[31,32] (chapter 8, section 8.5). Protein S16 binds along with these. The addition of the six proteins S4, S7, S8, S15, S16, and S17 (step I in the scheme of fig. 78) is a prerequisite for the transition of the intermediate ribonucleoprotein from a less compact to a more compact state (step II in fig. 78)[16]. Apparently, it is this transition that requires an elevated temperature during self-assembly[30]. As a result of this transition, the 16S RNA almost reaches the maximal compact state of its overall folding which is characteristic of this RNA within the mature 30S ribosomal subunit (see chapter 7, section 7.4).

Proteins S20, S6–S18, S5, S9, S11, S12, S13, and S19 may enter the complex concurrently with the aforementioned proteins, even before the transition of the complex to a more compact state[30]. However, fig. 78, which shows the sequence and interdependence of protein binding in the course of *E. coli* 30S subunit reconstitution, presents the incorporation of these proteins into the ribonucleoprotein as step III of self-assembly, since these proteins, even when bound, are not strictly necessary for the transition into a more compact state and can bind to the complex after this transition. Moreover, as shown in fig. 78, the binding of most proteins at this stage of self-assembly depends on the presence of the set of previously bound proteins[31–32]. For example, the binding of proteins S9 and S19 requires that protein S7 be bound with RNA. The attachment of proteins S6–S18 depends on the presence of protein S15 and, probably to a lesser extent, of protein S8. The binding of protein S11 requires the presence of proteins S6–S18. The attachment of protein S5 is induced by proteins S8 and S16. (It should be emphasized once again that steps II and III shown in the scheme of fig. 78 are not strictly sequential but appear to proceed concurrently. In other words, the compactization and binding of nine proteins do not greatly depend on each other and may proceed in parallel; *in vitro*, step II can be accomplished even after step III has been completed.)

Only after the ribonucleoprotein has undergone transition to compact conformation can the last set of proteins, consisting of S3,

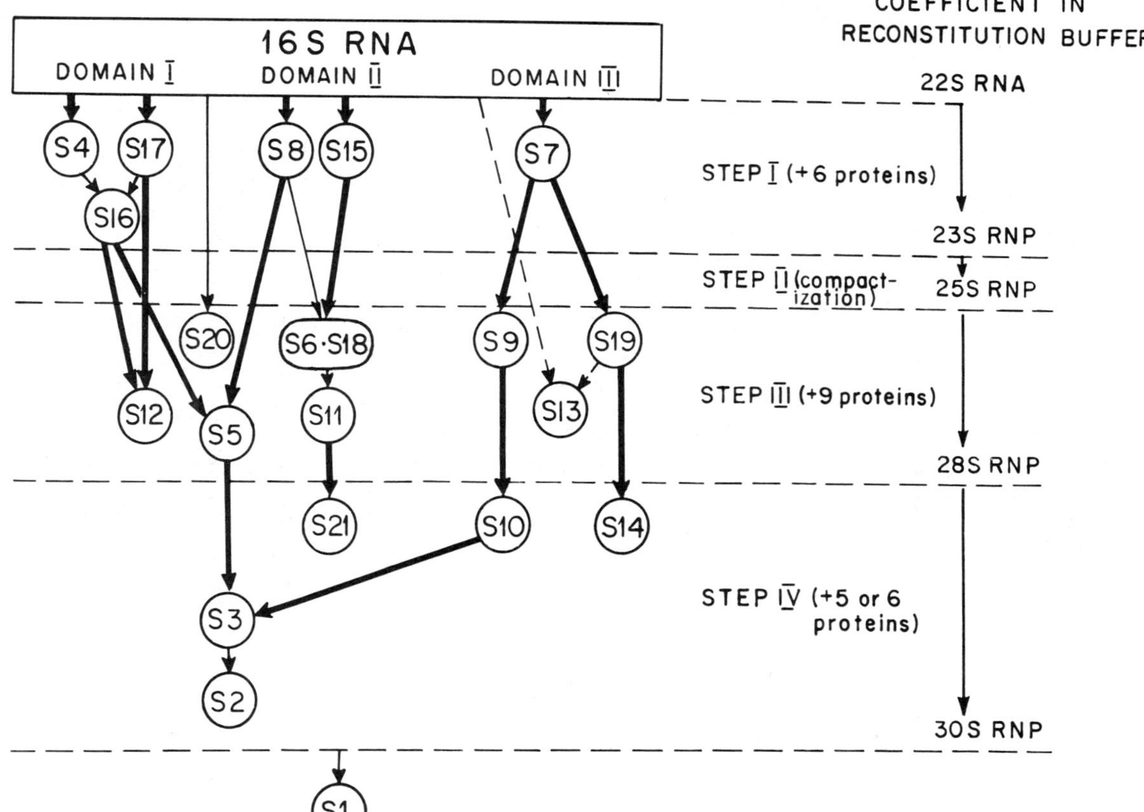

Figure 78 Scheme of self-assembly of the 30S ribosomal subunit from 16S RNA and 21 proteins ("assembly map"). Compiled from the results obtained by Nomura and co-workers[29–32]. The thick arrows from the RNA to a protein or from one protein to another symbolize the great dependence of the binding of the subsequent partner on the previous one; the thin arrows indicate a weak dependence. Some weak interactions have been omitted for the sake of clarity.

S10, S14, S21, as well as S2 and S1, be added to the complex (fig. 78, step IV); this step yields the completed biologically active 30S ribosomal subunit[30]. The incorporation of each of these proteins into the complex requires the presence of proteins bound at previous stages, as well as the final overall folding of the 16S RNA. The binding of protein S10 requires the presence of protein S9, the addition of protein S14 depends on protein S19, protein S3 may become incorporated only if proteins S5 and S10 are present, and the binding of protein S21 is stimulated by the presence of protein S11. The binding of protein S2 is affected by protein S3 and probably by the whole local structure of the ribonucleoprotein. The binding of the largest acidic protein, S1, also requires the correct folding of the ribonucleoprotein; however, it is difficult to determine which specific proteins are necessary for its addition.

An analysis of the complete scheme of 30S ribosomal subunit reconstitution demonstrates that the assembly of each structural lobe of the particle proceeds on the corresponding domain of 16S RNA more or less independently. Thus, proteins S4, S16, S17, S20, as well as S12, are assembled on the 5′-terminal domain (I), forming the subunit body. The middle domain (II) binds proteins S8, S15, S6–S18, as well as S11 and S21, yielding the assembled side bulge of the particle. The 3′-proximal domain (III) with protein S7 incorporates proteins S9, S13, and S19, followed by proteins S10 and S14, and forms the head of the 30S ribosomal subunit.

Interdomain and interlobe interactions should also receive some attention. The most characteristic cases are the addition of protein S5, which depends simultaneously on domains I and II with proteins contained therein; and the attachment of protein S3, which depends on all three domains of RNA and their corresponding proteins. It is likely that protein S5 finds its place somewhere on the boundary between the subunit body and its side bulge, while protein S3 is located at the junction of the head, body, and side bulge of the 30S ribosomal subunit. According to information on self-assembly and the evidence provided by chemical crosslinking data, which are not given here, the head protein S13 may be also in contact with the two other lobes of the ribosomal subunit.

A similar analysis of the *E. coli* 50S ribosomal subunit assembly from 23S RNA, 5S RNA, and 32 proteins revealing the interdependence of protein binding and the sequence of stages can also be conducted on the basis of the experimental data available[33].

There is every reason to assume that the assembly of ribosomes *in vivo* proceeds mainly via the route demonstrated in the course of their reconstitution *in vitro*.

References

1. F. C. Chao (1957), "Dissociation of macromolecular ribonucleoprotein of yeast," *Arch. Biochem. Biophys.* 70:426–431.

2. A. Tissières and J. D. Watson (1958), "Ribonucleoprotein particles from *E. coli*," *Nature* 182:778–780.

3. J. Marcot–Queiroz and R. Monier (1965), "Interactions between RNA's from *Escherichia coli* ribosomes," *J. Mol. Biol.* 14:490–505.

4. H. F. Noller (1979), "Structure and topography of ribosomal RNA, in *Ribosomes: Structure, function and genetics*, ed. G. Chambliss, G. R. Craven, J. Davies, K. Davis, L. Kahan, and M. Nomura, pp. 3–22 (Baltimore: University Park Press).

5. H. F. Noller, J. A. Kop, V. Wheaton, J. Brosius, R. R. Gutell, A. M. Kopylov, F. Dohme, W. Herr, D. A. Stahl, R. Gupta, and C. R. Woese (1981), "Secondary structure model for 23S ribosomal RNA," *Nucleic Acids Res.* 9:6167–6189.

6. E. Metspalu, M. Ustav, and R. Villems (1983), "5S RNA-protein complex is involved in ribosomal subunit association," *FEBS Letters* 153:125–127.

7. A. Tissières, D. Schlessinger, and F. Gros (1960), "Amino acid incorporation into proteins by *Escherichia coli* ribosomes," *Proc. Nat. Acad. Sci. U.S.A.* 46:1450–1463.

8. N. V. Belitsina and A. S. Spirin (1970), "Studies on the structure of ribosomes. IV. Participation of aminoacyl-transfer RNA and peptidyl-transfer RNA in the association of ribosomal subparticles," *J. Mol. Biol.* 52:45–55.

9. A. S. Spirin (1971), "On the equilibrium of the association-dissociation of ribosomal subparticles and on the existence of the so-called 60S intermediate (swollen 70S) during centrifugation of the equilibrium mixture," *FEBS Letters* 14:349–353.

10. A. S. Spirin (1974), "Structural transformations of ribosomes (dissociation, unfolding, and disassembly)," *FEBS Letters* 40:S38–S47.

11. A. S. Spirin and E. B. Lishnevskaya (1971), "Effect of nonionic agents on the stability of association of ribosomal subparticles," *FEBS Letters* 14:114–116.

12. V. D. Vasiliev, O. M. Selivanova, and V. E. Koteliansky (1978), "Specific self-packing of the ribosomal 16S RNA," *FEBS Letters* 95:273–276.

13. A. S. Spirin, N. A. Kisselev, R. S. Shakulov, and A. A. Bogdanov (1963), "On the structure of ribosomes: Reversible unfolding of the ribosomal particles into ribonucleoprotein strands and possible model of packing," *Biokhimiya* 28:920–930.

14. L. P. Gavrilova, D. A. Ivanov, and A. S. Spirin (1966), "Studies on the structure of ribosomes. III. Stepwise unfolding of 50S particles without loss of protein," *J. Mol. Biol.* 16:473–489.

15. R. F. Gesteland (1966), "Unfolding of *Escherichia coli* ribosomes by removal of magnesium," *J. Mol. Biol.* 18:356–371.

16. I. N. Serdyuk, S. C. Agalarov, S. E. Sedelnikova, A. S. Spirin, and R. P. May (1983)," On the shape and compactness of the isolated ribosomal 16S RNA and its complexes with ribosomal proteins," *J. Mol. Biol.* 169:409–425.

17. I. N. Serdyuk, S. Ch. Agalarov, G. M. Gongadze, A. T. Gudkov, S. E. Sedelnikova, R. P. May, and A. S. Spirin (1984), "On the shape and compactness of ribosomal RNAs and their complexes with proteins in solution," *Mol. Biol.* (*U.S.S.R.*) 18:244–262.

18. A. S. Spirin (1964), in *Macromolecular structure of ribonucleic acids*, pp. 161–204 (New York: Reinhold).

19. A. S. Spirin, N. V. Belitsina, and M. I. Lerman (1965), "Use of formaldehyde fixation for studies of ribonucleoprotein particles by caesium chloride density-gradient centrifugation," *J. Mol. Biol.* 14:611–615.

20. T. Itoh, E. Otaka, and S. Osawa (1968), "Release of ribosomal proteins from *Escherichia coli* ribosomes with high concentrations of lithium chloride," *J. Mol. Biol.* 33:109–122.

21. V. D. Vasiliev and V. E. Koteliansky (1977), "The 30S ribosomal subparticle retains its main morphological features after removal of half the proteins," *FEBS Letters* 76:125–128.

22. V. D. Vasiliev, V. E. Koteliansky, and G. V. Rezapkin (1977), "The complex of 16S RNA with proteins S4, S7, S8, S15 retains the main morphological features of the 30S ribosomal subparticle," *FEBS Letters* 79:170–174.

23. M. I. Lerman, A. S. Spirin, L. P. Gavrilova, and V. F. Golov (1966), "Studies on the structure of ribosomes. II. Stepwise dissociation of protein from ribosomes by caesium chloride and the re-assembly of ribosome-like particles," *J. Mol. Biol.* 15:268–281.

24. A. S. Spirin and N. V. Belitsina (1966), "Biological activity of the re-assembled ribosome-like particles," *J. Mol. Biol.* 15:282–283.

25. K. Hosokawa, R. Fujimura, and M. Nomura (1966), "Reconstitution of functioning active ribosomes from inactive subparticles and proteins," *Proc. Nat. Acad. Sci. U.S.A.* 55:198–204.

26. T. Staehelin and M. Meselson (1966), "*In vitro* recovery of ribosomes and of synthetic activity from synthetically inactive ribosomal subunits," *J. Mol. Biol.* 16:245–249.

27. P. Traub and M. Nomura (1968), "Structure and function of *E. coli* ribosomes. V. Reconstitution of functionally active 30S ribosomal particles from RNA and proteins," *Proc. Nat. Acad. Sci. U.S.A.* 59:777–784.

28. M. Nomura and V. Erdmann (1970), "Reconstitution of 50S ribosomal subunits from dissociated molecular components," *Nature* 228:744–748.

29. P. Traub and M. Nomura (1966), "Structure and function of *Escherichia coli* ribosomes. VI. Mechanism of assembly of 30S ribosomes studied *in vitro*," *J. Mol. Biol.* 40:391–413.

30. W. A. Held and M. Nomura (1973), "Rate-determining step in the reconstitution of *Escherichia coli* 30S ribosomal subunits," *Biochemistry* 12:3273–3281.

31. S. Mizushima and M. Nomura (1970), "Assembly mapping of 30S ribosomal proteins from *E. coli*," *Nature* 226:1214–1218.

32. W. A. Held, B. Ballon, S. Mizushima, and M. Nomura (1974), "Assembly mapping of 30S ribosomal proteins from *Escherichia coli*," *J. Biol. Chem.* 249:3103–3111.

33. K. H. Nierhaus (1979), "Analysis of the assembly and function of the 50S subunit from *Escherichia coli* ribosomes by reconstitution," in *Ribosomes: Structure, function, and genetics*, ed. G. Chambliss, G. R. Craven, J. Davies, K. Davis, L. Kahan, and M. Nomura, pp. 267–299 (Baltimore: University Park Press).

PART III

FUNCTIONING OF THE RIBOSOME

Chapter 11

Functional Activities and Functional Sites of the Ribosome

11-1 Working cycle of the ribosome

At any given time in the course of elongation, the ribosome is attached to the coding region of mRNA and retains the molecule of the peptidyl-tRNA (fig. 79). The peptidyl-tRNA is a nascent peptide chain bound through its C-terminus to the tRNA which has donated the last amino acid residue to the peptide. Such a ribosome can bind or may become capable of *binding the aminoacyl-tRNA* determined by the next mRNA codon (fig. 79(I)). The binding of the aminoacyl-tRNA results in the retained peptidyl-tRNA and the newly bound aminoacyl-tRNA being present on the ribosome simultaneously. Their side-by-side location and the catalytic activity of the ribosome are prerequisites of the *transpeptidation* reaction: the C-terminus of the peptidyl residue is transferred from the tRNA (to which it had previously been bound) to the amino group of the aminoacyl-tRNA (fig. 79(II)). As a result, the formation of a new peptidyl-tRNA with the peptide elongated by one amino acid residue at the C-end takes place; the other product of the reaction is the deacylated tRNA. In order to make the ribosome

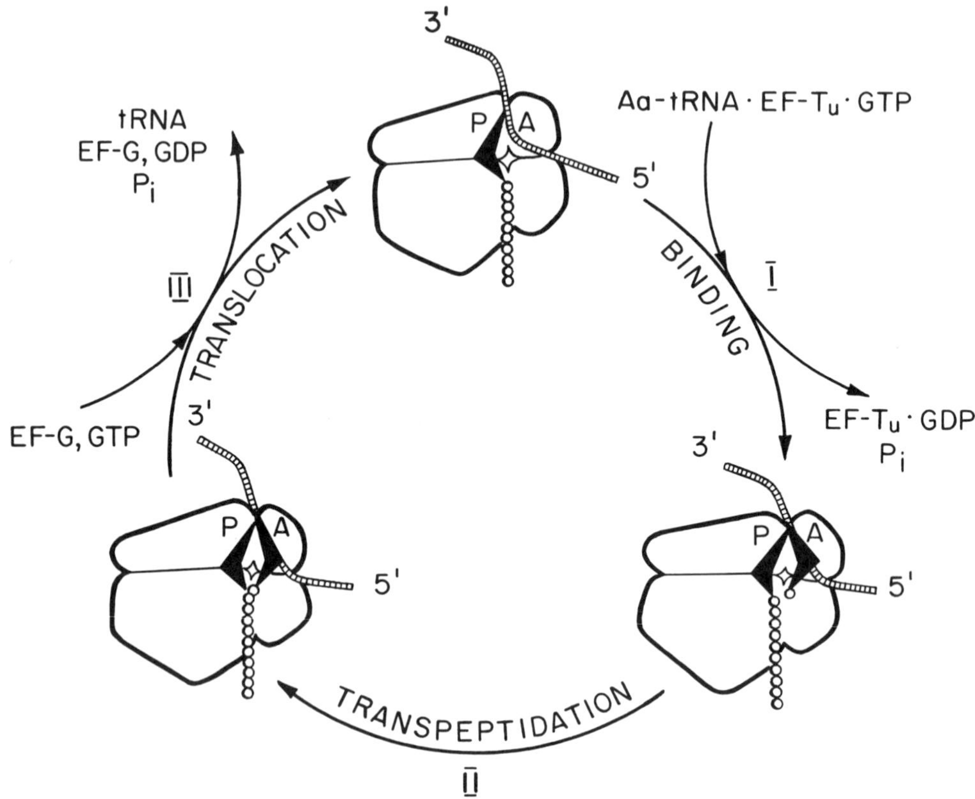

Figure 79 Elongation cycle of the ribosome[1,2].

competent to the binding of the next aminoacyl-tRNA, some intraribosomal ligand (tRNA and mRNA) displacements should be realized, resulting in the vacation of a place for the aminoacyl-tRNA and in the positioning of the next mRNA codon (fig. 79(III)); this step is called *translocation*.

Thus, the working cycle of the ribosome in the course of elongation consists of three steps: codon-dependent binding of aminoacyl-tRNA (step I), transpeptidation (step II), and translocation (step III). The binding of aminoacyl-tRNA requires the presence of the EF-T_u protein, and is accompanied by the hydrolysis of a GTP molecule. Transpeptidation is catalyzed by the ribosome itself. Translocation requires another protein, the EF-G, and is also accompanied by GTP hydrolysis.

According to the classical model[1,2], the ribosome has two sites of tRNA binding, the A-site and the P-site. At stage I the aminoacyl-tRNA in the complex with EF-T_u and GTP binds to the A-site and the vacant template codon located therein. At this time the peptidyl-tRNA is in the P-site. The binding of the aminoacyl-tRNA ends in GTP hydrolysis on the ribosome and the release of the EF-T_u·GDP complex and orthophosphate into solution. At stage II the aminoacyl-tRNA located in the A-site reacts with the peptidyl-tRNA in the P-site; this results in the peptide C-terminus being transferred to the aminoacyl-tRNA. Now, the elongated peptidyl-tRNA (its tRNA residue) is occupying the A-site while the deacylated tRNA formed in the reaction is located in the P-site. At stage III the ribosome interacts with EF-G and GTP which catalyze the displacement of the peptidyl-tRNA (its tRNA residue) along with the template codon from the A-site to the P-site, as well as the release of the deacylated tRNA from the P-site into solution. After completion of these events GTP undergoes hydrolysis; EF-G, GDP, and orthophosphate are released from the ribosome. This again leads to the situation whereby the peptidyl-tRNA is located in the P-site while the next template codon is located in the A-site; thus the A-site is ready to accept the next aminoacyl-tRNA molecule. Translation of the whole coding sequence of the template polynucleotide and corresponding polypeptide elongation on the ribosome are achieved by the repetition of such cycles. It should be pointed out that both the initiation and termination of translation are simply modifications of the ribosomal working elongation cycle outlined above (see chapters 16 and 18).

An analysis of the ribosomal working cycle demonstrates that the ribosome performs the following functions in the course of translation: (1) binding and retention of mRNA, (2) retention of peptidyl-tRNA, (3) binding of aminoacyl-tRNA, (4) binding of translation protein factors, (5) participation in the catalytic hydrolysis of GTP, (6) catalysis of transpeptidation, and (7) intraribosomal displacements referred to as translocation.

11-2 Binding functions

Binding and retention of the template polynucleotide (mRNA-binding site)

The ribosome has an intrinsic affinity to template polynucleotides. It has long been known that vacant ribosomes effectively bind polyuridylic acid (this fact may explain the wide use of poly(U) as a template in cell-free translation systems). It is likely that the absence of a stable secondary and tertiary structure in poly(U) is an important factor contributing to its effective binding with the ribosomes. In the case of mRNA from natural sources, there are definite preferential sites on the polynucleotide for binding vacant ribosomes (see below, as well as chapter 16). In contrast, the perfect continuous double helix of RNA cannot serve as a binding site for vacant ribosomes.

At the same time, in the course of translation (elongation) the ribosome passes along the entire coding sequence of mRNA and thus can transiently hold the template at any region of the sequence. Apparently the ribosome should unfold the translated template polynucleotide in such a way that the template section hold on the ribosome is devoid of its original secondary and tertiary structure. Codon-anticodon interactions with tRNA undoubtedly contribute to a retention of mRNA on the translating ribosome.

A translating ribosome bound to the template polynucleotide protects a rather long nucleotide sequence from external nucleases. Early experiments with poly(U) have demonstrated that the ribosome covers the 25-residue-long section, making it inaccessible to pancreatic ribonuclease[3]. In the case of natural mRNA, e.g. R17 or f2 phage RNA, the *Escherichia coli* ribosomes protect the 30- to 60-residue-long sections from nucleases[4,5]. More recently the functional 70S ribosome has been shown to protect a template polynucleotide region of about 45 to 50 nucleotide residues long[6,7]; within this region, however, there is a nuclease-accessible point that divides the protected region into the fragments of 27 to 30 residues and 15 to 20 residues long[7]. (From some observations it follows that this sensitive ribosome-covered point of the template is located immediately before the codon bound with the tRNA in the A-site of the ribosome; the protected 15 nucleotide long fragment seems to be the 3′-proximal part while the 30 nucleotide long fragment is the 5′-proximal section of the ribosome-bound template region[6].) In any case, all these results suggest that the mRNA-binding site of the ribosome is of a considerable size; apparently it extends for at least 100 Å, i.e. about half of the ribosome.

The first problem regarding the localization of the functional sites of the ribosome has to do with whether they are assigned to one of the two ribosomal subunits, or to both subunits together. In the simplest case the experimental solution of this problem is as follows. Ribosomes are dissociated to yield large and small subunits, the subunits are separated, and the tested ligand is added to each of them (in the presence of a sufficient concentration of magnesium ions, which is required to observe any binding to the ribosome). It has been demonstrated in this type of experiment that the isolated 30S subunit binds the template polynucleotide whereas the 50S subunit does not[8,9]

On the basis of this result, it is generally accepted that the mRNA-binding site of the ribosome is located only on the small (30S or 40S) subunit.

Several approaches have been used for identifying the ribosomal proteins that take part in the organization of the mRNA-binding site of 30S ribosomal subunits. Experiments on the partial disassembly of the subunits have demonstrated that the mRNA-binding capacity is lost when the following group of proteins has been removed from the particles: S1, S2, S3, S5, S9, S10, and S14; the reconstitution of the subunit without any one of these proteins, however, yields 30S particles active in the poly(U) binding[10]. The considerable length of the mRNA-binding site suggests a multicenter binding of mRNA to the particle, i.e. the participation of several binding points of the ribosome surface. If this is indeed the case, then the omission of any single binding point would not be critical to the function. On the other hand, the loss of the poly(U) binding capacity as a result of the dissociation of all the above-mentioned proteins may be due to a disruption in the overall structure of the corresponding ribosomal region, but not necessarily to the removal of the main components of the active site.

Another approach to identifying the proteins forming ribosomal functional sites makes use of affinity labelling[11–13]. With this technique a chemically active (fig. 80) or photoactivable group (fig. 81) is introduced into a corresponding ligand (e.g. mRNA, tRNA, translation factor, guanylic nucleotide, antibiotic) which specifically binds with the ribosome. This group attacks the ribosomal components located nearby and becomes crosslinked with them. Proteins crosslinked with the ligand may then be identified. Oligonucleotide mRNA analogs containing either bromoacetyl, or 4-[(N-2-chlorethyl-N-methyl)-amino]-benzylidene, or photoactivable arylazide groups (fig. 81(2)) at the 3′- or 5′-end, as well as poly(s^4U), a photoactivable thio-derivative of poly(U), have been used for studying the mRNA-binding site of the ribosome. Furthermore, it has become possible to achieve photo-activation of nonmodified synthetic and natural template oligo- and polynucleotides, e.g. poly(U) and MS2 RNA, in order to produce their crosslinking with the nearest neighbors in the ribosome. It is clear, however, that this approach does not allow the components directly forming the mRNA-binding site and the components located nearby to be distinguished. Of the proteins either belonging to the mRNA-binding site of the 30S ribosomal subunit or located nearby, proteins S1, S3, and S5 can be most reliably identified in this way, although at least a dozen other 30S ribosomal proteins, including S4, S9, and S18, have also been reported. On the basis of evidence for the interdomain or interlobe position of proteins S3 and S5 (see chapter 9, as well as

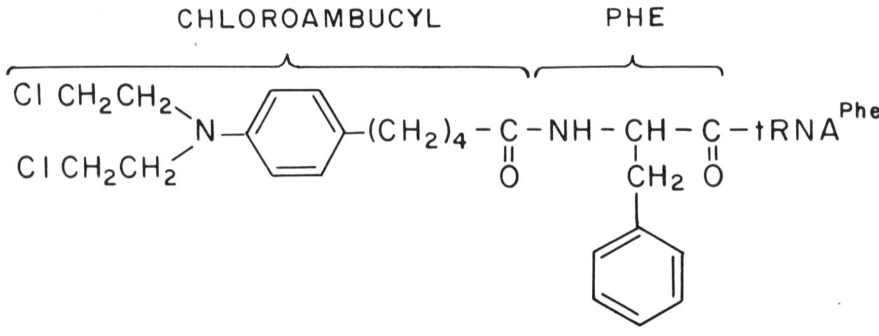

BROMOACETYL PHE

Br CH_2-C-NH-CH-C-$tRNA^{Phe}$

Figure 80 Chemically active analogs of aminoacyl-tRNA first used as affinity labels to probe the tRNA-binding sites of the ribosome[11,12].

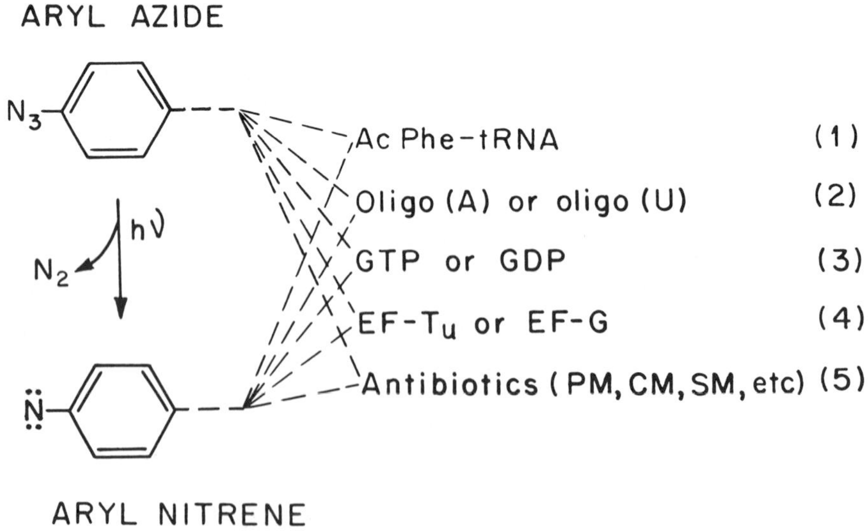

ARYL AZIDE

N_3

hν

N_2

Ac Phe-tRNA (1)

Oligo (A) or oligo (U) (2)

GTP or GDP (3)

EF-Tu or EF-G (4)

Antibiotics (PM, CM, SM, etc) (5)

ARYL NITRENE

Figure 81 Photoactivation of the arylazide group attached to one of the ribosome functional ligands (listed from 1 to 5 on the right). The nitrene radical formed attacks any group found in close proximity.

chapter 10, section 10.3), one may suggest that the mRNA-binding site is located in the region of the grooves dividing the head, body, and side bulge of the 30S ribosomal subunit.

Protein S1 is not strictly required for the binding and translation of poly(U), although it does stimulate these functions. However, it is indispensable for the binding and translation of mRNA from natural sources. Furthermore, its intrinsic capacity for forming complexes with polynucleotides has been detected. The binding of protein S1 with RNA results in the loosening or unfolding of the RNA secondary structure. Taking also into account that protein S1 neighbors mRNA on the ribosome, one may assume that it directly participates in the formation of the mRNA-binding site[14].

The chemically active analogs of mRNA can be crosslinked not only with the proteins of the 30S subunit but with 16S RNA as well, and some of these analogs are crosslinked preferentially with the 16S RNA. Again, it is not clear whether this reflects the involvement of ribosomal RNA in mRNA binding or whether it is merely a consequence of certain 16S RNA regions being in the neighborhood of mRNA-binding centers. The unpaired region of the 16S RNA between domain III and the long compound hairpin of the 3′-terminal sequence (helices 57 and 58; see chapter 7, fig. 41) has been found close to the mRNA codon present in the P-site, or even in direct contact with it[15]. Moreover, when the vacant ribosome binds with the natural mRNA during initiation, the mRNA polypurine tract appears to form a complementary complex with the polypyrimidine sequence which is located near the extreme 3′-end of the 16S RNA (see chapter 16, section 16.4). This pairing, of course, disappears when elongation begins.

In order to directly locate the mRNA-binding site on the morphologically visible surfaces of the ribosome, the immuno-electron microscopy studies of 30S subunits and 70S ribosomes bound with short poly(U) carrying a covalently linked hapten on either the 3′ or 5′-end have been performed. Using this approach the template polynucleotide ends have been detected in the region of the groove separating the head and the body of 30S subunit, mainly on its external (facing from the 50S subunit) side and near the side bulge (platform)[16].

Taking all the evidence into consideration, it appears that the mRNA chain binds to and passes along the 30S subunit somewhere on the boundary between its lobes or between the 16S RNA domains. It is likely that the binding site is located in the region of the groove that separates the head from the side bulge and the head from the body. This extended region seems to contain the 3′-end of the 16S RNA and proteins S3 and S5 (see chapter 9, sections 9.2 and 9.3, and

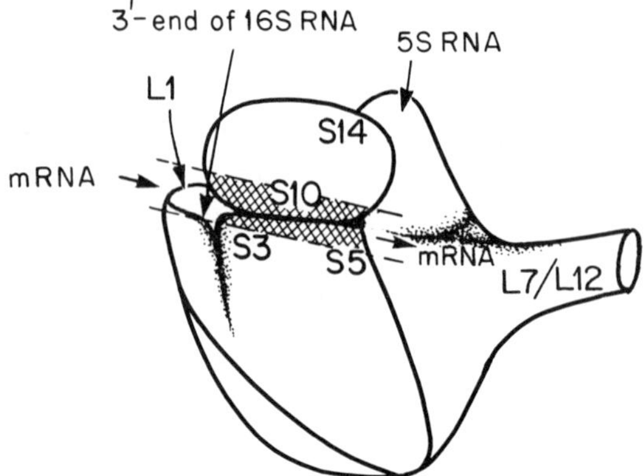

Figure 82 Tentative localization of the mRNA-binding site (crosshatched area) on the ribosome. Proteins S3, S5, and S10 are either involved in the formation of this site or located in the neighborhood.

chapter 10, section 10.3), as well as the stem of the 16S RNA domain III and protein S1. Crosslinks of chemically active mRNA analogs with components of the 50S ribosomal subunit provide evidence that proteins L1 and L7/L12 are located close to the mRNA[17]; these proteins are found on two different sides, directly opposite the groove separating the head from the side bulge and the head from the body of the 30S ribosomal subunit respectively (see chapter 9, section 9.2, and fig. 66).

On the basis of general considerations it may be assumed that the association of the template polynucleotide with the ribosome permits a slippage of the polynucleotide chain along the mRNA-binding site. This is an obvious requirement for the sequential reading of the mRNA chain in the course of translation. The proposed position of tRNA residues on the ribosome (see below) and the possible trajectories of their displacements during translocation (see chapter 14, section 14.8) suggest that the mRNA slips from the side with the side bulge adjacent to protein L1 to the opposite side of the 30S subunit, facing protein L7/L12, as indicated by short arrows in the scheme of fig. 82.

Retention of peptidyl-tRNA or deacylated tRNA (tRNA-binding P-site)

The ribosome possesses an intrinsic affinity to tRNA. A vacant ribosome can bind any tRNA or its derivative, e.g. aminoacyl-tRNA or peptidyl-tRNA, in the absence of a template polynucleotide. The presence of a template polynucleotide makes this binding specific: only the tRNA corresponding to the template codon will be bound. In the cases where tRNA or its derivative is accepted by the vacant ribosome, one of the two tRNA-binding sites is filled first. This is the same site that is occupied by the peptidyl-tRNA prior to transpeptidation or by the deacylated tRNA after transpeptidation in the translating ribosomes; it is called the P-site (see fig. 79). The affinity of tRNA for the P-site of vacant *E. coli* 70S ribosomes with a messenger polynucleotide is characterized by a binding constant of about 10^7 to $10^9\,M^{-1}$, depending on the Mg^{++} concentration, ionic strength and temperature (the affinity increases with increasing Mg^{++} concentration, decreasing ionic strength and decreasing temperature)[18–21].

The retention of tRNA in the P-site of the translating ribosome has an important feature. It is vital that the peptidyl-tRNA should not be exchangeable with the medium during translation. Correspondingly, the peptidyl-tRNA bound in the P-site of the translating ribosome should not be in equilibrium with exogenous tRNA, but rather occluded, i.e. its dissociation rate should be very low. In contrast, when the deacylated tRNA occupies the P-site after transpeptidation has taken place, the site becomes exchangeable and the tRNA may be released. It is not known whether the apparent nonequilibrium retention of the peptidyl-tRNA in the P-site of the translating ribosome is due to the contribution of the peptidyl residue that is anchored by the ribosomal particle during elongation, or whether this reflects the intrinsic property of the P-site in the posttranslocation state ribosome.

Experiments with separated ribosomal subunits have demonstrated that both the small and the large subunit possess a certain affinity for tRNA. The capacity for a codon-specific binding of tRNA, however, is found only for the small (30S or 40S) ribosomal subunit, an obvious result of the fact that only this subunit, but not the large one, can bind and hold the template polynucleotide. At the same time, after dissociation of the translating ribosomes the peptidyl-tRNA often remains bound to the large (50S or 60S) subunit. At present there is good reason to believe that both ribosomal subunits are involved in the formation of the tRNA-binding P-site.

Early experiments on the partial disassembly of the 30S subunit have shown that the removal of the group of proteins S1, S2, S3, S5, S9, S10, and S14 results in a loss not only of the mRNA-binding capacity but of tRNA-binding activity as well[10]. The recovery of tRNA-binding activity during reconstitution requires first and foremost proteins S10 and S14, as well as S3. In further experiments on the full reconstitution of the 30S subunits[22,23], proteins S10 and S14 were found to be most critical for the restoration of the tRNA-binding function, while proteins S3, S9, S11, and S19 did play a part although a lesser one. It should also be mentioned here that the absence of protein S1 did not affect the tRNA-binding capacity of ribosomes, and the omission of protein S2 had only a minor effect. In accordance with the above, the addition of proteins S3 and S14 to the preparations of ribosomes was shown to stimulate their tRNA-binding capacity[24]. This effect was probably due to the repair of the tRNA binding sites which were partially damaged in the course of particle isolation (these proteins can be partly washed off during the isolation procedure). It was still not clear, however, to what extent the above-mentioned proteins were involved in forming the P-site as opposed to the A-site of the ribosome.

The affinity labeling technique should provide more selective information, since chemically active or photoactivable tRNA analogs (figs. 80 and 81(1)) could be positioned specifically at the P-site, and then the proteins adjacent to tRNA could be identified following crosslinking. It would be even better to use direct crosslinking of the tRNA occupying the P-site with its immediate neighbors by means of the ultraviolet light-induced activation of nucleic acid bases. The initial results, although somewhat controversial (see reviews by Ofengand (1980) and Cooperman (1980)), have demonstrated that proteins S5, S9, S10, and S11 of the 30S ribosomal subunit are found in the immediate vicinity of some unpaired bases of the tRNA when it occupies the P-site. In addition, proteins L7/L12, L11, and L27 of the 50S ribosomal subunit also show evidence of contact with the tRNA polynucleotide chain. Crosslinks of the tRNA with the ribosomal RNA have been observed as well. Specifically, the anticodon loop of the tRNA accommodated in the P-site can be crosslinked by ultraviolet irradiation with the 16S RNA in the region of nucleotide residue 1400, on the border between domain III and the 3′-terminal part[15].

Unfortunately, no final conclusions about proteins of the tRNA-binding P-site and the localization of the site on the ribosomes can be made from the data available. Nevertheless, taking into account the information on the mRNA-binding site (see above), one may assume that the anticodon arm of the L-shaped tRNA molecule associates with the 30S subunit in the region of the groove separating the head from the body, somewhere between proteins S5 and S10; protein S3 will then also be located near this region. The acceptor stem of the tRNA occupying the P-site should be in contact with the 50S subunit because the peptidyl transferase center is there (see below); it is probable that such a protein of the 50S subunit as L7/L12, as well as protein L11 at the base of the L7/L12 stalk, is in the vicinity of the central part or corner of the L-shaped tRNA, while the tip (3′-end) of the acceptor arm is close to protein L27, located in the region of the

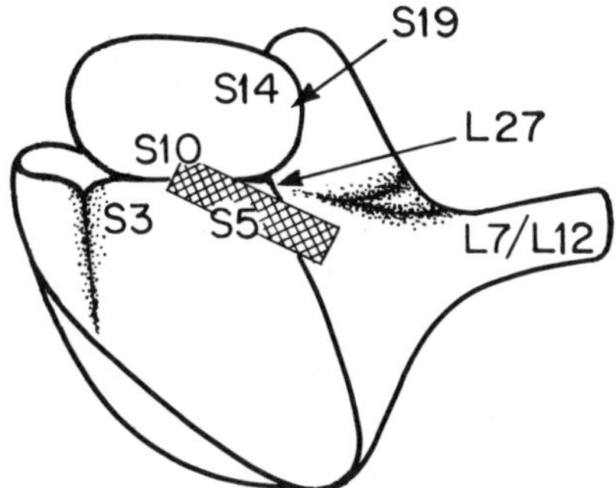

Figure 83 Tentative location of tRNA (crosshatched area) in the ribosomal P-site.

peptidyl transferase center, somewhere near the head of the 50S subunit (see below).

The likely position of the tRNA-binding P-site on the overlap projection of the 70S ribosome is shaded in fig. 83[25]. (It should again be emphasized that experimental arguments for this localization of the P-site are controversial and that the scheme therefore should be regarded as hypothetical.)

Binding of aminoacyl-tRNA (tRNA-binding A-site)

When the P-site is filled with tRNA (whether it is deacylated tRNA or peptidyl-tRNA or aminoacyl-tRNA makes no difference), the ribosome becomes capable of binding the second tRNA molecule. This binding takes place at another tRNA-binding site, called the A-site. Binding in the A-site is greatly stimulated by the template polynucleotide; in this case binding is codon specific, i.e. only tRNA corresponding to the codon placed in the site becomes bound. The affinity of tRNA toward the A-site is approximately one order of magnitude lower than toward the P-site[19–21]. Aminoacyl-tRNA has a somewhat greater affinity to the A-site than does the deacylated tRNA. In the course of normal translation, the binding of the aminoacyl-tRNA is specifically stimulated further by the EF-T_u protein.

It appears that the A-site, like the P-site, is formed by both ribosomal subunits. In any case, the tRNA anticodon should be placed in the immediate vicinity of the mRNA codon, i.e. on the small (30S or 40S) ribosomal subunit, whereas the acceptor end should interact with the peptidyl transferase center, i.e. with the large (50S or 60S) subunit.

There is even less information on the ribosomal components

involved in organizing the tRNA-binding A-site than there is on those of the P-site. Indeed, it is more difficult to experiment on the A-site. For one thing, it is impossible to test the A-site using a vacant ribosome because the binding to the A-site requires that the P-site be filled. Therefore, to test the A-site, the ribosome should be specially precharged with tRNA or its derivative which fills the P-site. On the other hand, the binding of aminoacyl-tRNA to the A-site when the P-site contains peptidyl-tRNA or N-blocked aminoacyl-tRNA results in immediate transpeptidation. Therefore, the problem arises as to how to freeze the required pretranspeptidation state. There are other difficulties with the A-site. One of the first clear-cut pieces of evidence on the involvement of a specific ribosomal component in the A-site has been obtained in experiments where the P-site of the poly(U)-programmed ribosome was previously occupied by N-acetyl-phenylalanyl-tRNA and then the A-site was loaded with phenylalanyl-tRNA carrying a *para*-azidophenacyl group attached to s^4U8 of the central region (the corner) of tRNA[26]. Photoactivation of such a complex resulted in a unique crosslink between the tRNA central region (corner) and protein S19. It follows from the stereochemistry of the affinity label and the position of the label on tRNA that the contact between the corner of the L-shaped tRNA and protein S19 is achieved on the right side of the tRNA molecule, when the molecule is viewed from the external side of the corner, anticodon up. If the approximate position of protein S19 on the head of the 30S subunit is assumed to be between proteins S13 and S14 (see fig. 67), and if the anticodon arm of the tRNA is placed on the 30S subunit and the acceptor stem on the 50S subunit, the L-shaped tRNA molecule may be arranged in the A-site as shown in the scheme of fig. 84[25]. In this

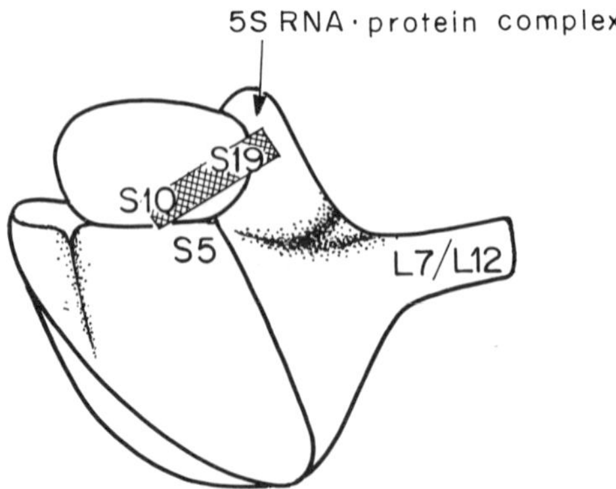

Figure 84 Tentative location of tRNA (crosshatched area) in the ribosomal A-site.

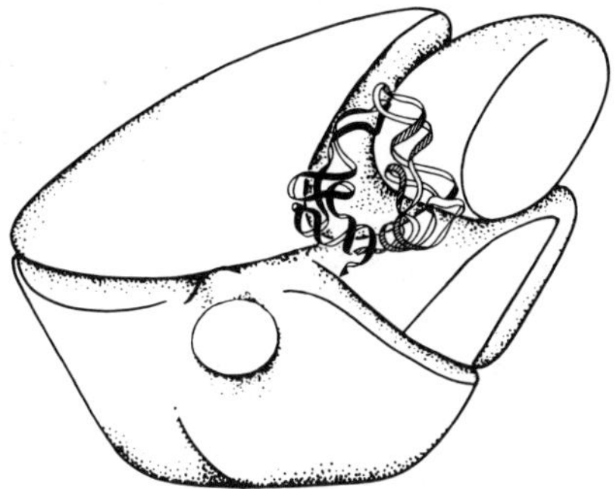

Figure 85 Contour drawing of the 70S ribosome model with the two tRNA molecules occupying A- and P-sites. (Redrawn from A. S. Spirin (1983), *FEBS Letters* 156 : 217–221, with permission.)

position, the corner of tRNA should be in contact with both the head of the 30S subunit and the head of the 50S subunit. In fact, there is indirect evidence that the 5S RNA-protein complex located in the head of the 50S subunit may be involved in the binding of tRNA to the A-site.

An examination of the electron microscopic images of the 70S ribosome or of its model reveals a pocket or cavity at the base of the L7/L12 stalk; this pocket or cavity is delimited by the noncovered concave surface of the 50S subunit and the lateral concave surface of the 30S subunit (see chapter 6, fig. 39). Both the groove separating the 30S subunit head from the body and the groove between the head and the body of the 50S subunit open out into this pocket or cavity. It is tempting to think that it is this pocket or cavity that accommodates the two tRNA-binding sites[25]. The model illustrating the accommodation of the two L-shaped tRNA molecules in this pocket is given in fig. 85. According to the model, the anticodons of the two tRNA molecules are located in the groove between the head and the body of the 30S subunit on its exposed (external) side, while the acceptor ends are found in the region of the groove dividing the head and the body of the 50S subunit on its interface side (this region is thought to bear the peptidyl transferase center; see section 11.3); the central parts of the two tRNA molecules fill the 70S ribosome pocket or cavity mentioned above.

It should be mentioned that an alternative model has been also proposed where the tRNA-binding A-site, as well as the P-site, are positioned on the other side of the ribosome, at the side bulge or platform of the 30S subunit[26,27].

Binding of translation factors and GTP (factor-binding site)

Elongation involves the periodic binding and release (once per cycle) of proteins EF-T_u and EF-G (or EF-1 and EF-2, respectively, in eucaryotes) by the translating ribosome. Each of these proteins is bound in the complex with GTP, and their release is the result of GTP hydrolysis. The binding and release cycle of EF-T_u takes place during aminoacyl-tRNA binding, whereas the binding and release of EF-G proceeds during the translocation stage. Also, initiation of translation involves the ribosomal binding of protein IF-2 (or eIF-2 in the case of eucaryotes) with GTP and the release of this protein as a result of GTP hydrolysis. Finally, in the course of the termination of translation, the ribosome binds and releases the RF protein; GTP takes part in this process as well. All these proteins interacting with ribosomes in the form of their GTP complexes appear to bind to the same region of the ribosomal particle. It is likely that their binding sites on the ribosome are either identical or at least overlapping. In any case, these proteins compete against each other for the binding site on the ribosome and cannot be present on it simultaneously.

Since all of the proteins mentioned are easily released from the ribosome after GTP hydrolysis, the study of their binding *in vitro* may be most conveniently performed if the GTP is replaced by a non-hydrolyzable analog, e.g. guanylyl methylene diphosphonate (GMP-PCP) or guanylyl imidodiphosphate (GMP-PNP) (fig. 86). The complex of the protein with such an analog interacts with the ribosome and is retained on it.

The study of the binding of protein EF-G to the 70S ribosomes has perhaps been the most thorough. EF-G with GMP-PCP may form a complex with both the translating and the vacant ribosome. EF-G with GTP also interacts with the translating ribosome and the vacant ribosome but it is not retained there because GTP undergoes hydrolysis and EF-G and GDP are released from the particle. In the presence of antibiotic fusidic acid (see chapter 14, fig. 108), however, EF-G preserves its affinity to the ribosome even after GTP has been cleaved[28]. The isolated 50S ribosomal subunit behaves in a manner similar to the complete ribosome: EF-G with GMP-PCP, as well as EF-G with GTP (or, to be more accurate, with the product of GTP cleavage) in the presence of fusidic acid forms a rather stable complex with the subunit[29,30]; the interaction of EF-G plus GTP with the 50S ribosomal subunits results in GTP cleavage and the release of EF-G and GDP. No appreciable interaction of EF-G with the isolated 30S ribosomal subunit has been detected. Thus, it may be concluded that the site responsible for EF-G binding is formed mainly by the 50S ribosomal subunit.

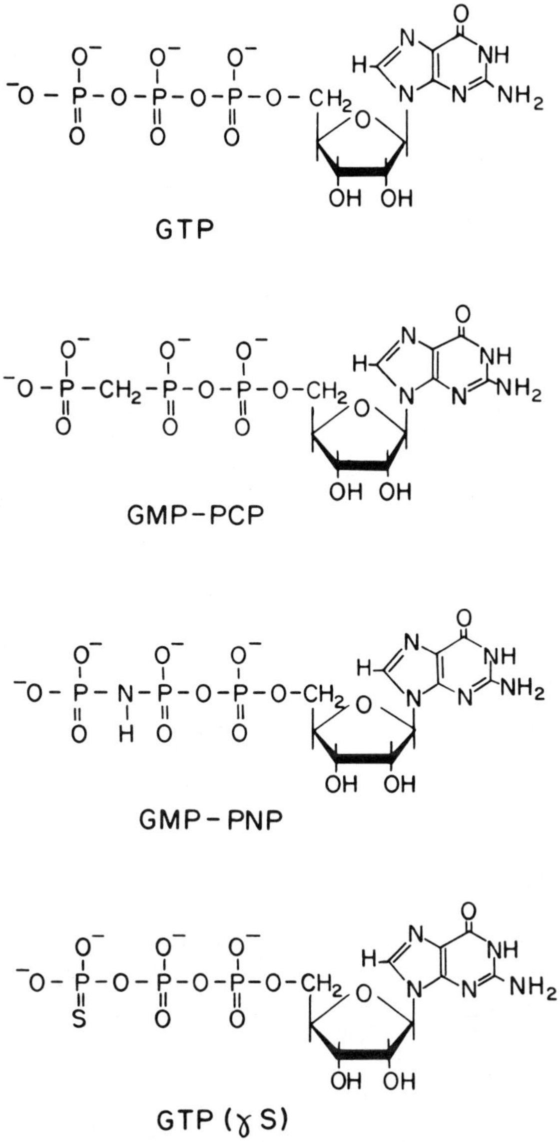

GTP

GMP–PCP

GMP–PNP

GTP (γ S)

Figure 86 GTP and its nonhydrolyzable and slowly hydrolyzable analogs used for studying the functions of the translation factors and the ribosome. GTP, guanosine 5′-triphosphate; GMP-PCP, nonhydrolyzable analog 5′-guanylyl methylene diphosphonate; GMP-PNP, very slowly hydrolyzable analog 5′-guanylyl imidodiphosphate; GTP(γS), slowly hydrolyzable analog guanosine 5′-(γ-thio) triphosphate.

In order to identify the 50S subunit proteins forming the factor-binding site, antibodies against various ribosomal proteins have been used. It has been found that antibodies against protein L7/L12 inhibit binding of EF-G, whereas antibodies against a wide variety of other ribosomal proteins do not affect this function[31]. Initially, these experiments led to the conclusion that protein L7/L12 forms the EF-G-binding site. Indeed, the selective removal of protein L7/L12 from the 50S subunit, achieved by treatment with a 0.5-M NH_4Cl-ethanol mixture, resulted in a markedly decreased binding of EF-G to the ribosome[32]. However, the affinity did not disappear; it simply decreased. It could be demonstrated that even when protein L7/L12 is completely removed from the ribosome, EF-G is still capable, though far less effectively, of binding and of performing its functions in GTP hydrolysis and translocation[33,34]. Therefore, it can be concluded that protein L7/L12 is only auxiliary in EF-G binding; because of the absence of other proteins involved, the ribosomal RNA should be the main binding component[34]. This has been confirmed in direct experiments: a specific affinity of a certain region of the 23S RNA molecule to EF-G has been detected[35].

Thiostrepton (an antibiotic preventing the binding of EF-G and EF-T_u to 50S ribosomal subunits[36]) is known to bind to the subunit in the region of protein L11 and the L11-protected 23S RNA sequence 1052–1112[37]; this sequence is organized in a compound hairpin with a large side loop and located in domain II (see chapter 7, fig. 44). It seems likely that this region of the 23S RNA molecule is either involved in, or close to, the binding site of EF-G on the 50S ribosomal subunit.

It should be pointed out that although it appears that the 50S ribosomal subunit is largely responsible for the recognition and binding of EF-G, the EF-G also comes into contact with the 30S subunit. Specifically, protein S12 can be crosslinked easily by a disulfide bond with the EF-G if the complex between the EF-G and 70S ribosomes is subjected to oxidation[38]. This result indicates that EF-G is located somewhere near the interface of the ribosomal subunits.

The position of an EF-G attachment site on the 50S subunit has been directly determined using immuno-electron microscopy[39]. For this purpose, a photoactivable arylazide derivative of EF-G (fig. 81(4)) was prepared and specifically bound to the particle in the presence of GTP and fusidic acid. Then, a covalent crosslink between EF-G and its attachment site was obtained by irradiation. Subunits with such a covalently linked protein (EF-G) were treated by antibodies against EF-G, and the complexes studied by electron microscopy. Figure 87 shows that the site of antibody attachment is near the base of the

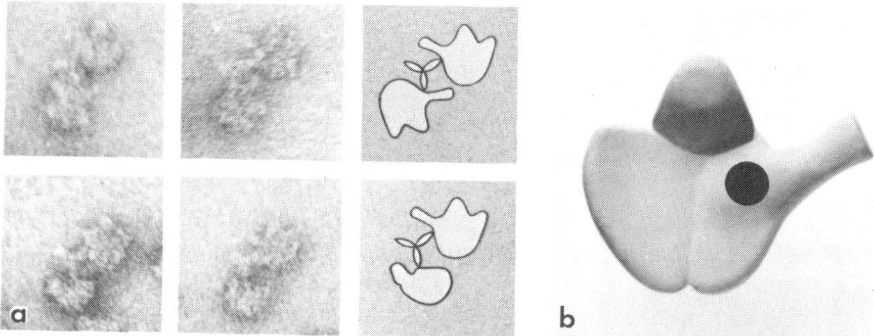

Figure 87 (a) Electron micrographs of 50S ribosomal subunits with EF-G reacted with antibodies against EF-G. (b) Photograph of the 50S subunit model with the approximate localization of EF-G on the subunit indicated (filled circle)[39]. (Courtesy of Dr. V. D. Vasiliev.)

L7/L12 stalk on the interface side of the 50S ribosomal subunit. It may thus be concluded that the EF-G-binding site is located in this region.

The other elongation factor, EF-T_u, is delivered to the ribosome as a complex with the aminoacyl-tRNA and GTP (Aa-tRNA·EF-T_u·GTP). EF-T_u interacts with the ribosome when the complex is bound. It can be shown that it binds to the 50S ribosomal subunit. The presence of EF-G on the 50S subunit prevents EF-T_u from interacting with the ribosome, which leads to the conclusion that the EF-T_u-binding site either coincides with or overlaps the EF-G-binding site (see review by Möller (1974)). As in the case of EF-G, the antibodies directed against protein L7/L12, and only these antibodies, inhibit the interaction between EF-T_u and the ribosome. The removal of protein L7/L12 strongly reduces the interaction between EF-T_u and the ribosome but does not stop it. EF-T_u may perform its functions on ribosomes without protein L7/L12 though far less effectively. Therefore, it appears that EF-T_u also binds to the 23S RNA[34].

The initiation factor IF-2, and the termination factor RF, also compete with EF-T_u and with EF-G for the binding site. Their interaction with the ribosome again depends on the presence of protein L7/L12, but it is not completely prevented if protein L7/L12 is absent from the ribosome. All of this information supports the assumption that the binding of all translation factors, using GTP as an effector, has many common features and that the ribosomes possesses a single factor-binding site near the L7/L12 stalk on the 50S ribosomal subunit.

11-3 Catalytic functions

GTPase

In all cases GTP must be bound to a translation factor, such as EF-T_u, EF-G, IF-2, RF-3, or their eucaryotic analogs, *prior* to factor binding to the ribosome. Therefore, it is apparent that the GTP-binding center is located on the factor protein itself. However, hydrolysis of the bound GTP into GDP and orthophosphate takes place *after* the factor is bound to the ribosome. In other words, GTPase activity requires both the factor and the ribosome in a complex. It is the attachment of the GTP-containing factor to the ribosome that results in GTP hydrolysis. It would be logical to suggest that the GTPase center is formed by a combinations of both the factor and the ribosome.

It is still not known, however, whether this is indeed the case. Doubts about the validity of this assumption stem from two groups of facts. First, a number of experiments with photoactivable GTP analogs (fig. 81(3)) have demonstrated that if a chemical crosslink between GTP and surrounding groups is induced after the EF-G·GTP·ribosome complex has been formed, then EF-G, but not the ribosomal components, is attacked preferentially, regardless of which moiety of GTP carries the photoactivable group[40]. Second, the antibiotic kirromycin (see chapter 12, fig. 95) has been shown to induce an intrinsic GTPase activity of EF-T_u, in the absence of ribosomes[41]. It is therefore now generally held that the attachment of the factor to the ribosome results in an activation of the intrinsic GTPase center of the factor, while the ribosome does not possess either a preexisting GTPase center or any indispensable complement of the GTPase center of the factor.

However, despite this evidence, it is still possible that the GTPase center of the factor is not complete and that some ribosomal components, including ribosomal RNA, participate directly in the completion of its functional state.

Peptidyl transferase

Peptidyl transferase activity is the main catalytic function of the ribosome responsible for synthesizing the polypeptide chain. In the translating ribosome, transpeptidation proceeds between the peptidyl-tRNA and the aminoacyl-tRNA:

$$\text{Pept}(n)\text{-tRNA}' + \text{Aa-tRNA}'' \rightarrow \text{tRNA}' + \text{Pept}(n+1)\text{-tRNA}''.$$

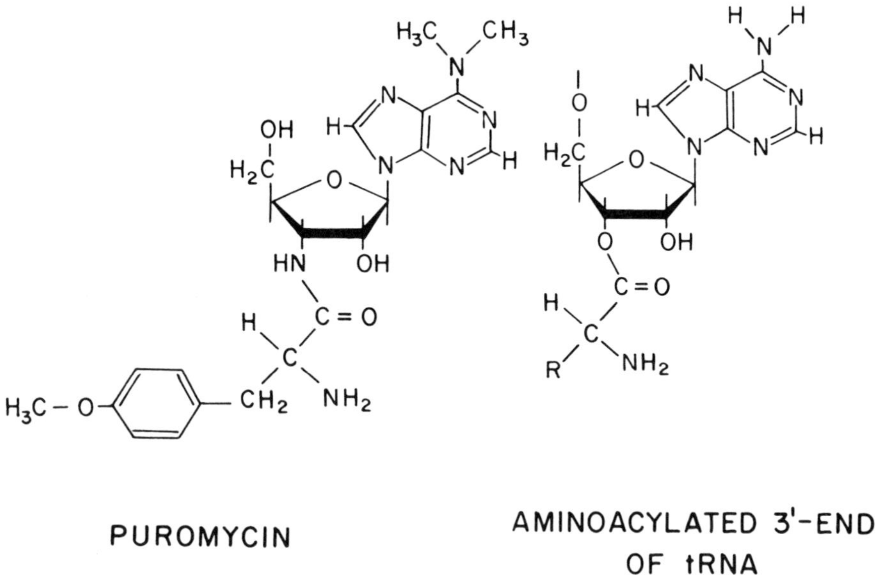

PUROMYCIN

AMINOACYLATED 3'-END OF tRNA

Figure 88 Puromycin (*left*) as an analog of aminoacylated 3′-terminal adenosine of aminoacyl-tRNA (*right*).

In this reaction the peptidyl-tRNA serves as a donor substrate, and the aminoacyl-tRNA as an acceptor substrate.

However, the ribosome catalyzes transpeptidation not only between these natural substrates. The antibiotic puromycin has turned out to be an excellent low-molecular-mass acceptor substrate in the reaction[42–45]. By its chemical nature, it is an analog of the aminoacylated 3′-terminal adenosine of the aminoacyl-tRNA molecule[46] (fig. 88); a dimethylated NH_2-group at position 6 of the adenine residue, a methylated hydroxyl of the tyrosine residue, and the amide bond between ribose and the aminoacyl residue instead of the ester bond are its characteristic features. The addition of puromycin to the translating ribosomes results in a reaction between the antibiotic as an acceptor substrate and peptidyl-tRNA as a donor substrate in the peptidyl transferase center of the ribosome:

$$\text{Pept-tRNA} + \text{PM} \rightarrow \text{tRNA} + \text{Pept-PM}.$$

In this way the peptide becomes transferred, not to the aminoacyl-tRNA, but to a low-molecular-mass compound which is not retained in the ribosome; as a result, the peptidyl puromycin is released from the ribosome. Thus puromycin leads to the abortion of the growing peptide.

The use of puromycin has played an important part in studies on the ribosome peptidyl transferase center. For example, its use has made possible the identification of the ribosomal subunit bearing the peptidyl transferase center. The isolated 50S subunit can retain peptidyl-tRNA, as well as show some labile interaction with the 3′-terminal fragments of the N-blocked aminoacyl-tRNA, which serve as donor substrates. The peptidyl transferase reaction occurs when puromycin is added to the 50S subunits carrying peptidyl-tRNA or its analogs[45,47] Hence, it can be concluded that the peptidyl transferase center is located entirely on the 50S subunit (or the 60S subunit in the case of eucaryotes). The small (30S or 40S) ribosomal subunit does not contribute to the catalysis of the reaction at all.

Not only the intact peptidyl-tRNA and aminoacyl-tRNA molecules but also the 3′-terminal fragments of the tRNAs with adjacent aminoacyl residues have been shown to serve as substrates of the ribosomal peptidyl transferase center[47,48]. As regards the acceptor substrate, this may become clear merely from looking at the puromycin structure. Numerous other aminoacyl adenosines, e.g. A-Phe, A-Tyr, A-Lys, A-Met, and A-Ala, can also serve as good acceptor substrates, although pA-Aa and, in particular, CpA-Aa are even more active. N-blocked derivatives, e.g. pA-(Ac-Aa) and pA-(F-Aa), and even more so CpCpA(Ac-Aa) and CpCpA(F-Aa), can play the part of donor substrates. Therefore, if aminoacyl oligonucleotides such as CpCpA-(F-Met) and puromycin or A-Phe are added to the isolated vacant 50S ribosomal subunits, the particle will work as a normal enzyme catalyzing transpeptidation between low-molecular-mass substrates:

$$\text{CpCpA-(F-Met)} + \text{A-Phe} \rightarrow \text{CpCpA}_{\text{OH}} + \text{A-(F-Met-Phe)}.$$

Neither product is retained by the 50S subunit, and both of them appear immediately in solution.

Naturally, there have been many attempts at isolating the "enzyme" from the 50S subunit, i.e. at finding the ribosomal protein responsible for catalyzing transpeptidation. However, none of these attempts have proved to be successful; the isolated proteins have not showed the presence of such activity. It has been concluded that the "enzyme" probably consists of several proteins tightly integrated in the 50S ribosomal subunit and that it undergoes disruption in the course of protein isolation.

Various analogs of peptidyl-tRNA and aminoacyl-tRNA, e.g. different N-bromoacetyl-aminoacyl-tRNA, N^{ε}-bromoacetyl-lysyl-tRNA, N-(*para*-nitrophenoxycarbonyl)-phenylalanyl-tRNA, and N-(2-nitro-4-azidophenyl)-glycyl-phenylalanyl-tRNA, which carry a chemically active or photoactivable group on the aminoacyl residue at the

3′-end of tRNA (fig. 80 and 81(1)), have been used as affinity labels for identifying proteins located in the region of the peptidyl transferase center. Most intense crosslinks have been observed in such experiments with proteins L2 and L27; proteins L6, L11, L16, L18, and L33 have also been reported as crosslinkable neighbors of the substrates of the peptidyl transferase center (see review by Cooperman (1980)).

In experiments on the partial disassembly and reconstitution of 50S ribosomal subunits, proteins L6, L11, and L16 have been found to be critical for peptidyl transferase activity. Adding them to the LiCl-stripped derivatives of the 50S subunit, which are devoid of more than a third of the initial proteins, led to a recovery in activity. It became clear later, however, that the addition of a large amount of protein L16 by itself also restored peptidyl transferase activity; it was found that proteins L6 and L11 simply facilitated the binding of protein L16 to the particle (see review by Nierhaus (1980)). Furthermore, there is evidence adding weight to the theory that protein L16 plays a key role in organizing the peptidyl transferase center: a chemical modification of its histidine residue resulted in the inactivation of the peptidyl transferase[49]. Nevertheless, it was eventually found that even an integrated protein L16 is neither a peptidyl transferase nor its active site: under modified ionic conditions, provided other ribosomal proteins are present, 50S subunits devoid of protein L16 showed noticeable peptidyl transferase activity[50].

At the same time, experiments on the affinity labeling of the peptidyl transferase center by active aminoacyl-tRNA or peptidyl-tRNA analogs repeatedly demonstrated that although proteins are frequently found as crosslinkable neighbors of these analogs, the ribosomal 23S RNA is still the preferred target. In many experiments, including the pioneer study of Knorre, Girshovich, and associates[11] (who were the first to use the principle of affinity labeling for investigating functional sites of the ribosome), crosslinks mainly with 23S RNA have been obtained. Crosslinking of the photoactivated label-carrying acceptor end of tRNA with position 2584 of the 23S RNA[51], as well as nucleotide replacements at positions equivalent to 2447–2504 in the 23S RNA accompanying the mutations of the peptidyl transferase center[52], suggest that the peptidyl transferase center is located in the region of the evolutionarily conservative sequence 2450–2600 of domain V (see chapter 7, fig. 44). It cannot be excluded that the peptidyl transferase center is organized preferentially by the 23S RNA rather than by proteins. A more cautious opinion, however, is that both the 23S RNA and some proteins of the 50S subunit are involved in organizing the peptidyl transferase center.

Localization of the peptidyl transferase center on the morphologically visible surfaces of the 50S subunit can be done from knowing the proteins which are complexed with the sequence 2450–2600 of the 23S RNA domain V, provided the positions of these proteins are known from immuno-electron microscopy. The protein L6 has been found to be crosslinkable with this sequence[53] and at the same time it has been detected by immuno-electron microscopy at the base of the L7/L12 stalk[54]. Indeed, according to chemical crosslinking with bifunctional reagents (see review by Traut *et al.*, 1980), the protein L6 is a neighbor of proteins L7/L12, L10, and L11 situated at the stalk and its base.

On the other hand, the peptidyl transferase center should neighbor with the head (central protuberance) of the 50S subunit, since the region of the 23S RNA that binds the 5S RNA-protein complex (sequence 2250–2350) and the region involved in forming the peptidyl transferase center (sequence 2450–2600) are the closely located parts of the same domain (V) (chapter 7, fig. 44). (The third, i.e., the 5′-proximal, part of the same 23S RNA domain binds protein L1, which is located on the side lobe of the 50S subunit.) This is confirmed by the fact that protein L18, which is a constituent of the 5S RNA-protein complex and consequently the head of the subunit, can be found among the components reacting with the active analogs of peptidyl transferase center substrates. Other proteins reacting with active substrate analogs, e.g. L2, L11, and L16, are neighbors of the 5S RNA-protein complex, as follows from their crosslinking with either protein L5 or protein L25 (see review by Traut *et al.* 1980). Immuno-electron microscopy observations where antibodies against puromycin covalently fixed in the peptidyl transferase center are used have also indicated its position near the head (central protuberance) of the 50S subunit but rather on the other side of the head, i.e., from the side of the L1 ridge[55,56] (although it cannot be excluded that the latter is a

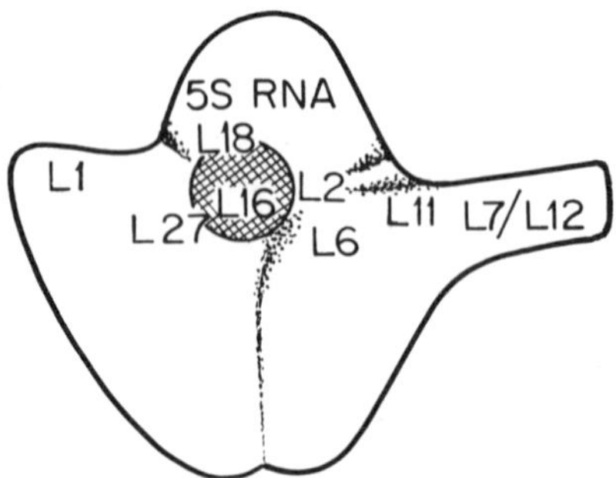

Figure 89 Tentative localization of the peptidyl transferase center (crosshatched region) on the contact surface (interface side) of the 50S ribosomal subunit.

result of a distortion of the antibody position due to specific orientation of the 50S subunit with its interface to the substrate). On the whole, it can be stated that the peptidyl transferase center is located on the 50S subunit, somewhere on its interface (concave) side, perhaps near the head (central protuberance), and probably in the region of the groove separating the head from the rest of the body.

The probable position of the peptidyl transferase center on the 50S ribosomal subunit is shown schematically in fig. 89.

11-4 Function of ligand displacements (translocation)

Transpeptidation in the course of the ribosome working cycle is followed by simultaneous displacements of the three large ligands: mRNA, peptidyl-tRNA, and deacylated tRNA. This may be defined as the mechanical function of the ribosome. Neither isolated ribosomal subunit is capable of even partially performing this function. It is likely that the mechanical function requires the ribosome to be constructed of two subunits.

In the search for the molecular mechanisms responsible for vectorial displacements of large ligands, attention should first be paid to a possible large-block mobility within the ribosome. Since the ribosome consists of two subunits, which are relatively loosely associated in the absence of ligands, it is possible, in principle, that the subunits are capable of moving relative to each other while functioning, either by locking and unlocking[57], or by relative sliding[58]. There is an experimental result demonstrating a small reduction in the compactness of the ribosome in the course of translocation[59,60]; this may be evidence in favor of the unlocking-locking mechanism.

Another possible mobile element of the ribosome is the L7/L12 stalk of the 50S ribosomal subunit. A considerable amount of information suggests that the L7/L12 stalk is involved directly in the functions of protein translation factors, and particularly in the EF-G-catalyzed translocation. It would come as no surprise that the mobility of the L7/L12 stalk[61] played a part in the ligand displacements during translocation, as well as, perhaps, in the course of aminoacyl-tRNA delivery into the ribosome.

The possibility of some interdomain (interlobe) mobility within ribosomal subunits, especially in the small subunit that seems to be more labile and changeable, cannot be excluded either.

References

1. J. D. Watson (1964), "The synthesis of proteins upon ribosomes," *Bull. Soc. Chim. Biol.* 46:1399–1425.
2. F. Lipmann (1969), "Polypeptide chain elongation in protein biosynthesis," *Science* 164:1024–1031.
3. M. Takanami and G. Zubay (1964), "An estimate of the size of the ribosomal site for messenger RNA binding," *Proc. Nat. Acad. Sci. U.S.A.* 51:834–839.
4. M. Takanami, Y. Yan, and T. H. Jukes (1965), "Studies on the site of ribosomal binding of f2 bacteriophage RNA," *J. Mol. Biol.* 12:761–773.
5. J. A. Steitz (1969), "Nucleotide sequences of the ribosomal binding sites of bacteriophage R17 RNA," *Cold Spring Harbor Symp. Quant. Biol.* 34:621–630.
6. J. R. Patton and C.-B. Chae (1983), "A method for mapping RNA initiation, termination, splice, and protein binding sites. Ribosome binding sites on β-globin messenger RNA," *J. Biol. Chem.* 258:3991–3995.
7. P. L. Woolenzien, C. F. Hui, C. Kang, R. F. Murphy, and C. R. Cantor (1985), "RNA structure, free and on the ribosome, as revealed by chemical and enzymatic studies," in *Mechanisms of Protein Synthesis*, ed. E. Bermek, pp. 2–22 (Springer–Verlag, Berlin–Heidelberg).
8. M. Takanami and T. Okamoto (1963), "Interaction of ribosomes and synthetic polyribonucleotides," *J. Mol. Biol.* 7:323–333.
9. J. E. Dahlberg and R. Haselkorn (1967), "Studies on the binding of turnip yellow mosaic virus RNA to *Escherichia coli* ribosomes," *J. Mol. Biol.* 24:83–104.
10. P. Traub, K. Hosokawa, G. R. Craven, and M. Nomura (1967), "Structure and function of *E. coli* ribosomes. IV. Isolation and characterization of functionally active ribosomal proteins," *Proc. Nat. Acad. Sci. U.S.A.* 58:2430–2436.
11. E. S. Bochkareva, V. G. Budker, A. S. Girshovich, D. G. Knorre, and N. M. Teplova (1971), "An approach to specific labeling of ribosome in the region of peptidyltransferase center using N-acylaminoacyl tRNA with an active alkylating grouping," *FEBS Letters* 19:121–124.
12. M. Pellegrini, H. Oen, and C. Cantor (1972), "Covalent attachment of a peptidyl-transfer RNA analog to the 50S subunit of *Escherichia coli* ribosomes," *Proc. Nat. Acad. Sci. U.S.A.* 69:837–841.
13. A. P. Czernilofsky and E. Kuechler (1972), "Affinity label for the tRNA binding site on the *Escherichia coli* ribosome," *Biochim. Biophys. Acta* 272:667–671.
14. A.-R. Subramanian (1963), "Structure and function of ribosomal protein

S1," *Progress in nucleic acid research and molecular biology*, ed. W. Cohn, vol. 28, pp. 102–142 (New York: Academic Press).

15. B. H. Taylor, J. B. Prince, J. Ofengand, and R. A. Zimmermann (1981), "Nonanucleotide sequence from 16S ribonucleic acid at the peptidyl transfer ribonucleic acid binding site of the *Escherichia coli* ribosome," *Biochemistry* 20:7581–7588.
16. A. G. Evstafieva, I. N. Shatsky, A. A. Bogdanov, Y. P. Semenkov, and V. D. Vasiliev (1983), "Localization of 5′ and 3′-ends of the ribosome-bound segment of template polynucleotides by immune electron microscopy, *EMBO Journal* 2:799–804.
17. O. I. Gimautdinova, G. G. Karpova, D. G. Knorre, and N. D. Kobetz (1981), "The proteins of the messenger RNA binding site of *Escherichia coli* ribosomes," *Nucleic Acids Res.* 9:3465–3481.
18. V. Kirillov, K. S. Kemkhadze, E. M. Makarov, V. I. Makhno, V. B. Odintsov, and Y. P. Semenkov (1980), "Mechanism of codon-anticodon interaction in ribosomes. Codon-anticodon interaction of aminoacyl-tRNA at the ribosomal donor site," *FEBS Letters* 120:221–224.
19. S. V. Kirillov, E. M. Makarov, and Y. P. Semenkov (1983), "Quantitative study of interaction of deacylated tRNA with *Escherichia coli* ribosomes," *FEBS Letters* 157:91–94.
20. H.-J. Rheinberger, H. Sternbach, and K. H. Nierhaus (1981), "Three tRNA binding sites on *Escherichia coli* ribosomes," *Proc. Nat. Acad. Sci. U.S.A.* 78:5310–5314.
21. R. Lill, J. M. Robertson, and W. Wintermeyer (1984)," The tRNA binding sites of ribosomes from *Escherichia coli*," *Biochemistry* 23:6710–6717.
22. M. Nomura, S. Mizushima, M. Ozaki, P. Traub, and C. V. Lowry (1969), "Structure and function of ribosomes and their molecular components," *Cold Spring Harbor Symp. Quant. Biol.* 34:49–61.
23. P. Traub and M. Nomura (1969), "Studies on the assembly of ribosomes *in vitro*," *Cold Spring Harbor Symp. Quant. Biol.* 34:63–67.
24. L. L. Randall–Hazelbauer and C. G. Kurland (1972), "Identification of three 30S proteins contributing to the ribosomal A site," *Molec. Gen. Genetics* 115:234–242.
25. A. S. Spirin (1983), "Location of tRNA on the ribosome," *FEBS Letters* 156:217–221.
26. J. Ofengand, F.-L. Lin, L. Hsu, and M. Boublik (1981), "Three-dimensional arrangement of tRNA at the donor and acceptor sites of the *E. coli* ribosome," in *Molecular approaches to gene expression and protein structure*, ed. M. A. Q. Siddiqui, M. Krauskopf, and H. Weissbach, pp. 1–31 (New York: Academic Press).
27. F.-L. Lin, M. Boublik, and J. Ofengand (1984), "Immunoelectron microscopic localization of the S19 site on the 30S ribosomal subunit which is crosslinked to A site bound transfer RNA," *J. Mol. Biol.* 172:41–55.

28. J. W. Bodley, F. J. Zieve, L. Lin, and S. T. Zieve (1970), "Studies on translocation. III. Conditions necessary for the formation and detection of a stable ribosome-G factor-guanosine diphosphate complex in the presence of fusidic acid," *J. Biol. Chem.* 245:5656–5661.

29. J. W. Bodley and L. Lin (1970), "Interaction of *E. coli* G factor with the 50S ribosomal subunit," *Nature* 227:60–62.

30. N. Brot, C. Spears, and H. Weissbach (1971), "The interaction of transfer factor G, ribosomes, and guanosine nucleotides in the presence of fusidic acid," *Arch. Biochem. Biophys.* 143:286–296.

31. J. J. Highland, J. W. Bodley, J. Gordon, R. Hasenbank, and G. Stöffler (1973), "Identity of the ribosomal proteins involved in the interaction with elongation factor G," *Proc. Nat. Acad. Sci. U.S.A.* 70:142–150.

32. E. Hamel, M. Koka, and T. Nakamoto (1972), "Requirement of an *E. coli* 50S ribosomal protein component for effective interaction of the ribosome with T and G factors and with guanosine triphosphate," *J. Biol. Chem.* 247:805–814.

33. B. R. Glick (1977), "The role of *Escherichia coli* ribosomal proteins L7 and L12 in peptide chain propagation," *FEBS Letters* 73:1–5.

34. V. E. Koteliansky, S. P. Domogatsky, A. T. Gudkov, and A. S. Spirin (1977), "Elongation factor-dependent reactions on ribosomes deprived of proteins L7 and L12," *FEBS Letters* 73:6–11.

35. A. S. Girshovich, E. S. Bochkareva, and A. T. Gudkov (1982), "Specific interaction of the elongation factor EF-G with the ribosomal 23S RNA from *Escherichia coli*," *FEBS Letters* 150:99–102.

36. E. Cundliffe (1980), "Antibiotics and prokaryotic ribosomes: Action, interaction, and resistance", in *Ribosomes: Structure, function, and genetics*, ed. G. Chambliss, G. R. Craven, J. Davies, K. Davis, L. Kahan, and M. Nomura, pp. 555 to 581 (Baltimore: University Park Press).

37. F. J. Schmidt, J. Thompson, K. Lee, J. Dijk, and E. Cundliffe (1981), "The binding site for ribosomal protein L11 within 23S ribosomal RNA of *Escherichia coli*," *J. Biol. Chem.* 256:12301–12305.

38. A. S. Girshovich, E. S. Bochkareva, and Y. A. Ovchinnikov (1981), "Elongation factor G and protein S12 are the nearest neighbors in the *Escherichia coli* ribosome," *J. Mol. Biol.* 151:229–243.

39. A. S. Girshovich, T. V. Kurtskhalia, Y. A. Ovchinnikov, and V. D. Vasiliev (1981), "Localization of the elongation factor G on *Escherichia coli* ribosomes," *FEBS Letters* 130:54–59.

40. A. S. Girshovich, V. A. Pozdnyakov, and Y. A. Ovchinnikov (1976), "Localization of the GTP-binding site in the ribosome·elongation-factor-G·GTP complex," *Eur. J. Biochem.* 69:321–328.

41. G. Chinali, H. Wolf, and A. Parmeggiani (1977), "Effect of kirromycin on elongation factor T_u: Location of the catalytic center for ribosome·elongation-factor-T_u GTPase activity on the elongation factor," *Eur. J. Biochem.* 75:55–65.

42. A. Morris and R. Schweet (1961), "Release of soluble protein from reticulocyte ribosomes," *Biochim. Biophys. Acta* 47:415–416.

43. D. W. Allen and P. C. Zamečnik (1962), "The effect of puromycin on rabbit reticulocyte ribosomes," *Biochim. Biophys. Acta* 55:865–874.

44. D. Nathans (1944), "Puromycin inhibition of protein synthesis: Incorporation of puromycin into peptide chains," *Proc. Nat. Acad. Sci. U.S.A.* 51:585–592.

45. R. R. Traut and R. E. Monro (1964), "The puromycin reaction and its relation to protein synthesis," *J. Mol. Biol.* 10:63–72.

46. M. Yarmolinsky and G. de la Haba (1959), "Inhibition by puromycin of amino acid incorporation into protein," *Proc. Nat. Acad. Sci. U.S.A.* 45:1721–1729.

47. R. E. Monro (1967), "Catalysis of peptide bond formation by 50S ribosomal subunits from *Escherichia coli*," *J. Mol. Biol.* 26:147–151.

48. R. E. Monro, J. Cerná, and K. A. Marcker (1968), "Ribosome-catalyzed peptidyl transfer: Substrate specificity at the P-site," *Proc. Nat. Acad. Sci. U.S.A.* 61:1042–1049.

49. R. M. Baxter and N. D. Zahid (1978), "The modification of the peptidyl transferase activity of 50S ribosomal subunits, LiCl-split proteins, and L16 ribosomal protein by ethoxyformic anhydride," *Eur. J. Biochem.* 91:49–56.

50. T. Maimets, M. Ustav, and R. Villems (1983), "The role of protein L16 and its fragments in the peptidyltransferase activity of 50S ribosomal subunits," *Eur. J. Biochem.* 135:127–130.

51. A. Barta, G. Steiner, J. Brosius, H. F. Noller, and E. Kuechler (1984), "Identification of a site on 23S ribosomal RNA located at the peptidyl transferase center," *Proc. Nat. Acad. Sci. U.S.A.* 81:3607–3611.

52. H. F. Noller, J. A. Kap, V. Wheaton, J. Brosius, R. R. Gutell, A. M. Kopylov, F. Dohme, W. Herr, D. A. Stahl, R. Gupta, and C. R. Woese (1981), "Secondary structure model for 23S ribosomal RNA," *Nucleic Acids Res.* 9:6167–6189.

53. I. Wower, J. Wower, M. Meinke, and R. Brimacombe (1981), "The use of 2-iminothiolane as an RNA-protein crosslinking agent in *Escherichia coli* ribosomes, and the localization on 23S RNA of sites crosslinked to proteins L4, L6, L21, L23, L27 and L29," *Nucleic Acids Res.* 9:4285–4302.

54. M. Stöffler–Meilicke, B. Epe, K. G. Steinhäuser, P. Woolley, and G. Stöffler (1983), "Immunoelectron microscopy of ribosomes carrying a fluorescence label in a defined position. Location of proteins S17 and L6 in the ribosome of *Escherichia coli*," *FEBS Letters* 163:94–98.

55. R. Lührmann, R. Bald, M. Stöffler–Meilicke, and G. Stöffler (1981), "Localization of the puromycin binding site on the large ribosomal subunit of *Escherichia coli* by immunoelectron microscopy," *Proc. Nat. Acad. Sci. U.S.A.* 78:7276–7280.

56. H. M. Olson, P. G. Grant, B. S. Cooperman, and D. G. Glitz (1982),

"Immunoelectron microscopic localization of puromycin binding on the large subunit of the *Escherichia coli* ribosome," *J. Biol. Chem.* 257:2649–2656.

57. A. S. Spirin (1969), "A model of the functioning ribosome: Locking and unlocking of the ribosome subparticles," *Cold Spring Harbor Symp. Quant. Biol.* 34:197–207.

58. M. S. Bretscher (1968), "Translocation in protein synthesis: A hybrid structure model," *Nature* 218:675–677.

59. A. S. Spirin, V. I. Baranov, I. N. Serdyuk, and R. May (1984), "Change of ribosome compactness during translocation," *Dokl. Akad. Nauk S.S.S.R.* 274:1260–1266.

60. A. S. Spirin and I. N. Serdyuk (1985), "Studies on structural dynamics of the translating ribosome," in *Mechanisms of Protein Synthesis*", ed. E. Bermek, pp. 92–104 (Springer–Verlag, Berlin–Heidelberg).

61. A. T. Gudkov, G. M. Gongadze, V. N. Bushuev, and M. S. Okon (1982), "Proton magnetic resonance study of the ribosomal protein L7/L12 *in situ*," *FEBS Letters* 138:229–232.

Further reading

Cantor, C. R.; Pellegrini, M.; and Oen, H. (1974). Affinity labeling techniques for examining functional sites of ribosomes. In *Ribosomes* (M. Nomura, A. Tissières, and P. Lengyel, eds.), pp. 573–585. Cold Spring Harbor, N.Y.: Cold Spring Harbor Laboratory.

Cooperman, B. S. (1980). Functional sites on the *E. coli* ribosome as defined by affinity labeling. In *Ribosomes: Structure, function, and genetics* (G. Chambliss, G. R. Craven, J. Davies, K. Davis, L. Kahen, and M. Nomura, eds.), pp. 531–554. Baltimore: University Park Press.

Krayevsky, A. A., and Kukhanova, M. K. (1979). The peptidyltransferase center of ribosomes. Progress in nucleic acids research and molecular biology (W. E. Cohn, ed.), vol. 23, pp. 1–51. New York: Academic Press.

Kurland, C. G. (1974). Functional organization of the 30S ribosomal subunit. In *Ribosomes* (M. Nomura, A. Tissières, and P. Lengyel, eds.), pp. 309–331. Cold Spring Harbor, N.Y.: Cold Spring Harbor Laboratory.

Möller, W. (1974). The ribosomal components involved in EF-G and EF-T_u-dependent GTP hydrolysis. In *Ribosomes* (M. Nomura, A. Tissières, and P. Lengyel, eds.), pp. 711–731. Cold Spring Harbor, N.Y.: Cold Spring Harbor Laboratory.

Nierhaus, K. H. (1980). Analysis of the assembly and function of the 50S subunit from *Escherichia coli* ribosomes by reconstitution. In *Ribosomes:*

Structure, function, and genetics (G. Chambliss, G. R. Craven, J. Davies, K. Davis, L. Kahan, and M. Nomura, eds.), pp. 267–294. Baltimore: University Park Press.

Ofengand, J. (1980). The topography of tRNA binding sites on the ribosome. In *Ribosomes: Structure, function, and genetics* (G. Chambliss, G. R. Craven, J. Davies, K. Davis, L. Kahan, and M. Nomura, eds.) pp. 497–529. Baltimore: University Park Press.

Pellegrini, M., and Cantor, C. R. (1977). Affinity labeling of ribosomes. In *Molecular mechanisms of protein biosynthesis* (H. Weissbach and S. Pestka, eds.), pp. 203–244. New York: Academic Press.

Traut, R. R.; Lambert, J. M.; Boileau, G.; and Kenny, J. W. (1980). Protein topography of *Escherichia coli* ribosomal subunits as inferred from protein crosslinking. In *Ribosomes: Structure, function, and genetics* (G. Chambliss, G. R. Craven, J. Davies, K. Davis, L. Kahan, and M. Nomura, eds), pp. 89–110. Baltimore: University Park Press.

Chapter 12

Elongation I: Aminoacyl-tRNA Binding with the Ribosome

12-1 Codon-anticodon interaction

Analysis of the elongation cycle may conveniently begin at the point when the peptidyl-tRNA occupies the P-site of the translating ribosome while the A-site with the codon of the template polynucleotide positioned there is vacant (chapter 11, fig. 79(top)). Such a ribosome is capable of binding the next aminoacyl-tRNA molecule.

Although the binding of the aminoacyl-tRNA to the ribosomal A-site appears to involve several binding centers of the site and, correspondingly, several regions of the tRNA molecule, the specificity of the bound aminoacyl-tRNA depends exclusively on the template codon. In other words it is the codon that is responsible for selecting the corresponding aminoacyl-tRNA (cognate aminoacyl-tRNA), i.e. the tRNA carrying the aminoacyl residue coded by a given codon.

Adaptor hypothesis and its proof

According to Crick's adaptor hypothesis[1] the structure of the amino acid residue is irrelevant as far as the selection of aminoacyl-tRNA by the codon is concerned. The codon is capable of complementary interaction only with the tRNA residue, which plays the part of adaptor. Therefore, the amino acid residue attached to such an adaptor well in advance becomes automatically presented to the ribosome without participating in codon recognition.

Experimental proof of this postulate of the adaptor hypothesis was obtained in the brilliant experiments conducted by Lipmann's and Benzer's groups[2]. Cys-tRNA^{Cys} was catalytically reduced using Raney nickel; as a result, the cysteine residue was converted into alanine while still bound to the tRNA^{Cys}:

$$\text{II} \qquad H_2N\text{—}\underset{\underset{\displaystyle SH}{|}}{\underset{|}{\underset{\displaystyle CH_2}{CH}}}\text{—}C\begin{matrix}\nearrow O \\ \searrow O\text{—}tRNA^{Cys}\end{matrix} \xrightarrow{H} H_2N\text{—}\underset{\displaystyle CH_3}{\underset{|}{CH}}\text{—}C\begin{matrix}\nearrow O \\ \searrow O\text{—}tRNA^{Cys}\end{matrix}$$

When the Ala-tRNA^{Cys} was added to a cell-free system containing ribosomes programmed by statistical poly(U,G) copolymer coding for phenylalanine, leucine, valine, glycine, tryptophan, and cysteine, but not coding for alanine, the synthesis of an alanine-containing polypeptide was observed. In another experiment, where the Ala-tRNA^{Cys} was added to the reticulocyte cell-free system of globin synthesis programmed with endogenous mRNA, alanine residues were incorporated into the synthesized polypeptide chain at positions normally occuped by cysteine residues[3].

The concept of anticodon

Codon recognition proceeds by the pairing of the codon with a complementary nucleotide triplet present in the adaptor. Hence, selecting aminoacyl-tRNA should be governed by a complementarity between the codon and this triplet, which is termed the *anticodon*. Experimental proof of the decisive part played by codon-anticodon complementarity in binding aminoacyl-tRNA has been provided by studies of mutationally altered tRNA with nucleotide changes in positions 34–36 (chapter 3, fig. 22) corresponding to the anticodon. For instance, when the GUA anticodon of tRNA^{Tyr} of *Escherichia coli* is changed into CUA, the Tyr-tRNA$^{Tyr}_{CUA}$ no longer recognizes the tyrosine UAC codon but does recognize the termination codon UAG[4,5].

Thus, selecting the aminoacyl-tRNA for binding to the ribosomal A-site is the result of complementary codon-anticodon interaction, implying that the codon and anticodon should form a complex with parameters of the Watson–Crick double helix in the A-like form. Characteristic features of this complex include the antiparallel orientation of chains; the formation of standard A·U, U·A, G·C, and C·G base pairs held by hydrogen bonds; and the stacking interaction of bases. In the case of tyrosine tRNA, the initial anticodon and mutationally altered anticodon should be paired with the tyrosine codon and the termination codon, respectively, as follows:

5′		3′	5′		3′
	G·C			C·G	
	U·A			U·A	
	A·U			A·U	
3′		5′	3′		5′

Wobble hypothesis

Strict canonical base pairing, however, does not provide a general rule for the interaction between the first anticodon residue and the third residue of the codon. It has been noted that if an amino acid is coded by two, three, or four codons, the first two nucleotide residues of these codons are always identical; only the third position is different (chapter 2, figs. 2 and 3). Thus, a given amino acid is strictly coded by the two first codon positions but less strictly by the third position. On the other hand, it has been found that ribosomes programmed by different codons for the same amino acid may bind the same tRNA species; in other words, a tRNA can recognize more than one codon[6,7]. For example, the same $tRNA^{Phe}$ recognizes both UUU and UUC[6].

Upon analyzing this and other facts, Crick proposed his hypothesis about the pairing of the first nucleotide of the anticodon with the third residue of the codon; he suggested the possibility of base wobbling in this position[8]. This proposal implies that, in addition to standard A·U, U·A, G·C, and C·G pairing, as well as I·C pairing (I, a deaminated A derivative, pairs similarly with G), noncanonical pairs may form whose geometric parameters are *close* to the standard ones. Such pairs include A·G; G·A and I·A; G·U and I·U; U·G; U·U; and U·C and C·U (fig. 90).

The following characteristics of the codon dictionary (see chapter 3, fig. 3) should, however, be taken into account in order to fit the proposal to the facts: (1) U and C located in the third position of the codon are always equivalent according to the coding specificity; (2) A and G in the third codon position are often (but not always)

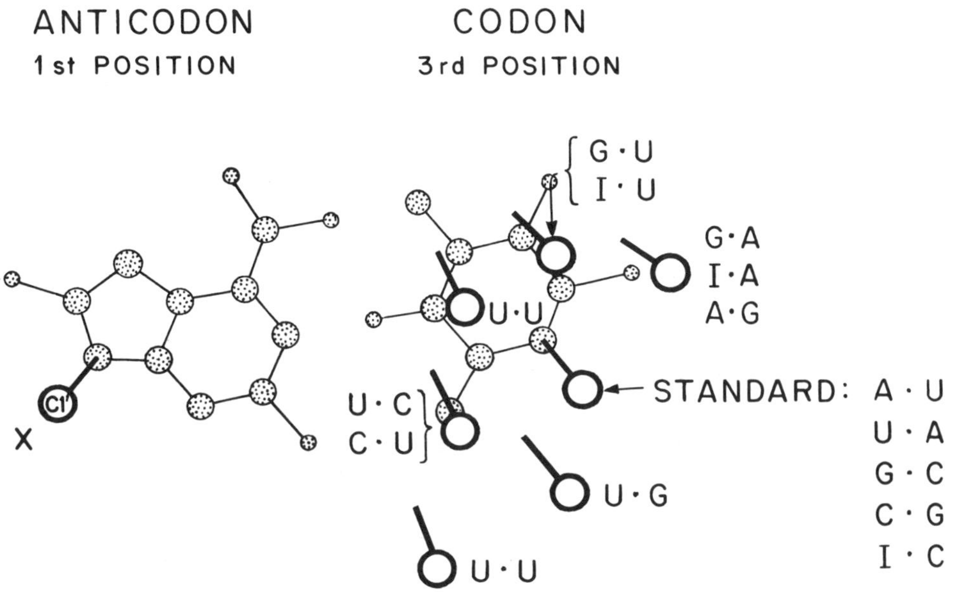

Figure 90 Relative positions and directions of N-glycosidic bonds of nucleoside residues in pairing the first base of the anticodon (*left*) with the third base of the codon (*right*)[8]. The heavy circles indicate the positions of C1′-atoms of the residues; the thick line sections designate N-glycosidic bonds. The standard A·U pair drawn in thinner circles and lines is shown for reference. If the C1′-atom of the anticodon residue is fixed in position X, the C1′-atom of the codon residue is found either in the standard position (in the cases of the standard pairs) or in positions deviated from the standard (in the cases of the wobble pairs).

equivalent; (3) pyrimidine nucleotides (U or C) and purine nucleotides (A or G), in the third position of the codon sometimes are not equivalent (i.e., can be distinguished).

On the basis of physical considerations, Crick excluded the possibility of A·G and G·A pairs formation since such an interaction would result in guanine NH_2-group dehydration, which is unfavorable from an energy standpoint. Furthermore, taking into account the characteristics of the code dictionary mentioned above, he also excluded the possibility of U·U, U·C, and C·U pairing. (If such pairing were allowed, pyrimidine and purine nucleotides in the third position would always be equivalent, i.e. not distinguished, a contradiction of the experimental data.) All of these considerations were summarized in following rules concerning pairing between the first nucleotide of the anticodon and the nucleotide at the third position of the codon:

A pairs with	U;
G pairs with	C
and	U;
I pairs with	C,
	U,
and	A;
U pairs with	A
and	G;
C pairs with	G.

By the time these rules were formulated, it was already known that I is often found in tRNA anticodons, while A in the first anticodon position is not detected and appears always to be converted into I by enzymatic deamination. The types of wobble base pairing proposed by Crick, compared to standard pairing, are given in fig. 91.

Following Crick's hypothesis, one may conclude that no amino acid can be coded by only one codon with A in the third position. Indeed, AUA together with AUU and AUG code for isoleucine; UUA and UUG code for leucine; but UGA does not code for any amino acid. This hypothesis predicts that all three isoleucine codons should be recognized by one tRNA with an IAU anticodon, and the two leucine codons mentioned above, by one tRNA which has an UAA anticodon:

III			
	5′ 3′	5′ 3′	5′ 3′
	I·U	I·C	I·A
tRNA^{Ile}	A·U	A·U	A·U
	U·A	U·A	U·A
	3′ 5′	3′ 5′	3′ 5′
	5′ 3′	5′ 3′	
	U·A	U·G	
tRNA^{Leu}	A·U	A·U	
	A·U	A·U	
	3′ 5′	3′ 5′	

Furthermore, according to the wobble hypothesis, I cannot be present in a tRNA anticodon when the corresponding amino acid is coded by two codons only. Hence, the hypothesis predicts that I could never be present in the first position of the anticodons in phenylalanine, tyrosine, cysteine, histidine, glutamic acid, glutamine, aspartic acid, and asparagine tRNAs.

Finally, when a given amino acid is coded by four codons, no fewer than two tRNAs with different anticodons should be present. One of these tRNA species may recognize U and C, while the other recog-

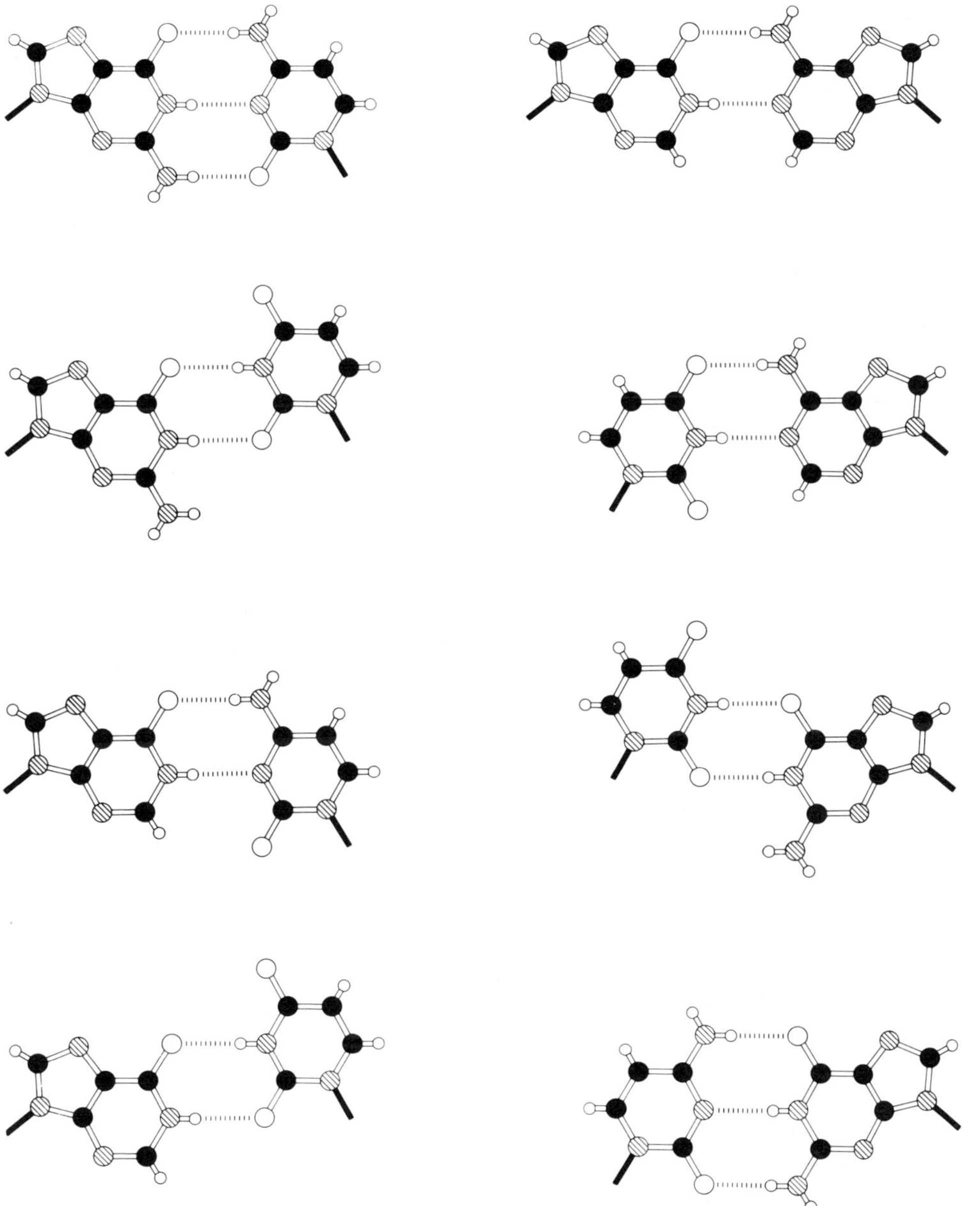

Figure 91 Base pairing between the first base of the anticodon and the third base of the codon: ball-and-stick drawings. *Left column, top to bottom*, G·C, G·U, I·C, and I·U. *Right column, top to bottom*, I·A, U·A, U·G, and C·G. Solid circles are carbons, hatched circles—nitrogens, large open circles—oxygens, and small open circles—hydrogens; filled sticks are N-glycosidic bonds between the base and ribose.

nizes A and G in the third codon position (if G and U are in the first position of their anticodons respectively); alternatively, one tRNA recognizes U, C, and A, while the other recognizes only G in the third codon position (if I and C are in the first position of their anticodons).

Wobbling at the first position of the anticodon and the third position of the codon, as well as most of the rules and predictions offered by this hypothesis, has been confirmed by all subsequent experimental information.

Corrections to wobbling rules

Some inconsistencies have been found, however, and so certain rules need to be amended. For example, three isoacceptor $tRNA^{Ser}$ species in yeast read codons UCU, UCC, UCA, and UCG; the anticodons are IGA, U*GA (U* is a modified U), and CGA. All of these tRNAs, including $tRNA^{Ser}_{U^*GA}$, are necessary to translation *in vivo*, suggesting that $tRNA^{Ser}_{IGA}$ is incapable of effectively recognizing a UCA codon in the cell[9]. In fact, it has yet to be proved that I pairs equally well with C, U, and A in the third codon position, and it is possible that a special isoacceptor tRNA may exist for codons ending with A.

On the other hand, conflicting information has been obtained, primarily *in vitro*, that I as well as G and modified U may pair in some cases with all four bases in the third position of the codon; in any case, the yeast $tRNA^{Val}$ with an IAC anticodon reads all four valine codons in a bacterial cell-free system[10] (the latter, however, may be a consequence of a miscodinglike mechanism induced by somewhat abnormal conditions *in vitro*; see section 12.5).

On the basis of such *in vitro* experiments, Lagerqvist[11] proposed the "two-out-of-three" hypothesis for codon-anticodon recognition. According to this hypothesis, in the case of eight unmixed codon families (leucine, valine, serine, proline, threonine, alanine, arginine, and glycine) only the first two nucleotides of the codons are recognized by the anticodons of their cognate tRNAs, whereas the third nucleotide of the codon does not participate in the recognition and could be any of the four. As regards the physical basis for such a codon-anticodon recognition mechanism, it has been suggested that pairing of the codons with anticodons in the first two positions is especially strong in the cases of the codon families indicated above; this, for example, could result from the participation of strong G·C and C·G pairs and the stabilizing contribution of the purine nucleotides at the second position of the anticodons.

Other inconsistencies have been found in studies on the mitochondrial genetic code and mitochondrial tRNA. It has been discovered that mitochondrial tRNAs containing unmodified U in the first position of the anticodon recognize all four codons with the same initial two-nucleotide sequence[12–14]. For example, $tRNA^{Val}$ with a UAC anticodon recognizes GUU, GUC, GUA, and GUG; $tRNA^{Ala}$ with UGC as an anticodon recognizes GCU, GCC, GCA, and GCG. In cases where the nucleotide in the first position of the anticodon is capable of recognizing only purine nucleotides in codons, U in the mitochondrial tRNA species is modified. In bacterial tRNA as well as in the cytoplasmic tRNA of eucaryotic cells, the U residue at the first position of the anticodon is, as a rule, modified. It appears that unmodified U in the first position of the anticodon could well be paired with all four bases, and the modifications are what reduces its interactions to pairing with purine nucleotides only, i.e. to Crick's rules. Perhaps certain characteristic modifications of U result in its interaction with A only, as is the case with yeast $tRNA^{Ser}$ with a U*GA anticodon[9].

It is also noteworthy that *E. coli* contains a minor $tRNA^{Ile}$ species that recognizes only the AUA codon for isoleucine; in the first position of the anticodon of this tRNA, an unidentified C derivative (but not U) has been found[15]. This again suggests that specialized anticodons recognizing only the codons terminating with A may exist.

Stereochemistry of codon-anticodon pairing

The structure of the anticodon loop described in chapter 3, section 3.2, features the helical arrangement of the chain section including the three anticodon nucleotides and two residues following them toward the 3′-end. The helix parameters are similar to those of a single chain within the standard RNA double helix. Its bases are stacked. The three bases forming the anticodon are oriented such that the groups responsible for pairing through the hydrogen bond formation are exposed (see chapter 3, fig. 20). The anticodon is thus ready to form a double-helical complex with the complementary sequence without its structure being significantly rearranged.

If the L-shaped tRNA molecule is viewed from the outer side of the corner, anticodon up, the groups of the three anticodon bases capable of pairing are turned more or less to the right from the plane containing both limbs of the molecule (see chapter 3, fig. 22). The first anticodon base (wobble position) is located at the very top of the anticodon arm, and the two other anticodon bases and the two subsequent bases of the loop descend from the first base helically, like a spiral staircase, from the left, downward and to the right (chapter 3, fig. 20, as well as fig. 22).

When the tRNA anticodon forms a complex with the mRNA, the

resulting segment of the double-stranded helix possesses standard (Watson–Crick) pairing downward from the top of the anticodon arm, and less strict (wobble) pairing at the top. The latter is in agreement with the greater steric freedom of the pair, located at the edge of the base stack, compared to the internal pairs of the helix. (In anticodon-codon pairing the third pair may also be considered to be internal since its conformational freedom is restricted by the adjacent stacked purine base of the anticodon loop, and probably also by the adjacent anticodon of the peptidyl-tRNA.) The mRNA codon interacting with the tRNA anticodon in the A-site should also acquire helical conformation. It is probable that such a conformation ready for an immediate interaction with the anticodon is induced by the corresponding mRNA-retaining center of the ribosome even before the aminoacyl-tRNA binding occurs.

The binding of aminoacyl-tRNA to the ribosomal A-site during elongation assumes that the P-site is occupied by the peptidyl-tRNA.

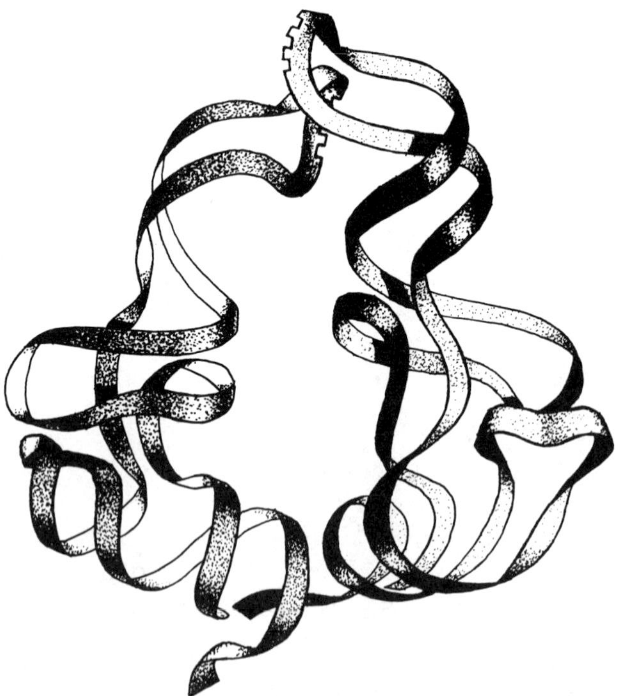

Figure 92 Mutual orientation of two ribosome-bound tRNA molecules represented as ribbon-drawn models: the anticodons are immediate neighbors on the mRNA chain, the acceptor ends are also brought together while the corners are arranged apart.

It has been demonstrated that the codon-anticodon interaction between the mRNA and the tRNA persists when the tRNA is in the P-site[16–18]. This implies that the triplet next to the codon located at the A-site (toward the 5′-end of the mRNA) is also involved in base pairing and has a helical conformation.

At the same time, the orientation of the two tRNA molecules on the ribosome should allow their aminoacyl ends to be brought into close proximity, thus providing for subsequent transpeptidation[19,20] (fig. 92). This implies strict requirements regarding the arrangement of the two codon-anticodon duplexes relative to each other. On the basis of steric considerations, it must be assumed that a flexible hinge exists between the two codon-anticodon duplexes in order that the two tRNA molecules are aligned so as to bring their acceptor ends together[20]. Such a flexible hinge may be realized by rotations around the bonds of the internucleotide C(3′)—O—P—O—C(5′)—C(4′) bridge connecting the two mRNA codons, which result in a kink being formed; in other words, the two codon-anticodon duplexes will not be coaxial and, therefore, will not be stacked with each other (fig. 93). Two L-shaped tRNA molecules can be positioned in such a way that the planes containing both limbs of the molecule meet at an angle, while the apexes of both limbs, i.e. the anticodons and the acceptor ends, are brought into immediate proximity (fig. 92).

12-2 Participation of the elongation factor (EF-T_u or EF-1) in aminoacyl-tRNA binding

In procaryotes the codon-dependent binding of the aminoacyl-tRNA to the ribosomal A-site is catalyzed by the protein referred to as elongation factor T_u (EF-T_u)[21–23]; in eucaryotes the corresponding factor is called EF-1[24,25]. The aminoacyl-tRNA comes to the ribosome as a complex with this protein and GTP[26].

EF-T_u and its interactions

EF-T_u is a labile protein consisting of two or three domains and has a molecular mass of about 45,000 daltons. The polypeptide chain of EF-T_u consists of 393 amino acid residues in the case of *E. coli*[27]. The dimensions of the EF-T_u molecule are 75 Å × 50 Å × 45 Å and thus not very different from those of tRNA[28]. Investigations of the three-dimensional structure of EF-T_u are in progress[28,29]. The GTP-binding

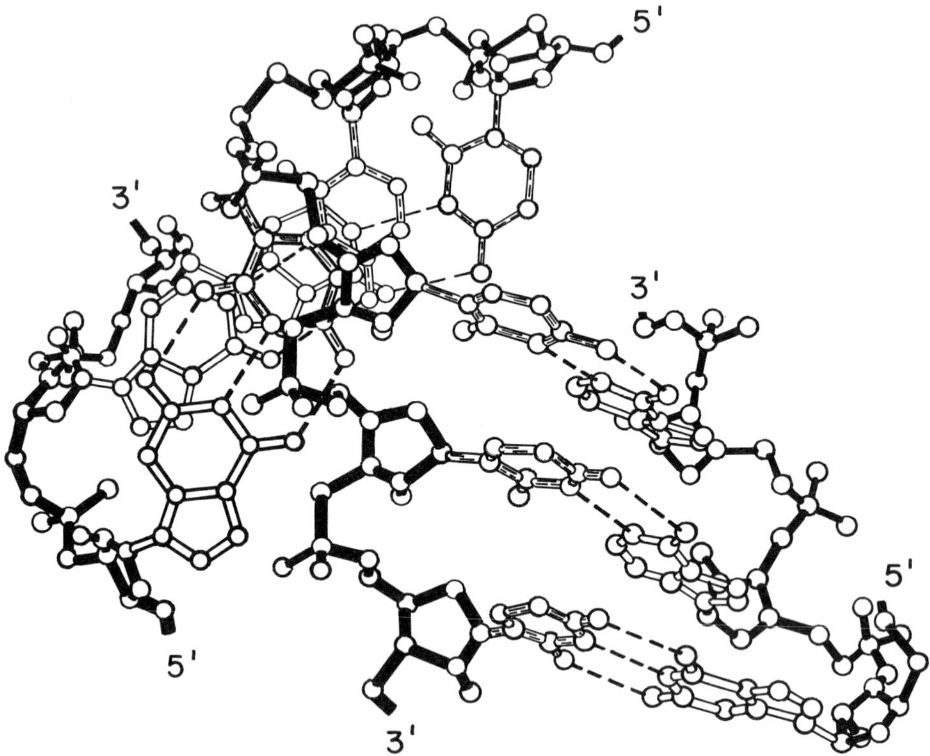

Figure 93 Double-helical structures formed by the two codon-anticodon complexes in the ribosome: ball-and-stick drawing (without hydrogens). The mRNA chain (six nucleotide residues) is seen in the middle; the acceptor tRNA anticodon is on the right and the donor tRNA anticodon is on the left. There exists a kink between two mRNA codons such that if the backbone of the 3′-side codon of the mRNA lies approximately on the plane of the figure, the chain direction of the 5′-codon is almost perpendicular to this plane. The broken lines designate hydrogen bonding between the codon and anticodon bases. (Kindly provided by Dr. V. I. Lim.)

site of the protein is located on the compact N-end-proximal domain (sequence 60–240) and contains a functionally important SH-group (Cys-137). The association between the aminoacyl-tRNA and EF-T_u appears to be of a multicenter nature and involves at least two domains of EF-T_u. Another SH-group present in the N-proximal domain (Cys-81) is important for the binding of the aminoacyl residue of the aminoacyl-tRNA molecule. There is evidence that EF-T_u bound to the aminoacyl-tRNA covers primarily the left side of the acceptor stem and the central part of the L-shaped tRNA molecule when the latter is viewed from the outer side of the corner, anticodon up[30].

Free EF-T_u can bind GTP as well as GDP, the affinity constant to GDP being one order of magnitude greater than that to GTP. The free EF-T_u, instead of the nucleotides, can interact with another protein which has a molecular mass of about 30,000 daltons and is referred to as the elongation factor T_s (EF-T_s). As a result, a complex between EF-T_u and EF-T_s—EF-T_u·T_s(which is sometimes called EF-T)—is formed. If EF-T_u is bound to GDP, the complex formation with EF-T_s results in GDP displacement, as follows:

$$\text{EF-T}_u\cdot\text{GDP} + \text{EF-T}_s \rightleftharpoons \text{EF-T}_u\cdot\text{T}_s + \text{GDP}.$$

The EF-T_u·T_s complex is capable of interacting with GTP, yielding EF-T_u·GTP and free EF-T_s in the presence of excess GTP:

$$\text{EF-T}_u\cdot\text{T}_s + \text{GTP} \rightleftharpoons \text{EF-T}_u\cdot\text{GTP} + \text{EF-T}_s.$$

Thus, EF-T_s is a factor which displaces GDP from the slowly dissociating EF-T_u·GDP complex and thereby catalyzes the exchange of GDP for GTP according to the following overall equation:

$$\text{EF-T}_u\cdot\text{GDP} + \text{GTP} \xrightleftharpoons{\text{EF-T}_s} \text{EF-T}_u\cdot\text{GTP} + \text{GDP}.$$

Free EF-T_u, as well as EF-T_u·GDP and EF-T_u·T_s complexes, has a low, if any, affinity to aminoacyl-tRNA. Only the addition of GTP induces a strong affinity between EF-T_u and aminoacyl-tRNA, resulting in a ternary complex being formed[31]:

$$\text{EF-T}_u\cdot\text{GTP} + \text{Aa-tRNA} \rightleftharpoons \text{Aa-tRNA}\cdot\text{EF-T}_u\cdot\text{GTP}.$$

EF-T_u complexed with GTP does not bind either deacylated or N-blocked aminoacyl-tRNA. The aminoacyl residue and, in particular, its free amino group are necessary to the formation of the ternary complex. The nature of the side group of the aminoacyl residue is of no decisive significance (though some variations in the affinity of aminoacyl-tRNA for EF-T_u·GTP complex depending on the side group have been reported[32]). It is thought that the addition of GTP induces a conformational rearrangement in the EF-T_u protein, resulting in the appearance of an affinity to aminoacyl-tRNA[33]. In contrast, GDP appears to fix EF-T_u in the conformation which has little affinity to aminoacyl-tRNA.

The eucaryotic EF-1 behaves in a similar manner[31]. Two forms of the factor have been described: EF-1_L (light) and EF-1_H (heavy). EF-1_L may be regarded as an analog of EF-T_u; it is a monomeric protein with a molecular mass of about 50,000 daltons and is capable of binding GTP or GDP and of forming the ternary complex Aa-tRNA·EF-1_L·GTP. EF-1_H appears to be an aggregated analog of the EF-T_u·T_s complex; it consists of EF-1_L (α-subunit) and two other proteins (β- and γ-subunits), with molecular masses of 30,000 and 55,000 daltons, respectively, which form a tight complex with each other. The β-subunit of EF-1_H seems to play the part of an EF-T_s analog, whereas the role of the larger γ-subunit has yet to be determined. *In vitro*, the β-subunit and the $\beta \cdot \gamma$-complex appear interchangeable with regard to function.

Binding the ternary complex with the ribosome

The content of EF-T_u (or EF-1) in the cell is high, and practically all cellular aminoacyl-tRNAs exist as ternary complexes with this protein and GTP. Both the tRNA and the protein moiety of the complex show affinity to the ribosome. As a result of codon-anticodon recognition, the complex that carries the tRNA corresponding to the template codon binds to the ribosome.

The selective binding of the ternary complex to the ribosome governed by the codon-anticodon interaction in the A-site induces hydrolysis of the bound GTP into GDP and orthophosphate; orthophosphate is released. At this point EF-T_u exists as a complex with GDP (EF-T_u·GDP) and, because this complex has no affinity to the ribosome or to the aminoacyl-tRNA, EF-T_u·GDP is released from the ribosome. Aminoacyl-tRNA remains bound in the A-site and becomes capable of reacting with the peptidyl-tRNA[31,33]. Thus the following sequence of reactions can be written:

IV (1) Aa-tRNA·EF-T_u·GTP + RS·mRNA $\rightarrow$ RS·mRNA·Aa-tRNA·EF-T_u·GTP;

(2) RS·mRNA·Aa-tRNA·EF-T_u·GTP + H_2O $\rightarrow$ RS·mRNA·Aa-tRNA·EF-T_u·GDP + P_i;

(3) RS·mRNA·Aa-tRNA·EF-T_u·GDP $\rightarrow$ RS·mRNA·Aa-tRNA + EF-T_u·GDP.

Role of GTP hydrolysis

What happens if GTP is replaced by its nonhydrolyzable analogs, e.g. guanylyl methylene diphosphonate (GMP-PCP) or guanylyl imidodiphosphate (GMP-PNP)? EF-T_u (as well as the eucaryotic EF-1) is capable of interacting with the analog and, as a result of the interaction, the protein acquires an affinity to aminoacyl-tRNA. The protein, GTP analog, and aminoacyl-tRNA yield a ternary complex which then binds to the ribosome. The next reaction (2), however, does not take place. Therefore, EF-T_u continues to retain its affinity to the aminoacyl-tRNA and the ribosome, and the aminoacyl-tRNA remains on the ribosome in the complex with EF-T_u and the nonhydrolyzable GTP analog. In this case the aminoacyl-tRNA cannot serve as the acceptor substrate in the transpeptidation reaction with the peptidyl-tRNA, and the next stage of the elongation cycle becomes blocked[33].

There are two explanations as to why the aminoacyl-tRNA complexed with EF-T_u is unable to participate in transpeptidation. According to one, the ternary aminoacyl-tRNA·EF-T_u·GTP or aminoacyl-tRNA·EF-T_u·GMP-PCP complex associates with the ribosome and binds to the template codon, but the tRNA cannot occupy the A-site proper before GTP is cleaved and EF-T_u is released. Therefore, a special ribosomal site responsible for such a preliminary codon-dependent binding of the aminoacyl-tRNA has been suggested; this site is called the *recognition site*, or R-site[35]. (However, since the tRNA located in the hypothetical R-site interacts with the template codon positioned in the A-site, the R-site cannot be considered to be an independent site but should, rather, overlap with the A-site.) GTP hydrolysis could then be postulated to provide the energy required for transferring the aminoacyl-tRNA from the R-site to the A-site. When a nonhydrolyzable GTP analog is used, the aminoacyl-tRNA gets stuck in the R-site and therefore cannot react with the peptidyl-tRNA.

Two sets of experimental information fail to support this hypothesized situation, however. One set demonstrates that the aminoacyl-tRNA bound to the ribosome in the presence of EF-T_u and a nonhydrolyzable GTP analog, such as GMP-PCP, may be rendered active in transpeptidation not by GTP hydrolysis but by simple washing away of EF-T_u and the GTP analog[36]. The other data set shows that aminoacyl-tRNA bound to the ribosome in the presence of EF-T_u and GTP does not become capable of reacting with the peptidyl-tRNA even after GTP hydrolysis if the EF-T_u remains bound on the ribosome. This latter evidence comes from experiments with kirromycin, an antibiotic that inhibits elongation. Kirromycin binds to EF-T_u and increases its affinity to the ribosome. As a result, all of the steps of binding the EF-T_u to GTP, to aminoacyl-tRNA, and eventually to the ribosome, proceed normally except that the EF-T_u·GDP complex is not released after GTP hydrolysis. As a consequence, the next step of the elongation cycle—transpeptidation—remains blocked[37].

The second explanation for the inability of aminoacyl-tRNA complexed with EF-T_u to participate in transpeptidation is based on the assumption that the EF-T_u, a bulky protein molecule bound to the aminoacyl moiety of the aminoacyl-tRNA in the A-site, simply screens the amino group from interacting with the peptidyl transferase center or with the ester group of the peptidyl-tRNA. The available experimental information does not contradict this simple explanation. If this is indeed the case, then the significance of GTP hydrolysis lies in destroying the effector (GTP) which has induced the affinity of EF-T_u to the ribosome and the aminoacyl-tRNA, thus allowing EF-T_u to be released.

Nevertheless, even with this explanation one cannot exclude the possibility that before EF-T_u is released, the aminoacyl-tRNA molecule is not fully accommodated in the A-site; for instance, the protein may prevent just its acceptor end from properly entering the peptidyl transferase center. (However, introduction of the concept of the R-site to designate the aminoacyl-tRNA position that greatly overlaps its final position in the A-site is not justified. In this text the A-site is defined in the structural sense as a ribosomal site where the aminoacyl-tRNA interacts with the mRNA codon adjacent to the codon holding the peptidyl-tRNA in the P-site. If there is a series of intermediate *overlapping* aminoacyl-tRNA positions, they should rather be considered as substates of the ligand within the same structural site.)

Thus, the codon-dependent binding of the aminoacyl-tRNA with the ribosomal A-site is catalyzed by EF-T_u protein plus GTP. Catalysis appears to be the result of the binding of EF-T_u together with the aminoacyl-tRNA, to the ribosome, and not of GTP hydrolysis. GTP hydrolysis may be required to destroy EF-T_u affinity to the ribosome and to facilitate its release, thereby allowing the next step of the elongation cycle (transpeptidation) to proceed.

12-3 "Nonenzymatic" (factor-free) binding of aminoacyl-tRNA

The codon-dependent binding of the aminoacyl-tRNA to the ribosomal A-site may be achieved even in the absence of EF-T_u[38,39] (see epigraph to chapter 16, section 16.6). Such factor-free binding requires a somewhat higher magnesium ion concentration in the medium (10 to 15 mM at 100 mM NH_4Cl or KCl) compared to the Mg^{2+} concentration for the EF-T_u-promoted binding (5 to 10 mM). The factor-free binding is functional: the bound aminoacyl-tRNA can participate in the transpeptidation reaction with the peptidyl-tRNA in the P-site[39,40]. The factor-free ("nonenzymatic") binding of the aminoacyl-tRNA differs from the EF-T_u-promoted ("enzymatic") binding in that it proceeds at a considerably slower rate.

Thus:

$$\text{RS·mRNA} + \text{Aa-tRNA} \xrightarrow[\text{slowly}]{\text{spontaneously}} \text{RS·mRNA·Aa-tRNA.}$$

Nevertheless, the eventual result of this reaction is similar to that observed for the fast binding of the ternary Aa-tRNA·EF-T_u·GTP complex followed by the subsequent cleavage of GTP and the release of EF-T_u:

$$\text{V} \quad \text{RS·mRNA} + \text{Aa-tRNA} \xrightarrow[\text{fast}]{+\text{EF-Tu}} \text{RS·mRNA·Aa-tRNA.}$$

(GTP, H_2O in; GDP, P_i out)

12-4 Inhibitors

Numerous specific inhibitors, predominantly antibiotics, are known to block different stages of the ribosomal elongation cycle[41]. Most of these act selectively either upon procaryotic (bacterial) or upon eucaryotic ribosomes. A classical example of specific inhibition of both EF-T_u-promoted and nonenzymatic binding of the aminoacyl-tRNA with the A-site of a bacterial 70S ribosome is provided by tetracyclines (fig. 94). Tetracyclines also inhibit the codon-dependent binding of the aminoacyl-tRNA with the isolated 30S subunit. This may be taken as evidence that it is the part of the A-site and not part of the P-site that shows tRNA-binding capacity on the isolated 30S subunit. In agreement with this, the specific binding site for tetracycline has been detected on a 30S subunit, although at greater concentrations the antibiotic can bind also to the 50S subunit, but with side effects. It is noteworthy that when the ternary Aa-tRNA·EF-T_u·GTP complex interacts with the 70S ribosome in the presence of tetracycline, GTP undergoes hydrolysis and EF-T_u·GDP is released but

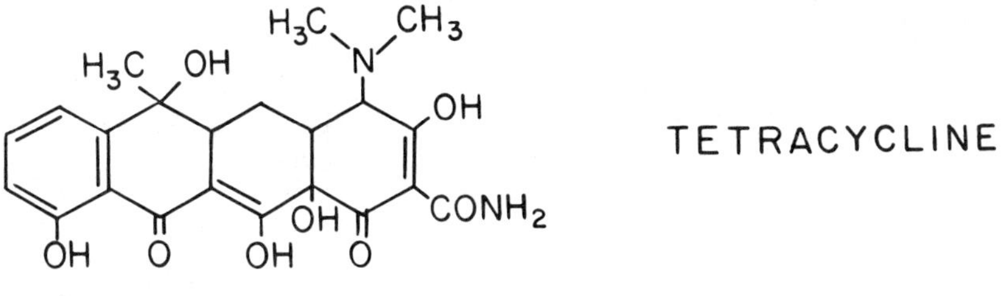

TETRACYCLINE

Figure 94 Tetracycline, an inhibitor of aminoacyl-tRNA binding with the ribosome.

aminoacyl-tRNA does not remain in a bound state. It is likely that tetracycline bound on a 30S ribosomal subunit somewhere in the region of the tRNA-binding A-site decreases the affinity of the site to tRNA, resulting in a poor retention of tRNA after EF-T_u release.

Although eucaryotic cell membranes are impermeable to tetracyclines, these antibiotics exert a strong inhibitory action in eucaryotic cell-free systems by suppressing the binding of the aminoacyl-tRNA to the 80S ribosomes.

There are several known inhibitors that suppress primarily the EF-T_u- or EF-1-promoted binding of the aminoacyl-tRNA to the ribosome by blocking the factor-binding site of the 50S or 60S ribosomal subunit. Antibiotics thiostrepton and related siomycin provide typical examples of such action. Large molecules of these antibiotics bind tightly to the 50S subunit in the region of protein L11 and prevent EF-T_u, as well as EF-G, from interacting with the factor-binding site[42]. Ribosomes lacking protein L11 are shown to be functionally active and insensitive to the antibiotic. It is remarkable that in certain mutants resistant to the antibiotic protein L11 is completely absent (rather than being simply altered).

Fusidic acid (see chapter 14, fig. 108) generally is considered to be a translocation inhibitor but in essence is an inhibitor of aminoacyl-tRNA binding. The antibiotic does not affect the ribosome but binds to EF-G. As a result, the protein acquires an increased affinity to the factor-binding site of the 50S ribosomal subunit and remains complexed with the site after translocation and GTP hydrolysis. The complex between EF-G·GDP and fusidic acid thus inhibits the interaction of EF-T_u and the aminoacyl end of tRNA with the 50S ribosomal subunit.

Finally, it is worthwhile to consider kirromycin as an antibiotic affecting EF-T_u; kirromycin (fig. 95) does not inhibit the binding of the aminoacyl-tRNA but rather suppresses the subsequent stage of the cycle. It affects EF-T_u just as fusidic acid affects EF-G: EF-T_u acquires a high affinity to both the ribosome and the aminoacyl-tRNA, with the result that the factor protein is not released after GTP hydrolysis and blocks the aminoacyl terminus from participating in transpeptidation[37]. It should be emphasized that the interaction of kirromycin with EF-T_u induces the intrinsic GTPase activity of the protein in the absence of the ribosome. These observations provided the basis for concluding that the GTPase center is present on the EF-T_u itself rather than on the ribosome, and thus the ribosome just activates the latent GTPase of EF-T_u.

Another group of antibiotics also affects the binding of the aminoacyl-tRNA to the ribosomal A-site, but the action is of a

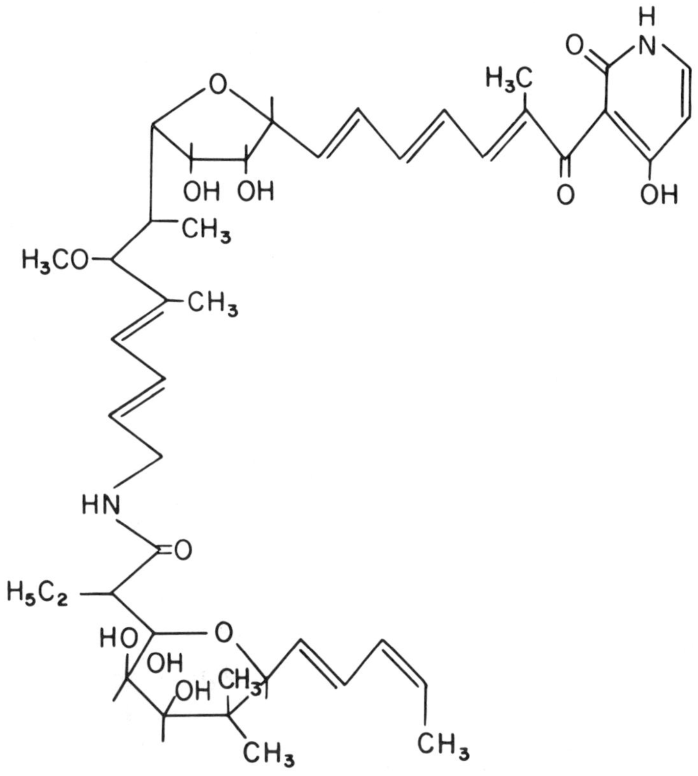

KIRROMYCIN
(MOCIMYCIN)

Figure 95 Kirromycin, an inhibitor affecting EF-T_u.

completely different nature. These antibiotics are the so-called aminoglycoside antibiotics, the principal members of which are streptomycin (fig. 96), neomycin, and kanamycin. Antibiotics in this group contribute to the ribosomal retention of the aminoacyl-tRNAs, which do not correspond to the codon positioned in the ribosomal A-site[43–45]. Such a *miscoding* results in synthesis of incorrect polypeptides characterized by a large number of errors, with a consequent cytotoxic or bactericidal effect on cells. Streptomycin specifically affects bacterial 70S ribosomes, and kanamycin and neomycin are known to induce miscoding with eucaryotic 80S ribosomes as well. The main site of the antibiotic binding to the ribosome appears to occur on the small (30S or 40S) ribosomal subunit, although the effect depends on the interaction of both subunits and is manifested only with the complete 70S or 80S ribosome.

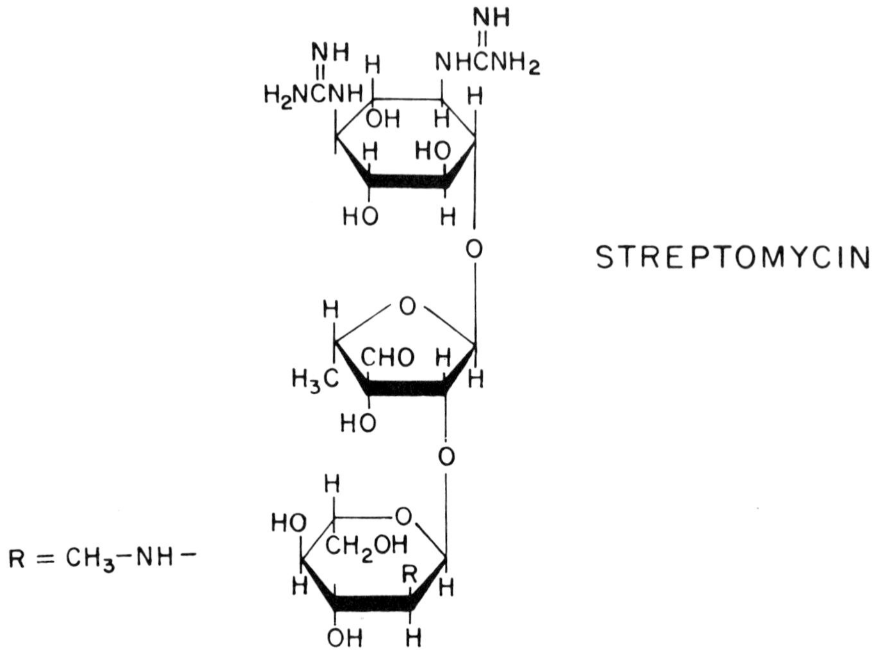

STREPTOMYCIN

R = CH_3-NH-

Figure 96 Streptomycin, an inhibitor stimulating misbinding of aminoacyl-tRNAs.

12-5 Miscoding

Miscoding resulting from the incorrect binding of aminoacyl-tRNA to the ribosome in the presence of streptomycin and other aminoglycoside antibiotics is not a qualitatively new aspect in the function of translation machinery. A certain low level of erroneous aminoacyl-tRNA binding to the ribosome is always present in translation, and the aminoglycoside antibiotics only increase the magnitude[43,44]. A particular template codon located in the ribosomal A-site may erroneously bind only a limited set of other (noncognate) aminoacyl-tRNAs of *related* coding specificity. Aminoglycoside antibiotics stimulate, as a rule, the binding of the same set of noncognate aminoacyl-tRNAs without contributing new ambiguities[45].

Misreading of poly(U)

In the simplest cell-free system of poly(U)-directed polyphenylalanine synthesis with *E. coli* ribosomes, one can detect a marked misincorporation of leucine and isoleucine[46,47], as well as of serine, tyrosine, and valine in decreasing levels[43,44,48–50]. The incorporation of these five amino acids, in particular of leucine and isoleucine, is greatly stimulated by streptomycin[43,44]. Figure 97 shows the correct codon-anticodon pairing of the template with tRNA^{Phe} and the known incorrect types of UUU codon pairing with anticodons of *E. coli* tRNA^{Ile}, tRNA^{Leu}, tRNA^{Ser}, tRNA^{Tyr}, and tRNA^{Val}. ($\text{tRNA}_1^{\text{Leu}}$ possessing anticodon CAG and tRNA^{Cys} possessing anticodon GCA appear incapable of appreciable binding with the poly(U)-programmed ribosome.)

```
tRNA^Phe       5' G · U 3'          tRNA^Ile       5' G · U 3'
                  A · U                               A · U
               3' A · U 5'                         3' U   U 5'

tRNA_2^Leu     5' N · U 3'          tRNA_3^Leu     5' G · U 3'
                  A · U                               A · U
               3' G   U 5'                         3' G   U 5'

tRNA_4^Leu     5' C   U 3'          tRNA_5^Leu     5' A*  U 3'
                  A · U                               A · U
               3' A · U 5'                         3' A · U 5'

tRNA_1^Ser     5' U^V· U 3'         tRNA^Tyr       5' G^Q· U 3'
                  G   U                               U   U
               3' A · U 5'                         3' A · U 5'

tRNA_1^Val     5' U^V· U 3'         tRNA_2^Val     5' G · U 3'
                  A · U                               A · U
               3' C   U 5'                         3' C   U 5'
```

Figure 97 Codon-anticodon pairing of poly(U) with cognate and noncognate tRNA species. U^V, 5′-carboxymethoxyuridine (cmo^5U or V); G^Q, queuosine (Q) (see fig. 15); A*, an unidentified derivative of A; N, an unidentified nucleoside.

It can be thought that two of the three nucleotide positions in the codon-anticodon complex should be paired (including noncanonical pairing at the wobble position) in order to provide perceptable binding of noncognate tRNA in the A site of the ribosome (it is assumed that bases N and V form a noncanonical pair with U at the wobble position when $tRNA_2^{Leu}$, $tRNA_1^{Ser}$ or $tRNA_1^{Val}$ are bound).

Principal types of mispairing

Analysis of miscoding in the cases of poly(U) and other template polynucleotides *in vitro*, as well as *in vivo*, including miscoding induced by various aminoglycoside antibiotics, has demonstrated that errors are largely due to G·U or U·G pairing (or juxtaposition without pairing), as well as to U·U pairing (or juxtaposition) at any position of the anticodon-codon duplex. Mispairing or juxtaposition of the U·C or C·U types are less common. Some rare errors are due to the juxtaposition of the C in the anticodon and the A of the codon. In exceptional cases errors involve the formation of A·G or G·A pairs or juxtapositions, as well as of C·C, A·A, and G·G juxtapositions[51].

It appears that all types of juxtapositions are possible in the wobble position, including I·G, G·A, G·G, U·U, U·C, C·A, C·U, and C·C. This probably explains why all of the isoacceptor tRNAs can recognize in the cell-free system all of the four codons of a given codon family, i.e. of codons that have the same initial two nucleotides (see section 12.1).

Factors contributing to miscoding

In addition to aminoglycoside antibiotics, a number of less specific factors including ionic conditions of the medium can increase the number of errors in the codon-dependent entry of the aminoacyl-tRNA into the translating ribosomes. Generally, all factors increasing the affinity of the tRNA to the ribosome result in increased miscoding. Increased Mg^{2+} concentration in the medium and the addition of diamines (e.g. putrescine) or polyamines (e.g. spermidine) increase the number of errors during cell-free translation[48,49]. Ethyl alcohol and other hydrophobic agents added in even low concentrations also increase miscoding[50]. Urea, in contrast, leads to a decrease in the miscoding level. As for general environmental factors, a lower temperature, decreased pH, and low ionic strength also contribute to a higher miscoding level[48,49].

Structural features of the tRNA itself also play a part in the accuracy of the codon-dependent binding of tRNA in the A-site. In particular, the structure of the D-hairpin may be important. For example, the mutational alteration of G to A in the D arm of $tRNA^{Trp}$ stimulates the pairing of this tRNA (which has a CCA anticodon) with the noncognate codons UGA or UGU[51]. In the absence of ribosomes, the affinity of the CCA anticodon to both the complementary triplet and the noncognate UGA triplet does not change as a result of such a mutation. Therefore, there is every reason to believe that the alteration of the D arm increases the affinity of tRNA to the A-site. This is consistent with the model wherein the L-shaped tRNA molecule positioned in the A-site interacts with the ribosome mainly at its right side, when the tRNA is viewed with the anticodon up, from the outer side of the corner (meanwhile the left side of the tRNA interacts with the EF-T_u; see chapter 11).

Furthermore, the ribosome structure plays an important role in the accuracy of aminoacyl-tRNA selection. Gorini and associates[45,52] were the first to demonstrate that certain mutations leading to alterations in ribosomal components may either decrease or increase the level of miscoding. It has been found that certain mutational alterations in protein S12 resulting in resistance to streptomycin (*str*A-mutations) confer greater fidelity on the ribosome in the codon-dependent selection of aminoacyl-tRNA. In contrast, the so-called *ram*-mutations (ribosome ambiguity mutations) involving protein S4 make the ribosome less selective and increase the level of miscoding. Specific mutational alterations in protein S5 also decrease the selectivity of the ribosome. Characteristically, all the known types of mutational alterations affecting the fidelity of codon-anticodon recognition in the ribosomal A-site concern only 30S ribosomal subunit components. There is no doubt, therefore, that the part of the A-site located on the 30S subunit is instrumental to the fidelity of codon-anticodon recognition. It may well be that the 30S subunit components forming the A-site and located nearby—particularly the tightly-clustered group of proteins S4, S5, and S12—are vital not only to the strength of tRNA retention but also may define the degree of structural rigidity/flexibility of the tRNA anticodon or mRNA codon positioned on the ribosome.

The environmental factors listed above, as well as intrinsic structural factors, affect the extent of the selectivity of the mRNA-programmed ribosome with respect to tRNA. At the same time, the miscoding level in the system depends not only on selectivity as defined by the intrinsic properties of the components under given conditions, but on the ratio of the components as well. It is apparent that as the ratio of the concentration of noncognate tRNAs to that of cognate tRNA increases, the probability of misbinding becomes higher[51]. It is for this reason that, for example, when poly(U) is translated in the cell-free system and the cognate phenylalanyl-tRNA is depleted as a result of synthesis, the incorporation of leucine, isoleucine, and other incorrect amino acids into the polypeptide tends to increase. In an extreme case the system with poly(U)-programmed ribosomes may be supplied only with leucyl-tRNA, and then pure polyleucine will be synthesized on poly(U) as a template (although, of course, at a markedly slower rate than with polyphenylalanine synthesis)[53]. In natural systems the amount of amino acids misincorporated into the synthesized polypeptide may depend greatly on the concentrations of different aminoacyl-tRNA species. Thus, cell starvation for an amino acid results in amino acids with near-coding specifities extensively replacing this amino acid in the polypeptide chains[54–56].

Miscoding level *in vivo* under normal conditions

Since miscoding depends largely on a number of environmental and structural factors, it is clear that its level in the cell-free systems varies greatly. Therefore, it is important to estimate the natural miscoding level in normal living cells that are not subjected to extreme conditions and do not carry mutations affecting the protein-synthesizing machinery. For many reasons, however, it is difficult to make this estimate. First, errors may occur at stages preceding aminoacyl-tRNA binding; they may involve a misacylation of tRNA, for example. Such errors result in the overestimation of the level of miscoding. Second, the wrong product may be eliminated at stages subsequent to aminoacyl-tRNA binding and the amino acid misincorporation into the peptide chain. This elimination may occur either as a premature release of the nascent peptide or as a hydrolysis of the wrong protein product. Third, if the completed wrong product possesses properties other than those of the normal product, it could remain unintegrated into the structures where its normal counterpart is present, or it could be lost during the procedures normally used for isolating the corresponding normal protein. These circumstances result in the miscoding level being underestimated, sometimes to a considerable degree.

Furthermore, the data on the misincorporation of one amino acid do not permit judgments about the misincorporation pattern of other amino acids. In particular, miscoding within one codon group (i.e. among codons differing only in the third nucleotide) is more probable than other replacements. It follows from the code dictionary (chapter 2, fig. 3) that the most probable amino acid replacements, resulting from miscoding, would be the following: Phe ↔ Leu, Cys ↔ Trp, His ↔ Gln, Ile ↔ Met, Asn ↔ Lys, Ser ↔ Arg, and Asp ↔ Glu. It is expected that the replacements from left to right (mispairing with U or C in the third codon position) are more probable than replacements from right to left (mispairing with A or G in the third codon position). Indeed, lysine substitutions for asparagine are far more frequent than replacements of lysine; similarly, histidine is often replaced by glutamine, whereas glutamine is rarely replaced[57].

Several attempts have been made at estimating the level of miscoding *in vivo*. Loftfield's classical estimates provided data on the rate of misincorporation of valine instead of threonine in the α-chains of rabbit globin[58,59]; a value of about $2–6 \times 10^{-4}$ was obtained. The frequency of cysteine misincorporation, probably instead of arginine, into the completed (folded and assembled) *E. coli* flagellin which normally does not contain cysteine was of the same order of magnitude (approximately 10^{-4} per codon)[60]. It was not clear, however, to what extent this value reflected the level of mistranslation on ribosomes (see the first paragraph of this section). Later more direct estimates of the miscoding level *in vivo* were done. An average frequency of translational errors of about 4×10^{-4} per codon in *E. coli* was calculated from the experiments where amino acid substitutions were indicated by a change of the electric charge of a protein due to the appearance or loss of a charged residue (the experiments were done in the presence of streptomycin stimulating only translational errors, and the values obtained were reduced to the error values for the normal cell growth by taking into account the independently determined 10-fold stimulation of the error rate by streptomycin)[61].

Some codons, however, can be misread much more frequently than the above estimated rate. For instance, in the experiments on the *in vivo* translation of MS2 coat protein mRNA the frequency of misreading of the asparagine codon AAU leading to the substitution of the positively charged lysine for the uncharged asparagine was about 5×10^{-3} (though the frequency of misreading of the asparagine codon AAC was nearly an order of magnitude less, being about 2×10^{-4})[62]. The consequence is that even under conditions of normal growth, up to 1–3% of the molecules of MS2 phage coat protein synthesized in *E. coli* have substitutions of lysine for asparagine as a result of translational errors.

It has been recently demonstrated that the high rates of translational errors *in vivo* can be observed not only in the cases of amino acid substitutions within the same codon group due to mispairing of a pyrimidine nucleotide at the third position of a codon. The tryptophan codon UGG in the ribosomal protein S6 mRNA of *E. coli* permits cysteine to be incorporated instead of tryptophan with a frequency of $3–4 \times 10^{-3}$ (Ref. 63); evidently, for some reason, a high rate of mispairing of a purine nucleotide (G) at the third codon position takes place here. Cysteine substitutes for arginine in the ribosomal protein L7/L12 also with a high frequency of 10^{-3} (Ref. 63), though in this case the codons of the two amino acids belong to different codon groups.

It should be mentioned that the most recent estimates of the miscoding level *in vivo* have given even higher values, from 10^{-3} to 10^{-1} errors per codon[64], thus suggesting a revision of previous estimates. If this were the case for the majority of codons, however, no correct protein molecules would be virtually synthesized during translation. At present, the most reasonable estimate, consistent with the bulk of the experimental data available, is an average value of miscoding *in vivo* from 10^{-4} to 10^{-3} errors per codon, perhaps with a few exceptional deviations for some codons to 10^{-2}–10^{-1}.

The level of miscoding is usually much higher in cell-free systems; the replacement frequency may be as great as 10^{-2} and, within the same codon group (Phe → Leu), even 10^{-1} per codon. However, under controlled ionic conditions (in a low Mg^{2+} concentration and with optimal proportions of other components), a level of miscoding approaching the values of 10^{-4} to 10^{-3} errors per codon may be attained[65] (see also Ref. 77).

It is important to note that under certain conditions the level of miscoding *in vivo* can be greatly increased, both in bacterial and in animal cells. This can be achieved by starving for certain amino acids, as well as by adding ethyl alcohol and some other agents to the medium. As already mentioned miscoding in the cell increases in response to aminoglycoside antibiotics.

A certain miscoding level may be of great biological significance and therefore is maintained in evolution. Bacterial mutants with a low miscoding level (streptomycin-resistant mutants) can be obtained, but this level is always higher in wild strains that have been isolated from nature and are more adapted to survival. There is no doubt that, in certain circumstances, miscoding contributes to survival, e.g. in cases of mutations that would otherwise be lethal. Thus, misbinding of an aminoacyl-tRNA to the termination codon produced by the mutation of a sense codon may ensure that a functional protein molecule is completed. The result of this "deception for the sake of salvation" is that the nonsense mutant survives. Similarly, certain mutants with the point amino acid replacements in important proteins, otherwise lethal, may survive through miscoding. In addition, the cell may permanently employ an infrequent misbinding of aminoacyl-tRNA to the normal termination codon; this results in an mRNA read-through beyond its usual coding region (see chapter 18, section 18.1) and therefore in a synthesis of small amounts of longer polypeptides yielding functionally differing proteins, which may be necessary to some processes in the cell. It is also possible that a cell occasionally replaces an amino acid in a similar manner in order to synthesize some needed variant of a given protein in small amounts.

In any event, too great a coding accuracy during translation would probably restrict the cellular flexibility necessary for survival.

Kinetic mechanisms of miscoding and miscoding correction

It follows from what has been said above that miscoding depends primarily on the affinity of anticodon to codon. Affinity is strongest when the anticodon is complementary to the codon, and therefore the cognate tRNA binds to the codon preferentially. Affinity is lower, but still exists, in the case of partial complementarity and, thus, some noncognate ("near-cognate") tRNA species may bind to the codon as well.

Experiments on different polynucleotide complexes have demonstrated that the difference in affinity constants, in the case of complementary or partially complementary pairing, is largely determined by the differences in the lifetimes of the complexes, i.e. in the rates of dissociation, whereas the rates of formation do not differ greatly, if they differ at all[66,67]. All this suggests that the probability of a wrong anticodon-codon complex being formed is just as high as the probability of a correct one, but the wrong complex decays faster, i.e. has a shorter lifetime.

An interesting model, making use of the pair complementary interaction between different tRNAs through their anticodons, has been suggested for measuring the values of the lifetimes of anticodon complexes with complementary and partially complementary sequences[67–70]. For example, $tRNA^{Phe}$ with the GAA anticodon on one hand, and $tRNA^{Lys}$ with the s^2UUU anticodon or $tRNA^{Glu}$ with the s^2UUC anticodon on the other, can pair through their anticodons according to all the rules of codon-anticodon interaction:

	5′		3′	
		G·C		
$tRNA^{Phe}$		A·U		$tRNA^{Glu}$
		A·U*		
	3′		5′	

	5′		3′	
		G·U		
$tRNA^{Phe}$		A·U		$tRNA^{Lys}$
		A·U*		
	3′		5′	

Thus, the anticodon of $tRNA^{Glu}$ or $tRNA^{Lys}$ can be regarded as a model of the ribosome-fixed codon (UUC or UUU, respectively) for binding $tRNA^{Phe}$. It has been demonstrated that the fully complementary anticodon-anticodon complex $tRNA^{Phe} \cdot tRNA^{Glu}$ has a half-life at least two orders of magnitude greater than that of the complexes simulating misbinding, e.g. $tRNA_2^{Leu} \cdot tRNA^{Lys}$, $tRNA_3^{Leu} \cdot tRNA^{Lys}$, and $tRNA_4^{Leu} \cdot tRNA^{Lys}$ (see fig. 97). (The complex $tRNA_5^{Leu} \cdot tRNA^{Lys}$, however, has a half-life just one order of magnitude lower than that of the fully complementary complex).

Although codon-anticodon affinity is necessary for the misbinding of aminoacyl-tRNA on the ribosome, this does not imply that the level of misbinding should be proportional to the affinity constant. The fact that the different affinities of tRNA species to a particular codon are largely due to differences in the complex dissociation rates makes kinetic parameters of the ribosome critical to the level of miscoding[71–75].

If reversible codon-dependent tRNA binding in the elongation cycle is followed by an irreversible or slowly reversible stage, e.g. the locking of aminoacyl-tRNA in the A-site or transpeptidation, then the rate of the irreversible process will affect the level of miscoding: the higher the rate of the subsequent irreversible (or slowly reversible) stage the greater the level of miscoding[71,72]. In fact, because the times of complex formation of the codon with the cognate tRNA and the noncognate tRNA are roughly the same the discrimination between the tRNAs is based only on that the wrong complex decays faster. However, if the rate of the next stage of the cycle is very fast and is comparable to the rate at which the wrong complex decays, the probability is high that the noncognate tRNA will be drawn into the cycle. Clearly, the higher the rate of the following stage, then the more the tRNA binding conditions differ from the equilibrium ones, and the greater the probability of the noncognate tRNAs being captured. Conversely, if the subsequent irreversible stage proceeds at a slower rate, the conditions will approach equilibrium and the discrimination between cognate and noncognate tRNA species will depend to a larger extent on the difference in their affinity constant toward the codon. As a means of reducing the miscoding level, the ribosome may possess a special mechanism of *kinetic delay* at the stage of reversible codon-dependent aminoacyl-tRNA binding. EF-T_u could play this role: this protein bound with GTP blocks the subsequent step of the cycle until GTP is hydrolyzed, while the ternary complex remains reversibly associated with the ribosome.

On the other hand, discrimination between the cognate and noncognate tRNA on the ribosome may be even further amplified than follows from the simple difference in affinity constants. If in the process of aminoacyl-tRNA binding there are two successive reversible phases which are separated by a virtually irreversible step, the aminoacyl-tRNA will have two independent chances to dissociate. In this case the overall dissociation probability is equal to the product of the dissociation probabilities at the two stages, i.e. discrimination between the cognate and noncognate tRNA will be significantly amplified compared to the difference in affinity constants[73,74]. This is a mechanism of *kinetic correction*, or "*proofreading*." Again, EF-T_u could perform the role: the first stage of reversible binding could include the binding of the ternary Aa-tRNA·EF-T_u·GTP complex prior to GTP hydrolysis; GTP hydrolysis would represent an irreversible separating step; and the aminoacyl-tRNA would then have another, independent chance of dissociating from the complex with the codon at the stage of EF-T_u·GDP dissociation from the ribosome:

VI RS* + Aa-tRNA·EF-T_u·GTP $\overset{1}{\rightleftharpoons}$

RS*·Aa-tRNA·EF-T_u·GTP

↙ 2 (GTP hydrolysis)

RS*·Aa-tRNA·EF-T_u·GDP + P_i

3 ⇅ (left) 3′ ⇅ (right)

RS*·Aa-tRNA + EF-T_u·GDP   RS*·EF-T_u·GDP + Aa-tRNA

4 ↓ (tRNA capture)   ⇅ 3″

RS*·Aa-tRNA†   RS* + EF-T_u·GDP

↓

Continuation of
the elongation cycle

(In this scheme RS* refers to the translating ribosome with mRNA and peptidyl-tRNA; Aa-tRNA† is an aminoacyl-tRNA, irreversibly captured either as a result of transpeptidation or due to its occlusion within the pretranspeptidation ribosome; see below.) In the case of the cognate aminoacyl-tRNA, the rate of the side reaction 3′ (decay of a codon-anticodon complex prior to EF-T_u·GDP release) will be low compared to the main reaction 3 (release of EF-T_u·GDP), i.e. $k_3 \gg k_3'$. In the case of a less stable complex with the noncognate tRNA, the side-reaction 3′ proceeds much faster and competes with the main reaction 3, i.e. $k_3' > k_3$. Thus, discrimination between the cognate and noncognate tRNA is accomplished twice: on the basis of differences in the rates at which the codon-anticodon complex decays in reaction 1 (difference in k_{-1}) and on the basis of the difference in the rates at which the codon-anticodon complex decays in reaction 3′ (difference in $k_{3'}$). GTP hydrolysis, which is an apparently irreversible process, serves to separate these two stages and, in this sense, the released free energy dissipating into heat increases translation fidelity.

Of course, this scheme is hypothetical and is presented here only to highlight the possibility of such a kinetic proofreading. Meanwhile, there are experimental observations which in fact demonstrate that EF-T_u can reduce the number of errors in translation[76,77]. In addition, there are independent data on the existence of two different phases in aminoacyl-tRNA binding at which the codon-anticodon complexes may decay[78–80].

12-6 Sequence of events and molecular mechanisms

Scanning of tRNA species

In the course of elongation, different aminoacyl-tRNA species are present in the solution surrounding the ribosome. The aminoacyl-tRNA corresponding to the template codon positioned in the A-site should be selected from this mixture by the ribosome. To achieve this end, a rapid scanning of different aminoacyl-tRNA (Aa-tRNA·EF-T_u·GTP complexes) should be performed and only the complex recognized by the ribosome as a codon cognate should remain bound to the ribosome. Such scanning assumes multiple collisions between the codon fixed on the ribosome and the anticodons of different tRNAs. It may be asked whether this is achieved by random diffusional collisions between nonoriented tRNAs and the ribosome or whether there is a rapid formation and decay (a run-through) of short-lived intermediate complexes between the tRNAs and the ribosome, where collisions between the anticodons and the codon occur in the proper orientation.

Studies on fast kinetics in the course of tRNA binding with the ribosome (achieved by measuring changes in the intensity and polarization of fluorescence of the proflavine label attached to the anticodon loop or to the D loop of tRNA)[81] favor the second alternative. Indeed, the kinetics of tRNA binding with the ribosome have been shown to be multiphasic. The first rapid phase was interpreted to be the formation of an intermediate short-lived complex which does not depend on the template codon (fig. 98(1)). The intermediate short-lived complex detected in these experiments

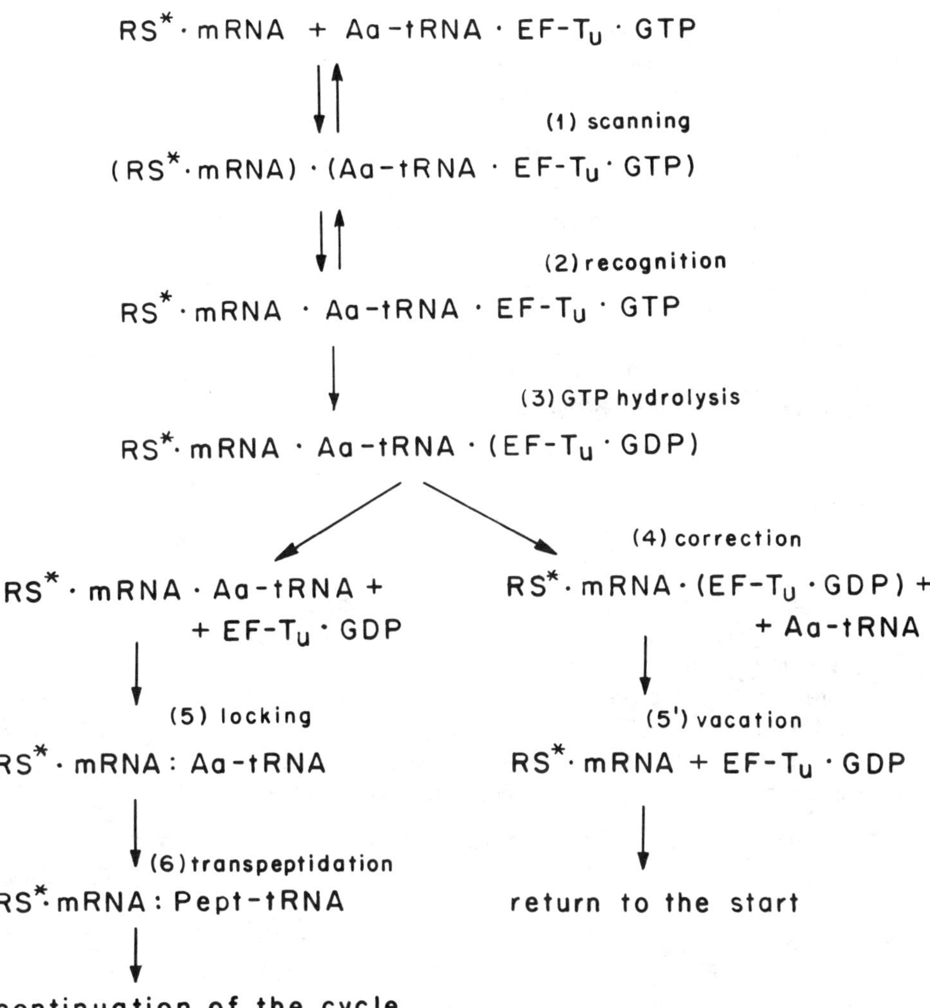

Figure 98 Pathway of aminoacyl-tRNA entry into the elongating ribosome. RS* is an elongating ribosome carrying peptidyl-tRNA in the P-site. Points connect the participants of a noncovalent reversible complex; a colon means occlusion (locking) of a ligand in the ribosome. The bracketed parts of complexes are presumed to be in loose and transient association with the rest of a complex.

results from a collision between the tRNA and the ribosome followed by a specific orientation of the tRNA. The formation of this complex does not depend either on the presence of the codon in the A-site or on the nature of the codon, if it is present. In other words, the codon-anticodon interaction is not used in the complex formation. The complex can be formed both in the presence and in the absence of EF-T_u suggesting that this protein does not participate in the catalysis of the first rapid phase of binding.

It can be assumed that the rapid formation and decay of such short-lived complexes of the ribosome with tRNA is used for the scanning (run-through) of anticodons. Only when the anticodon fits the codon does this result in recognition which induces the rearrangement of the complex and thus leads to the transition to the next binding phase.

It is likely that the tRNA within the intermediate short-lived complex is transiently bound to the ribosomal site overlapping the A-site. The tRNA-binding subsite of the small ribosomal subunit appears to take part in this transient tRNA binding.

Recognition of anticodon

Studies on the multiphasic kinetics of tRNA binding have demonstrated that when the P-site is occupied and only when the cognate codon is present in the A-site, the above rapid phase is followed by a slow phase during which the intermediate short-lived complex rearranges into a stable complex[81]. As a result, the tRNA becomes specifically bound in the ribosomal A-site. Without EF-T_u, i.e. in the case of nonenzymatic tRNA binding, this process is very slow; it is this process that is greatly accelerated by EF-T_u with GTP:

$$\text{RS·mRNA} + \text{Aa-tRNA} \xrightleftharpoons{\text{fast}} \text{(RS·mRNA)·Aa-tRNA} \xrightleftharpoons{\text{very slow}} \text{RS·mRNA·Aa-tRNA};$$

$$\text{RS·mRNA} + \text{Aa-tRNA·EF-T}_u\text{·GTP} \xrightleftharpoons{\text{fast}} \text{(RS·mRNA)·(Aa-tRNA·EF-T}_u\text{·GTP)} \xrightleftharpoons{\text{moderate}} \text{RS·mRNA·Aa-tRNA} + \text{EF-T}_u\text{·GDP} + \text{P}_i.$$

The process accelerated by EF-T_u and GTP seems to proceed through several steps. It begins with codon-anticodon recognition. Only when the anticodon fits the codon sterically may tRNA fill the A-site. As tRNA is bound within the ternary complex, its entry is accompanied by an interaction between EF-T_u and the factor-binding site of the 50S subunit; this auxiliary interaction may contribute to the tRNA entry. Experiments with nonhydrolyzable or slowly hydrolyzable GTP analogs have demonstrated that, at this stage of codon-dependent binding, the aminoacyl-tRNA or, more accurately, the ternary aminoacyl-tRNA·EF-T_u·GTP complex, is reversibly associated with the ribosome and, therefore, if GTP hydrolysis is not very fast, equilibrium can be approached. Thus, this stage provides the first step of aminoacyl-tRNA selection according to its coding specificity (fig. 98(2)).

If the interaction underlying the formation of the intermediate short-lived complex continues after codon-anticodon recognition, the aminoacyl-tRNA will then be bound to the 70S ribosome by at least three points: (1) by the anticodon through the mRNA codon with the 30S ribosomal subunit, (2) by the acceptor arm through EF-T_u with the 50S ribosomal subunit, and (3) by an unknown point of the initial interaction (possibly through the D arm interacting with the head of the 30S ribosomal subunit).

GTP hydrolysis

In this state, hydrolysis of the EF-T_u-bound GTP is induced (fig. 98(3)). Hydrolysis seems to be induced by the EF-T_u interacting with the factor-binding site of the 50S subunit. The hydrolysis of GTP, an apparently irreversible step (due to a great difference in the thermodynamic potentials of GTP and its hydrolysis products), divides the aminoacyl-tRNA binding process into two phases which are not connected by equilibrium. The shorter the time interval between tRNA recognition and GTP hydrolysis, the lower the selectivity of the recognition phase. If hydrolysis is artificially delayed, e.g. by using a slowly hydrolyzable GTP analog such as guanosine 5′-(γ-thio)-triphosphate (see chapter 11, fig. 86), the discrimination between the cognate tRNA and near-cognate tRNA (e.g. $tRNA^{Phe}$ and $tRNA_2^{Leu}$) increases to a ratio of 10^4 in favor of the cognate species[82].

When the EF-T_u-bound GTP is hydrolyzed, no intermediate phosphorylated products are detected. The transfer of phosphate seems to proceed directly to a water molecule. Thus, it appears that there is no material (biochemical) coupling of the GTP hydrolysis with an energy-consuming process. All the free energy of GTP hydrolysis seems to dissipate directly into heat. (Meanwhile, it should be remembered that the free energy of the hydrolysis of the factor-bound GTP should not equal that of the unbound GTP in solution.)

GTP hydrolysis results in the disappearance of or a drastic decrease in EF-T_u affinity to the aminoacyl-tRNA and the ribosome. This is believed to be the result of some conformational rearrangement of EF-T_u when it becomes bound to GDP instead of to GTP.

Correction of aminoacyl-tRNA selection

GTP hydrolysis leads to an intermediate ribosomal complex in which EF-T_u no longer has affinity to the aminoacyl-tRNA. Consequently, the aminoacyl-tRNA has lost its attachment to the ribosome through EF-T_u. It now has another opportunity to dissociate, and if its dissociation rate is particularly high, as is the case for noncognate tRNA, it may be released prior to EF-T_u·GDP leaving the ribosome (fig. 98(4), right path). However, if the rate of its dissociation is not that great, because of a correct codon-anticodon interaction, EF-T_u·GDP will be released earlier (fig. 98(4), left path) and the aminoacyl terminus will be captured by the 50S ribosomal subunit. Thus, after GTP is hydrolyzed, the ribosome is presented with a new opportunity to correcting the result of aminoacyl-tRNA selection made prior to GTP hydrolysis.

There is an evidence that streptomycin inhibits just the correction (proofreading) step of the codon-directed selection of aminoacyl-tRNA by the ribosome[78,79]; the prevention of the release of (noncognate) aminoacyl-tRNA after GTP hydrolysis is a likely explanation of the antibiotic action. Though other aminoglycoside antibiotics, such as kanamycin and neomycin, have been reported to possess more complicated mechanisms of their action, the correction step seems to be one of their main targets either[83]. Mutational alterations of bacterial ribosomes, such as observed in streptomycin-resistant and ribosome-ambiguity mutants, may either affect both the initial discrimination of aminoacyl-tRNA and the correction stage[79], or, in some cases, selectively perturb the proofreading function[83].

Locking of aminoacyl-tRNA in the A-site

A number of independent experiments suggest that after GTP hydrolysis and the release of EF-T_u·GDP, the aminoacyl-tRNA becomes locked in the A-site even before transpeptidation[81,84,85]. This is the final phase of aminoacyl-tRNA binding to the A-site, i.e. "full A-site binding"[84] (fig. 98(5)).

Various hypotheses on the mechanism of such a locking can be construed, but it is difficult to imagine this locking taking place without a conformational rearrangement in the tRNA or the ribosome. One hypothesis suggests that the A-site is not strictly fixed or rigid in the 70S ribosome but can be adjusted in the course of codon-dependent aminoacyl-tRNA binding. For example, the release of EF-T_u, which is a large molecule, could allow a tighter interaction (attraction) between the 50S ribosomal subunit and the aminoacyl terminus of aminoacyl-tRNA as well as the 30S ribosomal subunit, resulting in aminoacyl-tRNA locking.

General scheme

In the course of enzymatic binding to the mRNA-programmed ribosome, the aminoacyl-tRNA passes through a series of consecutive

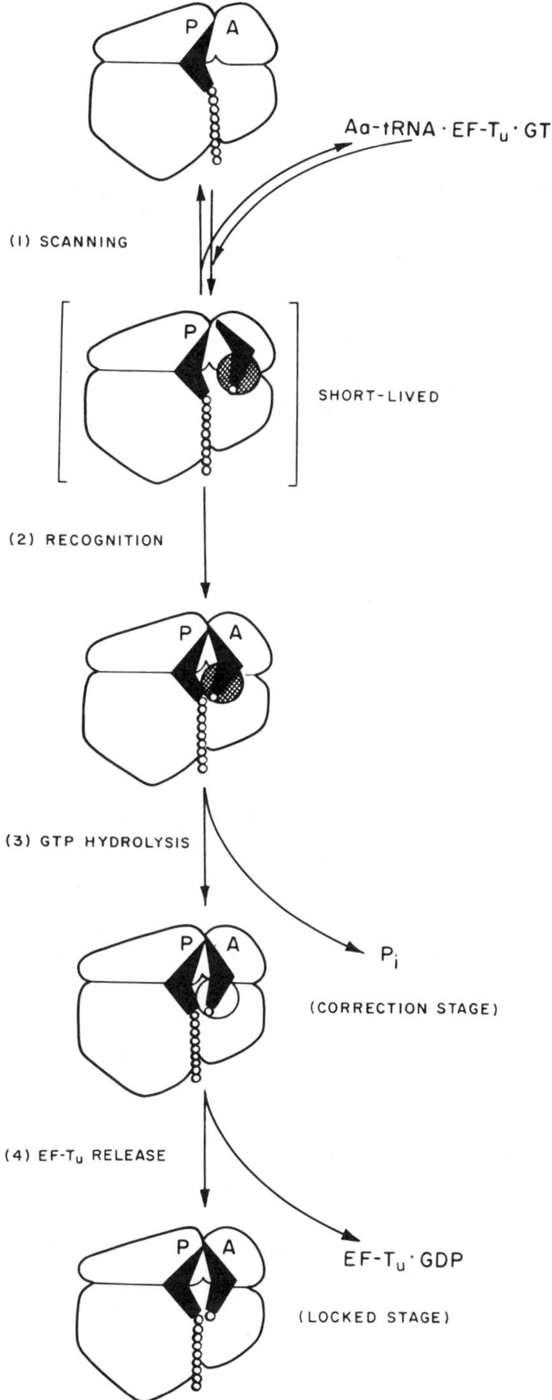

Figure 99 Sequence of events during aminoacyl-tRNA binding with the ribosome.

phases (fig. 98). Whether these phases involve binding to the same A-site, or whether the A-site is better defined as the final locked position of the aminoacyl-tRNA while the preceding phases are considered as corresponding to different overlapping sites, is largely a question of terminology. In any case, during transition from one phase to another, certain contacts between the aminoacyl-tRNA and the ribosome may change somewhat, but there is no redistribution of tRNA to other nonoverlapping sites. In this text the complete set of various possible overlapping positions of the aminoacyl-tRNA in the course of its binding with the ribosome is considered in terms of one A-site, perhaps with different states of binding therein.

The general scheme of the sequence of events described above is given in fig. 99.

References

1. F. H. C. Crick (1958), "On protein synthesis," *Symp. Soc. Exptl. Biol.* 12:138–163.

2. F. Chapeville, F. Lipmann, G. von Ehrenstein, B. Weisblum, W. J. Ray, and S. Benzer (1962), "On the role of soluble ribonucleic acid in coding for amino acids," *Proc. Nat. Acad. Sci. U.S.A.* 48:1086–1092.

3. G. von Ehrenstein, B. Weisblum, and S. Benzer (1963), "The function of sRNA as amino acid adaptor in the synthesis of hemoglobin," *Proc. Nat. Acad. Sci. U.S.A.* 49:669–675.

4. J. D. Smith, J. N. Abelson, B. F. C. Clark, H. M. Goodman, and S. Brenner (1966), "Studies on *amber* suppressor tRNA," *Cold Spring Harbor Symp. Quant. Biol.* 31:479–485.

5. H. M. Goodman, J. Abelson, A. Landy, S. Brenner, and J. D. Smith (1968), "Amber suppression: A nucleotide change in the anticodon of a tyrosine transfer RNA," *Nature* 217:1019–1024.

6. M. R. Bernfield and M. W. Nirenberg (1965), "RNA codewords and protein synthesis: The nucleotide sequences of multiple codewords for phenylalanine, serine, leucine, and proline," *Science* 147:479–484.

7. D. Söll, J. Cherayil, D. S. Jones, R. D. Faulkner, A. Hampel, R. M. Bock, and H. G. Khorana (1966), "sRNA specificity for codon recognition as studied by the ribosomal binding technique," *Cold Spring Harbor Symp. Quant. Biol.* 31:51–61.

8. F. H. C. Crick (1966), "Codon-anticodon pairing: The wobble hypothesis," *J. Mol. Biol.* 19:548–555.

9. P. Minz, U. Leupold, P. Agris, and J. Kohli (1981), "*In vivo* decoding rules

in *Schizosaccharomyces pombe* are at variance with *in vitro* data," *Nature* 294:187–188.

10. S. K. Mitra, F. Lustig, B. Akesson, and U. Lagerkvist (1977), "Codon:anticodon recognition in the valine codon family," *J. Biol. Chem.* 225:471–478.
11. U. Lagerkvist (1978), "'Two out of three': An alternative method for codon reading," *Proc. Nat. Acad. Sci. U.S.A.* 75:1759–1762.
12. J. E. Heckman, J. Sarnoff, B. Alzner–DeWeerd, S. Yin, and U. L. RajBhandary (1980), "Novel features in the genetic code and codon reading patterns in *Neurospora crassa* mitochondria based on sequences of six mitochondrial tRNAs," *Proc. Nat. Acad. Sci. U.S.A.* 77:3159–3163.
13. B. G. Barrell, S. Anderson, A. T. Bankier, M. H. L. de Bruijn, E. Chen, A. R. Coulson, J. Drouin, I. C. Eperon, D. P. Nierlich, B. A. Roe, F. Sanger, P. H. Schreier, A. J. H. Smith, R. Staden, and I. G. Young (1980), "Different patterns of codon recognition by mammalian mitochondrial tRNAs," *Proc. Nat. Acad. Sci. U.S.A.* 77:3164–3166.
14. S. G. Bonitz, R. Berlani, G. Coruzzi, M. Li, G. Macino, F. G. Nobrega, M. P. Nobrega, B. E. Thalenfeld, and A. Tzagoloff (1980), "Codon recognition rules in yeast mitochondria," *Proc. Nat. Acad. Sci. U.S.A.* 77:3167–3170.
15. Y. Kuchino, S. Watanabe, F. Harada, and S. Nishimura (1980), "Primary sequence of AUA-specific isoleucine tRNA from *E. coli*," *Biochemistry* 19:2085–2089.
16. N. de Groot, A. Panet, and Y. Lapidot (1971), "The binding of purified Phe-tRNA and peptidyl-$tRNA^{Phe}$ to *Escherichia coli* ribosomes," *Eur. J. Biochem.* 23:523–527.
17. V. B. Odinzov and S. V. Kirillov (1978), "Interaction of N-acetyl-phenylalanyl-$tRNA^{Phe}$ with 70S ribosomes of *Escherichia coli*," *Nucleic Acids Res.* 5:3871–3879.
18. J. Ofengand and R. Liou (1981), "Correct codon-anticodon base pairing at the 5′-anticodon position blocks covalent crosslinking between transfer ribonucleic acid and 16S RNA at the ribosomal P site," *Biochemistry* 20:552–559.
19. A. Rich (1974), "How transfer RNA may move inside the ribosome," in *Ribosomes*, ed. M. Nomura, A. Tissières, and P. Lengyel, pp. 871–884 (Cold Spring Harbor, N.Y.: Cold Spring Harbor Laboratory).
20. M. Sundaralingam, T. Brennan, N. Yathindra, and T. Ichikawa (1975), "Stereochemistry of messenger RNA (codon)-transfer RNA (anticodon) interaction on the ribosome during peptide bond formation," in *Structure and conformation of nucleic acids and protein-nucleic acid interactions*, ed. M. Sundaralingam and S. T. Rao, pp. 101–115 (Baltimore: University Park Press).
21. J. E. Allende, R. Munro, and F. Lipmann (1974), "Resolution of the *E. coli* aminoacyl sRNA transfer factor into complementary fractions," *Proc. Nat. Acad. Sci. U.S.A.* 51:1211–1216.

22. Y. Nishizuka and F. Lipmann (1966), "Comparison of guanosine triphosphate split and polypeptide synthesis with a purified *E. coli* system," *Proc. Nat. Acad. Sci. U.S.A.* 55:212–219.

23. J. Lucas-Lenard and F. Lipmann (1966), "Separation of three microbial amino acid polymerization factors," *Proc. Nat. Acad. Sci. U.S.A.* 55:1562–1566.

24. J. M. Fessenden and K. Moldave (1963), "Studies on amino acyl transfer from soluble RNA to ribosomes: Resolution of two soluble transferring activities," *J. Biol. Chem.* 238:1479–1484.

25. R. Arlinghaus, G. Favelukes, and R. Schweet (1963), "A ribosome-bound intermediate in polypeptide synthesis," *Biochem. Biophys. Res. Commun.* 11:92–96.

26. J. Gordon (1968), "A step-wise reaction yielding a complex between a supernatant fraction from *E. coli*, guanosine 5′-triphosphate, and aminoacyl-tRNA," *Proc. Nat. Acad. Sci. U.S.A.* 59:179–183.

27. K. Arai, B. F. C. Clark, L. Duffy, M. D. Jones, Y. Kaziro, R. A. Laursen, J. L'Italien, D. L. Miller, S. Nagarkatti, S. Nakamura, K. M. Nielsen, T. E. Petersen, K. Takahashi, and M. Wade (1980), "Primary structure of elongation factor T_u from *Escherichia coli*," *Proc. Nat. Acad. Sci. U.S.A.* 77:1326–1330.

28. K. Marikawa, T. F. M. LaCour, J. Nyborg, K. M. Rasmussen, D. L. Miller, and B. F. C. Clark (1978), "High resolution X-ray crystallographic analysis of a modified form of the elongation factor T_u: guanosine diphosphate complex," *J. Mol. Biol.* 125:325–338.

29. J. R. Rubin, K. Morikawa, J. Nyborg, T. F. M. LaCour, B. F. C. Clark, and D. L. Miller (1981), "Structural features of the GDP binding site of elongation factor T_u from *Escherichia coli* as determined by X-ray diffraction," *FEBS Letters* 129:177–179.

30. F. P. Wilkman, G. E. Siboska, H. U. Petersen, and B. F. C. Clark (1982), "The site of interaction of aminoacyl-tRNA with elongation factor T_u," *EMBO Journal* 1:1095–1100.

31. D. L. Miller and H. Weissbach (1977), "Factors involved in the transfer of aminoacyl-tRNA to the ribosome," in *Molecular mechanisms of protein biosynthesis*, ed. H. Weissbach and S. Pestka, pp. 323–313 (New York: Academic Press).

32. A. Pingoud and C. Urbanke (1980), "Aminoacyl transfer ribonucleic acid binding site of the bacterial elongation factor T_u," *Biochemistry* 19:2108–2112.

33. Y. Kaziro (1978), "The role of guanosine 5′-triphosphate in polypeptide chain elongation," *Biochim. Biophys. Acta* 505:95–127.

34. A. E. Johnson, R. H. Fairclough, and C. R. Cantor (1977), "Some approaches for the study of ribosome-tRNA interactions," in *Nucleic acid-protein recognition*, ed. H. J. Vogel, pp. 469–490 (New York: Academic Press).

35. J. A. Lake (1977), "Aminoacyl-tRNA binding at the recognition site is the first step of the elongation cycle of protein synthesis," *Proc. Nat. Acad. Sci. U.S.A.* 74:1903–1907.

36. H. Yokosawa, M. Kawakita, K. Arai, N. Inoue–Yokosawa, and Y. Kaziro (1975), "Binding of aminoacyl-tRNA to ribosomes promoted by elongation factor T_u: Further studies on the role of GTP hydrolysis," *J. Biochem. (Japan)* 77:719–728.

37. H. Wolf, G. Chinali, and A. Parmeggiani (1977), "Mechanism of the inhibition of protein synthesis by kirromycin: Role of elongation factor T_u and ribosomes," *Eur. J. Biochem.* 75:67–75.

38. A. Kaji and H. Kaji (1963), "Specific interaction of soluble RNA and polyribonucleic acid induced polysomes," *Biochem. Biophys. Res. Commun.* 13:186–192.

39. T. Nakamoto, T. W. Conway, J. E. Allende, G. J. Spyrides, and F. Lipmann (1963), "Formation of peptide bonds. I. Peptide formation from aminoacyl-sRNA," *Cold Spring Harbor Symp. Quant. Biol.* 28:227–231.

40. M. E. Gottesman (1967), "Reaction of ribosome-bound peptidyl transfer ribonucleic acid with aminoacyl transfer ribonucleic acid or puromycin," *J. Biol. Chem.* 242:5564–5571.

41. S. Pestka (1977), "Inhibitors of protein synthesis," in *Molecular mechanisms of protein biosynthesis*, ed. H. Weissbach and S. Pestka, pp. 467–555 (New York: Academic Press).

42. E. Cundliffe (1980), "Antibiotics and prokaryotic ribosomes: Action, interaction, and resistance," in *Ribosomes: Structure, function, and genetics*, ed. G. Chambliss, G. R. Craven, J. Davies, K. Davis, L. Kahan, and M. Nomura, pp. 555–581 (Baltimore: University Park Press).

43. J. E. Davies, W. Gilbert, and L. Gorini (1964), "Streptomycin, suppression, and the code," *Proc. Nat. Acad. Sci. U.S.A.* 51:883–890.

44. J. E. Davies, L. Gorini, and B. D. Davis (1965), "Misreading of RNA codewords induced by aminoglycoside antibiotics," *J. Mol. Pharmacol.* 1:93–106.

45. L. Gorini (1974), "Streptomycin and misreading of the genetic code," in *Ribosomes*, ed. M. Nomura, A. Tissières, and P. Lengyel, pp. 791–803 (Cold Spring Harbor, N.Y.: Cold Spring Harbor Laboratory).

46. J. H. Matthaei, O. W. Jones, R. G. Martin, and M. W. Nirenberg (1962), "Characteristics and composition of RNA coding units," *Proc. Nat. Acad. Sci. U.S.A.* 48:666–676.

47. M. S. Bretscher and M. Grunberg–Manago (1962), "Polyribonucleotide-directed protein synthesis using an *E. coli* cell-free system," *Nature* 195:283–284.

48. W. Szer and S. Ochoa (1964), "Complexing ability and coding properties of synthetic polyribonucleotides," *J. Mol. Biol.* 8:823–834.

49. S. M. Friedman and I. B. Weinstein (1964), "Lack of fidelity in the

translation of synthetic polyribonucleotides," *Proc. Nat. Acad. Sci. U.S.A.* 52:988–996.

50. A. G. So and E. W. Davie (1964), "The effects of organic solvents on protein biosynthesis and their influence on the amino acid code," *Biochemistry* 3:1165–1169.

51. R. H. Buckingham and H. Grosjean (1985), "The accuracy of messenger RNA: tRNA recognition on the ribosome," in *Accuracy in Molecular Processes*," ed. Galas, Kirkwood, and Rosenberger. Chapman et Hall, Publishers. Technical and Medical Publishing Division, London.

52. L. Gorini (1971), "Ribosomal discrimination of tRNAs," *Nature New Biol.* 234:264.

53. L. P. Gavrilova and N. M. Rutkevitch (1980), "Ribosomal synthesis of polyleucine on polyuridylic acid as a template," *FEBS Letters* 120:135–140.

54. P. H. O'Farrell (1978), "The suppression of defective translation by ppGpp and its role in the stringent response," *Cell* 14:545–557.

55. J. Parker, J. W. Pollard, J. D. Friesen, and C. P. Stanners (1978), "Shuttering: High-level mistranslation in animal and bacterial cells," *Proc. Nat. Acad. Sci. U.S.A.* 75:1091–1095.

56. J. Gallant and D. Foley (1979), "On the causes and prevention of mistranslation," in *Ribosomes: Structure, function, and genetics*, ed. G. Chambliss, G. R. Craven, J. Davies, K. Davis, L. Kahan, and M. Nomura, pp. 615–638 (Baltimore: Univeristy Park Press).

57. J. Parker and J. D. Friesen (1980), "'Two out of three' codon reading leading to mistranslation *in vivo*," *Mol. Gen. Genetics* 177:439–445.

58. R. B. Loftfield and D. Vanderjagt (1972), "The frequency of errors in protein biosynthesis," *Biochem. J.* 128:1353–1356.

59. S. F. Coons, L. F. Smith, and R. B. Loftfield (1979), "The nature of amino acid errors in *in vivo* biosynthesis of rabbit hemoglobin," *Fed. Proc.* 38:328.

60. P. Edelman and J. Gallant (1977), "Mistranslation in *E. coli*," *Cell* 10:131–137.

61. N. Ellis and J. Gallant (1982), "An estimate of the global error frequency in translation", *Mol. Gen. Genetics* 188:169–172.

62. J. Parker, T. C. Johnston, P. T. Borgia, G. Holtz, E. Remaut, and W. Fiers (1983), "Codon usage and mistranslation. *In vivo* basal level misreading of the MS2 coat protein message," *J. Biol. Chem.* 258:10007–10012.

63. F. Bouadloun, D. Donner, and C. G. Kurland (1983), "Codon-specific missense errors *in vivo*," *EMBO J.* 2:1351–1356.

64. K. Khazaie, J. H. Buchanan, and R. F. Rosenberger (1984), "The accuracy of $Q\beta$ RNA translation. 1. Errors during the synthesis of $Q\beta$ proteins by intact *Escherichia coli* cells," *Eur. J. Biochem.* 144:485–489.

65. P. C. Jelenc and C. G. Kurland (1979), "Nucleoside triphosphate regenera-

tion decreases the frequency of translation errors," *Proc. Nat. Acad. Sci. U.S.A.* 76:3174–3178.

66. M. E. Craig, D. M. Crothers, and P. Doty (1971), "Relaxation kinetics of dimer formation by self complementary oligonucleotides," *J. Mol. Biol.* 62:383–401.

67. H. Grosjean, D. Söll, and D. M. Crothers (1976), "Studies of the complex between tRNA with complementary anticodons. I. Origins of enhanced affinity between complementary triplets," *J. Mol. Biol.* 103:499–519.

68. J. Eisinger and N. Gross (1974), "The anticodon-anticodon complex," *J. Mol. Biol.* 88:165–174.

69. H. Grosjean, S. deHenau, and D. M. Crothers (1978), "On the physical basis for ambiguity in genetic coding interactions," *Proc. Nat. Acad. Sci. U.S.A.* 75:610–614.

70. H. Grosjean and H. Chantrenne (1980), "On codon-anticodon interactions," in *Chemical recognition in biology*, ed. F. Chapeville and A.-L. Haenni, pp. 347–367 (Berlin and Heidelberg: Springer–Verlag).

71. J. Ninio (1974), "A semi-quantitative treatment of missense and nonsense suppression in the *str*A and *ram* ribosomal mutants of *Escherichia coli*: Evaluation of some molecular parameters of translation *in vivo*," *J. Mol. Biol.* 84:297–313.

72. V. S. Schwartz and V. N. Lysikov (1974), "Physical mechanisms of the 'ribosomal screen,'" *Dokl. Akad. Nauk SSSR* 217:1446–1448.

73. J. J. Hopfield (1974), "Kinetic proofreading: A new mechanism for reducing errors in biosynthetic processes requiring high specificity," *Proc. Nat. Acad. Sci. U.S.A.* 71:4135–4139.

74. J. J. Hopfield and T. Yamana (1980), "The fidelity of protein synthesis," in *Ribosomes: Structure, function, and genetics*," ed. G. Chambliss, G. R. Craven, J. Davies, K. Davis, L. Kahan, and M. Nomura, pp. 585–596 (Baltimore: University Park Press).

75. C. G. Kurland (1980), "On the accuracy of elongation," in *Ribosomes: Structure, function, and genetics*, ed. G. Chambliss, G. R. Craven, J. Davies, K. Davis, L. Kahan, and M. Nomura, pp. 597–614 (Baltimore: University Park Press).

76. R. C. Thompson and P. J. Stone (1977), "Proofreading of the codon-anticodon interaction on ribosomes," *Proc. Nat. Acad. Sci. U.S.A.* 74:198–202.

77. L. P. Gavrilova, I. N. Perminova, and A. S. Spirin (1981), "Elongation factor T_u can reduce translation errors in poly(U)-directed cell-free systems," *J. Mol. Biol.* 149:69–78.

78. J. L. Yates (1979), "Role of ribosomal protein S12 in discrimination of aminoacyl-tRNA," *J. Biol. Chem.* 254:11550–11554.

79. R. C. Thompson, D. B. Dix, R. B. Gerson, and A. M. Karim (1981), "Effect of Mg^{2+} concentration, polyamines, streptomycin, and mutations in

ribosomal proteins on the accuracy of the two-step selection of aminoacyl-tRNAs in protein biosynthesis," *J. Biol. Chem.* 256:6676–6681.

80. T. Ruusala, M. Ehrenberg, and C. G. Kurland (1982), "Is there proofreading during polypeptide synthesis?" *EMBO Journal* 1:741–745.

81. W. Wintermeyer and J. M. Robertson (1982), "Transient kinetics of transfer ribonucleic acid binding to the ribosomal A and P sites: Observation of a common intermediate complex," *Biochemistry* 21:2246–2252.

82. R. C. Thompson and Amr M. Karim (1982), "The accuracy of protein biosynthesis is limited by its speed: High fidelity selection by ribosomes of aminoacyl-tRNA ternary complexes containing GTP[γS]," *Proc. Nat. Acad. Sci. U.S.A.* 79:4922–4926.

83. P. C. Jelenc and C. G. Kurland (1984), "Multiple effects of kanamycin on translational accuracy," *Mol. Gen. Genetics* 194:195–199.

84. R. C. Thompson, D. B. Dix, R. B. Gerson, and Amr M. Karim (1981), "A GTPase reaction accompanying the rejection of Leu-$tRNA_2$ by UUU-programmed ribosomes," *J. Biol. Chem.* 256:81–86.

85. Yu. P. Semenkov, E. M. Makarov, V. I. Makhno, and S. V. Kirillov (1982), "Kinetic aspects of tetracycline action on the acceptor (A) site of *Escherichia coli* ribosomes," *FEBS Letters* 144:125–129.

Further reading

Bermek, E. (1978). Mechanisms in polypeptide chain elongation on ribosomes. Progress in nucleic acid research and molecular biology (W. E. Cohn, ed.), vol. 21, pp. 63–100. New York: Academic Press.

Clark, B. F. C. (1980). Structure of tRNA during protein biosynthesis. In *Ribosomes: Structure, function, and genetics* (G. Chambliss, G. R. Craven, J. Davies, K. Davis, L. Kahan, and M. Nomura, eds.), pp. 413–444. Baltimore: University Park Press.

Gale, E. F.; Cundliffe, E.; Reynolds, P. E.; Richmond, M. H.; and Waring, M. J. (1982). Antibiotic Inhibitors of Ribosome Function, *The Molecular Basis of Antibiotic Action*, Second Edition, pp. 402–547. London: J. Wiley & Sons.

Gallant, J., and Foley, D. (1980). On the causes and prevention of mistranslation. In *Ribosomes: Structure, functions, and genetics* (G. Chambliss, G. R. Craven, J. Davies, K. Davis, L. Kahan, and M. Nomura, eds.), pp. 615–638. Baltimore: University Park Press.

Hardesty, B.; Culp, W.; and McKeehan, W. (1969). The sequence of reactions leading to the synthesis of a peptide bond on reticulocyte ribosomes. *Cold Spring Harbor Symp. Quant. Biol.* 34:331–344.

Kurland, C. G. (1977). Aspects of ribosome structure and function. In

Molecular mechanisms of protein biosynthesis (H. Weissbach and S. Pestka, eds.), pp. 81–116. New York: Academic Press.

Lockwood, A. H.; Hattman, S.; and Maitra, U. (1969). The nature of T-factor-guanine nucleotide complexes. *Cold Spring Harbor Symp. Quant. Biol.* 34:433–436.

Lucas-Lenard, J.; Tao, P.; and Haenni, A.-L. (1969). Further studies on bacterial polypeptide elongation. *Cold Spring Harbor Symp. Quant. Biol.* 34:455–462.

Ofengand, J. (1977). tRNA and aminoacyl-tRNA synthetases. In *Molecular mechanisms of protein biosynthesis* (H. Weissbach and S. Pestka, eds.), pp. 7–79. New York: Academic Press.

Parmeggiani, A., and Gottschalk, E. M. (1969). Isolation and some properties of the amino acid polymerization factors from *Escherichia coli*. *Cold Spring Harbor Symp. Quant. Biol.* 34:377–384.

Ravel, J. M.; Shorey, R. L.; Garner, C. W.; Dawkins, R. C.; and Shive, W. (1969). The role of an aminoacyl-tRNA-GTP-protein complex in polypeptide synthesis. *Cold Spring Harbor Symp. Quant. Biol.* 34:321–330.

Skoultchi, A.; Ono, Y.; Waterson, J.; and Lengyel, P. (1969). Peptide chain elongation. *Cold Spring Harbor Symp. Quant. Biol.* 34:437–454.

Weissbach, H. (1980). Soluble factors in protein synthesis. In *Ribosomes: Structure, function, and genetics* (G. Chambliss, G. R. Craven, J. Davies, K. Davis, L. Kahan, and M. Nomura, eds.), pp. 377–411. Baltimore: University Park Press.

Weissbach, H.; Brot, N.; Miller, D.; Rossman, M.; and Ertel, R. (1969). Interaction of guanosine triphosphate with *E. coli* soluble transfer factors. *Cold Spring Harbor Symp. Quant. Biol.* 34:419–431.

Chapter 13

Elongation II: Transpeptidation (Peptide Bond Formation)

13-1 Chemistry of the reaction

After the peptidyl-tRNA has occupied the P-site and the aminoacyl-tRNA has been fully bound at the A-site, the 3′-ends of the two tRNA residues are found close to each other in the region of the peptidyl transferase center of the 50S ribosomal subunit. This is followed by a nucleophilic attack *on the carbonyl* of the ester bond between the peptide residue and the tRNA moiety of the peptidyl-tRNA molecule *by the amino group* of the aminoacyl-tRNA molecule. As a result, an amide (peptide) bond is formed between the peptide residue and the aminoacyl-tRNA molecule (fig. 100). The peptidyl-tRNA and the aminoacyl-tRNA are the substrates of this reaction. The peptidyl-tRNA is a *donor* substrate, and the aminoacyl-tRNA plays the part of an *acceptor* substrate. The products of the reaction are the deacylated tRNA in the P-site and the peptidyl-tRNA with the peptide moiety elongated by one aminoacyl residue attached to the tRNA in the A-site, as is shown schematically in fig. 101.

It has already been mentioned that the 3′-terminal fragments of the

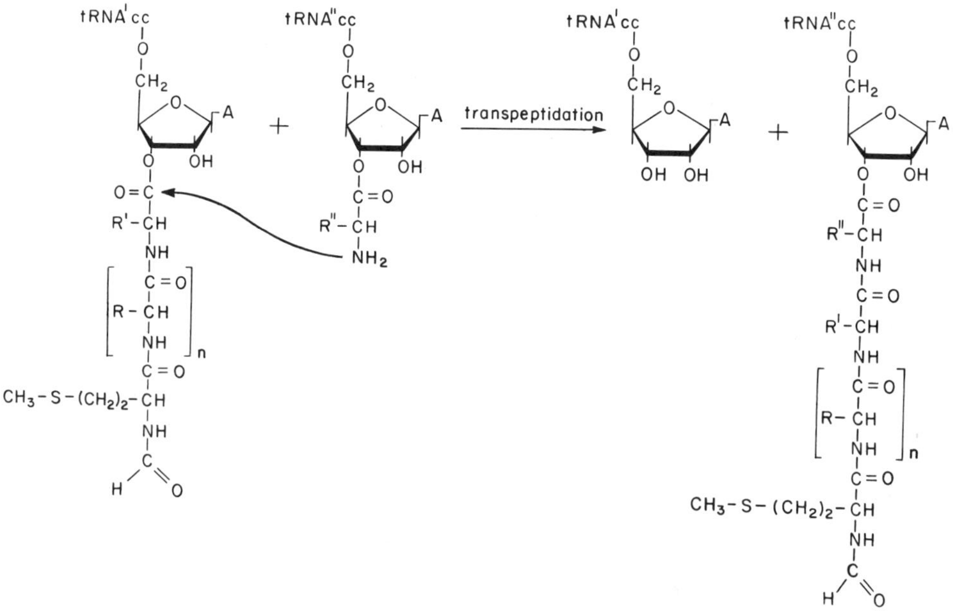

Figure 100 Transpeptidation reaction catalyzed by the ribosome.

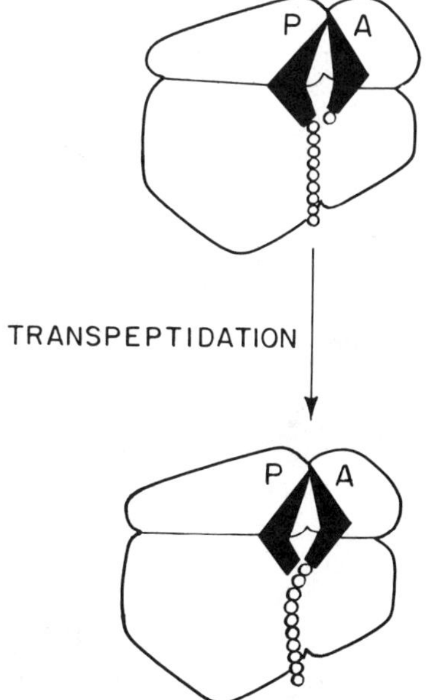

Figure 101 Scheme of transpeptidation on the ribosome.

aminoacyl-tRNA[1], the aminoacyl esters of the adenosine[2,3], as well as puromycin[4,5], can play the part of acceptor (or nucleophilic) substrate in the reaction with the peptidyl-tRNA, instead of the normal aminoacyl-tRNA molecule.

The reaction proceeds only with the NH_2-group occupying the α-position of an aminoacyl residue. The amino acid residue should be in L-configuration. The aminoacyl residue accommodated in the peptidyl transferase center must be attached to the 3′-position (not to the 2′-position) of the terminal adenosine of tRNA or another acceptor substrate. The latter requirement needs special comments. In solution, the aminoacyl residue can migrate between the 2′- and 3′-positions of the ribose residue in the tRNA terminal adenosine (see chapter 3, fig. 27). Binding the aminoacyl-tRNA to the EF-T_u seems to be possible with an aminoacyl residue in either of these positions. Both the ternary complexes Aa-(2′)tRNA·EF-T_u·GTP and Aa-(3′)tRNA·EF-T_u·GTP can bind to the ribosome. The migration of the aminoacyl residue between positions 2′ and 3′ may also be possible after GTP hydrolysis and the loss of the affinity to EF-T_u (the correction stage). It appears, however, that the aminoacyl residue is fixed in the 3′-position by the peptidyl transferase center[6].

The free 2′-hydroxyl of the acceptor substrate ribose is not essential for transpeptidation; if it is substituted (e.g., methylated) or is absent (2′-deoxyderivatives), the acceptor activity of the substrate is retained. The 2′-hydroxyl is, however, indispensable to the activity of the donor substrate.

The transpeptidation step in the ribosomal elongation cycle seems to be faster than either aminoacyl-tRNA binding or translocation. It appears that under no circumstances could transpeptidation be a limiting step of the cycle.

The catalytic mechanism of ribosomal transpeptidation has been the subject of numerous studies and discussions[7,8]. There is evidence that the histidine imidazole of some ribosomal protein may be involved in the catalysis. However, all attempts at isolating an acyl-enzyme (acyl-ribosome) intermediate of the kind found in proteinases catalyzing hydrolysis and transpeptidation have been unsuccessful. Perhaps such an intermediate should not exist in principle, because peptidyl-tRNA is already an activated macromolecular acyl-derivative which may be regarded as a functional analog of the acyl-enzyme group. Therefore, many investigators believe that transpeptidation in the ribosome is catalyzed simply by an appropriate spatial orientation and alignment of the aminoacyl-tRNA and peptidyl-tRNA reacting groups without the catalytic involvement of special nucleophile groups of the peptidyl transferase center. This hypothesis is strongly supported by experiments showing a low specificity of the ribosomal peptidyl transferase center with respect to the types of bonds formed in it. Indeed, if a hydroxyacyl residue (HO—CHR—CO—), rather than an aminoacyl residue (H_2N—CHR—CO—), is attached to the tRNA or its 3′-end analog, the hydroxy-derivative serves as a good acceptor substrate, and the ester bond is produced by the ribosome[9,10]. Similarly, when the acceptor substrate is a thioacyl derivative the ribosomal peptidyl transferase center catalyzes the formation of a thioester bond.[11] Moreover, the donor substrate can also be derived, and the ribosome is capable of catalyzing the attack on the thioester group of the donor by the amino group of the acceptor, thus forming a thioamide bond.[12] Finally, the phosphinoester analog of the donor has been shown to react with aminoacyl-tRNA in the ribosomal peptidyl transferase center, with the formation of an unnatural phosphinoamide bond.[13] The latter fact is especially important, since the geometry of the attacked group and the transition state of the reaction (trigonal bipyramid) are significantly different from those in the case of a normal donor (see section 13.4); the nucleophilic catalysis is hardly compatible with the realization of the reactions of both types by the same peptidyl transferase center.

At the same time, however, the ribosomal peptidyl transferase under certain conditions is capable of using water and low-molecular-weight alcohols, e.g., methanol and ethanol, as acceptor substrates, thus performing hydrolysis[14] or alcoholysis[15] of the peptidyl-tRNA (see chapter 18). These observations cannot be explained easily within the framework of the purely orientational mechanism of ribosomal peptidyl transferase action. Most reasonable is the hypothesis that the orientation effect of the peptidyl transferase center is indeed instrumental to transpeptidation on the ribosome, but that this effect can be aided by the contribution of specific microenvironments facilitating the reaction. It may well be that some groups located near the substrates properly oriented on the ribosome pull a proton away from the NH_2-group of the acceptor, thereby increasing its nucleophilic nature[16], on the one hand, and contribute to protonation of the carbonyl oxygen, thereby increasing the electrophilic properties of the attacked carbon of the donor ester group, on the other:

```
VII             O  R″   H
                ‖  |    |
tRNA″—O—C—CH—N—H----> (X)
                      ˙˙
                     /
              δ+ <--
tRNA′—O—C—CH—NH—···
                ‖  |
                O  R′
                ↑
                ¦
                H
                |
               (Y)
```

Whatever the case may be, it is certain that the reaction proceeds according to the mechanism of the S_n2 nucleophilic substitution through the so called *tetrahedral intermediate* (see section 13.4):

VIII

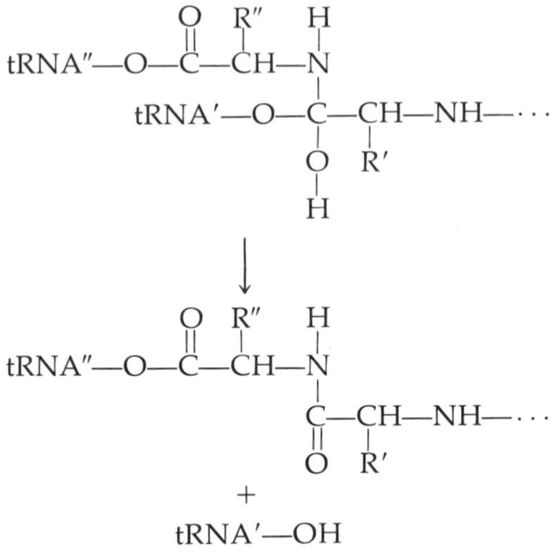

$$
\begin{array}{l}
\text{tRNA}''\text{—O—}\overset{\overset{\displaystyle O}{\|}}{C}\text{—}\overset{\overset{\displaystyle R''}{|}}{CH}\text{—}\overset{\overset{\displaystyle H}{|}}{N} \\
\qquad\qquad\qquad\qquad\quad | \\
\qquad\qquad \text{tRNA}'\text{—O—}\underset{\underset{\underset{\displaystyle H}{|}}{\displaystyle O}}{\overset{|}{\underset{|}{C}}}\text{—}\underset{\underset{\displaystyle R'}{|}}{CH}\text{—NH—}\cdots
\end{array}
$$

$$\downarrow$$

$$
\begin{array}{l}
\text{tRNA}''\text{—O—}\overset{\overset{\displaystyle O}{\|}}{C}\text{—}\overset{\overset{\displaystyle R''}{|}}{CH}\text{—}\overset{\overset{\displaystyle H}{|}}{N} \\
\qquad\qquad\qquad\qquad\quad | \\
\qquad\qquad\qquad\qquad\quad \underset{\underset{\displaystyle O}{\|}}{C}\text{—}\underset{\underset{\displaystyle R'}{|}}{CH}\text{—NH—}\cdots \\
\qquad\qquad\quad + \\
\qquad\qquad \text{tRNA}'\text{—OH}
\end{array}
$$

13-2 Energy balance of the reaction

The standard free energy of hydrolysis of the ester bond between the tRNA and the carbonyl group of the aminoacyl or peptidyl residue $\Delta G^{0\prime}$ is equal to about -7 to -8 kcal/mole. The standard free energy of hydrolysis of the peptide bond in a polypeptide of infinite length is approximately -0.5 kcal/mole. Thus, if the reaction substrates come directly from solution and the reaction products are released into solution, the net gain of free energy due to transpeptidation under standard conditions should be about -6.5 to -7.5 kcal/mole:

$$\text{Pept(n)-tRNA}' + \text{Aa-tRNA}'' \rightarrow \text{tRNA}' + \text{Pept(n + 1)-tRNA}''(-7 \mp 0.5\ \text{kcal}).$$

Such a formal calculation is usually put forward as an argument for ribosomal transpeptidation being fully supplied with free energy and therefore thermodynamically spontaneous.

It should be pointed out, however, that such an approach is valid only in the evaluation of ribosome-catalyzed model reactions between low-molecular-mass substrates[17,18], e.g.:

$$\text{F-Met-(3}')\text{ACC(5}') + \text{Phe-A} \xrightarrow{\text{ribosome}} \text{(5}')\text{CCA(3}') + \text{F-Met-Phe-A}.$$

Here the substrates come to the peptidyl transferase center directly from solution and the products are immediately and spontaneously released into solution.

In the elongation cycle, one substrate is always associated with the ribosome while the other comes to the peptidyl transferase center from a prebound state; one of the reaction products is not released into solution until translation has been completed, while the other product is released not as a result of transpeptidation but only after the next step of the elongation cycle. All this makes it impossible to give even an approximate estimation of free-energy change in the course of transpeptidation proceeding in the normal ribosomal elongation cycle. Of course, the reaction is fast, suggesting that there is a significant decrease in the free energy of the ribosomal complex in the course of transpeptidation. This decrease, however, must be less than −7 kcal/mole under standard conditions. For one thing, the free energy of hydrolysis of the bound substrate, provided hydrolysis products are released, should be lower than that of the free substrate since some portion of the energy had to be released upon binding if binding was spontaneous. In addition, the free energy of substrate hydrolysis, provided the product remains bound, should be lower than that when the product leaves the complex since the bound state of the product is associated with the accumulation of free energy, if the release of the product is thermodynamically spontaneous in principle. It can be assumed that the free energy of −7 kcal/mole, which would be released in transpeptidation if the substrate and the products were free, is in fact partly distributed at the preceding stage of aminoacyl-tRNA binding and the subsequent stage of translocation, thus driving the entire elongation cycle.

13-3 Inhibitors

Numerous specific inhibitors of the peptidyl transferase reaction catalyzed by procaryotic or eucaryotic ribosomes have been described. All the inhibitors, as expected, affect the large (50S or 60S, respectively) ribosomal subunit and have an affinity to it[19–21]. Many antibiotics commonly used for treating bacterial infections are inhibitors of the peptidyl transferase of the procaryotic 70S ribosome.

Chloramphenicol. Chloramphenicol (also known as chloromycetin) is the most well known inhibitor of peptidyl transferase in

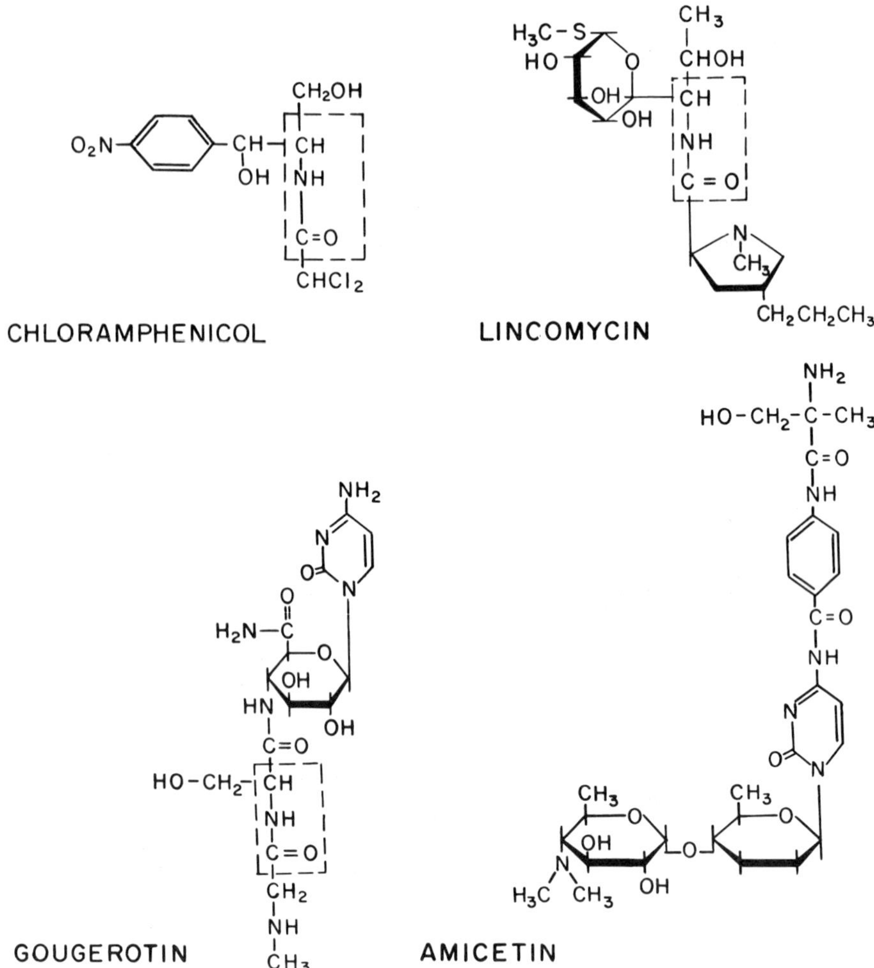

Figure 102 Antibiotics inhibiting the ribosomal peptidyl transferase center.

70S ribosomes (fig. 102). This is a broad-spectrum bacteriostatic antibiotic. It does not affect eucaryotic 80S ribosomes. Chemically, chloramphenicol is the analog of N-blocked amino alcohol with an aromatic radical. The dichloromethyl group is not strictly required for activity and can be changed for many moderately massive radicals. The aromatic nitro-group may also be changed for a number of other electronegative groups without resulting in any loss of antibiotic activity. The amide bond and stereochemistry of the —CO—NH-group with adjacent atoms are noteworthy with respect to the antibiotic action mechanism: this part of the molecule may imitate the peptide group with the adjacent C^{α}-atom and side radical.

Chloramphenicol binds loosely to the 70S ribosome or its isolated 50S subunit and can be washed away easily from the particles. Correspondingly, the action of this antibiotic is reversible. The antibiotic binds directly to the peptidyl transferase center, and seems to bind to the region of the center responsible for the interaction with the acceptor substrate; at any rate, the puromycin and 3′-terminal fragments of the aminoacyl-tRNAs compete with chloramphenicol for binding to the peptidyl transferase center. Various chemically active and photoactivable derivatives of chloramphenicol can be crosslinked covalently with the neighboring proteins in the ribosome; in particular, proteins L2, L16, L24, and L27 can be crosslinked with the antibiotic when it is bound in the peptidyl transferase center[8,21].

Chloramphenicol inhibits "natural" transpeptidation between the peptidyl-tRNA and aminoacyl-tRNA in the course of elongation, as well as the reaction of the peptidyl-tRNA or its analogs with puromycin. The simplest explanation for these effects is that chloramphenicol is an inactive analog of the acceptor substrate and after binding to the peptidyl transferase center competitively interferes with the interaction of true acceptors. However, there is also evidence for a noncompetitive mode of chloramphenicol inhibition, and it is possible that the bound antibiotic somehow inhibits a catalytic function of the peptidyl transferase center.

Lincomycin. Lincomycin (fig. 102), like chloramphenicol, affects only bacterial 70S ribosomes and is inactive toward eucaryotic 80S ribosomes. The peptidyl transferase center of the 50S ribosomal subunit is the target of antibiotic action and its binding site. Lincomycin competes with chloramphenicol for binding to the ribosome. It appears to inhibit competitively the interaction of the acceptor substrate with the peptidyl transferase center. The chemical structure of lincomycin, like that of chloramphenicol, shows the presence of an amide bond and a group simulating the peptide group adjacent to the C^{α}-atom of the amino acid residue (here, again, the alcohol hydroxyl group is present instead of the acidic one).

4-Aminohexose pyrimidine nucleoside antibiotics. This group of antibiotics includes such inhibitors of ribosomal transpeptidation as gougerotin, amicetin, blasticidin S, and bamicetin. The chemical structures of gougerotin and amicetin are given in fig. 102. All antibiotics in this group possess a nucleoside structure and may be regarded as analogs of the tRNA 3′-terminal adenosine. In addition, gougerotin (and blasticidin S) shows the presence of a structural motif traceable in chloramphenicol and lincomycin, e.g. the peptide group with the adjacent C^{α}-atom. Antibiotics of this group affect bacterial ribosomes although some of them, e.g. gougerotin and blasticidin S, can inhibit eucaryotic ribosomes as well. All these antibiotics bind to the 50S ribosomal subunit and inhibit the interaction between the acceptor substrates and the peptidyl transferase center of the ribosome. It is remarkable that the binding of the antibiotics stimulates the interaction of the low-molecular-mass analogs of the donor substrate with the peptidyl transferase center.

Anisomycin. This antibiotic inhibits transpeptidation with eucaryotic ribosomes only. It binds to the 60S subunit and serves as a competitive inhibitor of puromycin. It is apparent that anisomycin, like the antibiotics mentioned above, interferes with the interaction of the acceptor substrate with the peptidyl transferase center. Anisomycin is a powerful inhibitor and may block elongation completely, arresting the movement of ribosomes along the mRNA and thereby "freezing" the polyribosomes.

13-4 Stereochemistry

In order to understand both the molecular mechanism of the ribosome-catalyzed transpeptidation and the initial conformation of the peptide to be synthesized, it is vital to have knowledge of the conformation of the reacting substrates in the ribosomal peptidyl transferase center. Unfortunately, however, there is no direct evidence regarding the conformations of the tRNA 3′-termini and the adjacent aminoacyl residues at the reaction site.

As regards the acceptor substrate of the reaction, some information could be obtained by studying puromycin and its analogs. The conformation of puromycin in crystal has been solved by X-ray analysis[22] (fig. 103), and confirmed by studies on puromycin in solution[23]. Because puromycin is a good acceptor substrate in ribosomal transpeptidation, its structure may provide some information about the stereochemistry of aminoacyl and adenosine residues in the peptidyl transferase center. Furthermore, the puromycin analog with a more rigid conformation shown in fig. 104 can also serve as an acceptor substrate in transpeptidation and is even more active than puromycin[8]. In this case, the orientation of the purine ring with respect to the ribose is fixed, suggesting that the *anti* orientation is realized in the 3′-terminal adenosine of the acceptor substrate positioned in the peptidyl transferase center. Unfortunately, however, no reliable conclusion about the conformation acquired by the aminoacyl

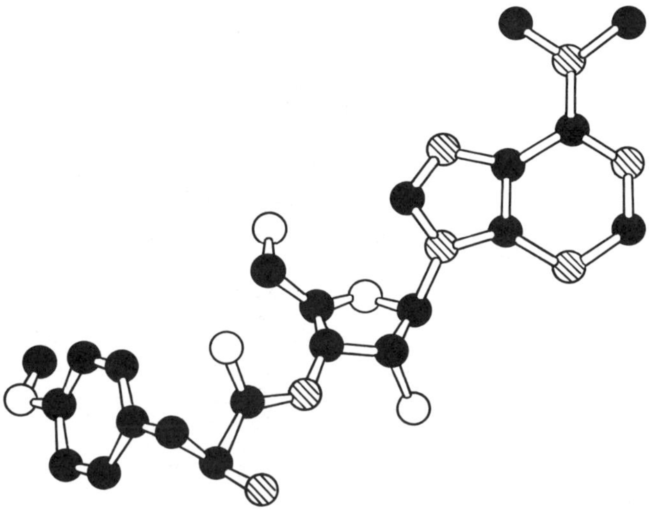

Figure 103 Skeletal model (without hydrogens) of puromycin based on X-ray analysis[22]. Filled circles are carbons, hatched circles—nitrogens, open circles—oxygen.

residue in the peptidyl transferase center can be reached using these data.

At the same time, on the basis of general considerations, one may propose that all types of aminoacyl residues of the acceptor substrate, on one hand, and C-terminal aminoacyl residues of the donor substrate, on the other, are positioned and presented to each other by

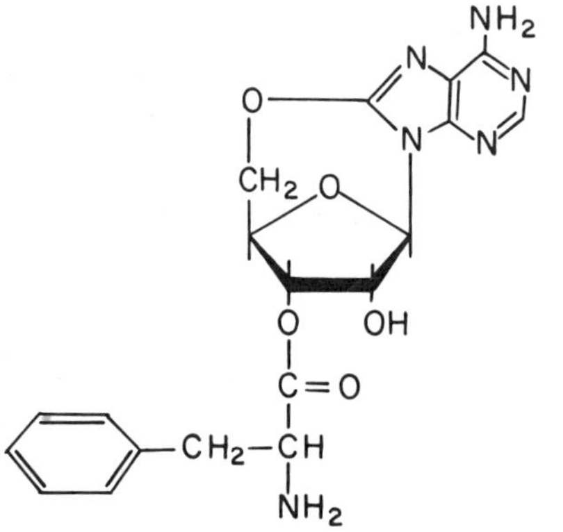

Figure 104 Active analog of puromycin with fixed *anti*-orientation of the purine ring[8]. (In the opposite *syn*-orientation, the ring is turned 180° around the N-glycosidic bond, so that the C8 of the ring is away from C5 of the ribose.)

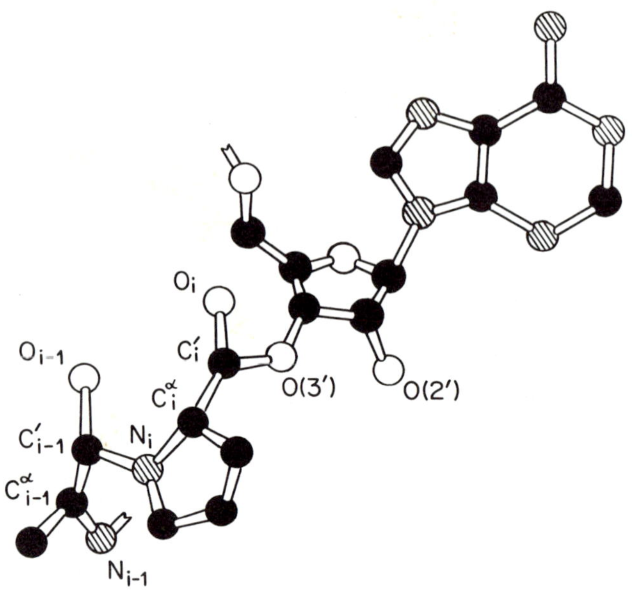

Figure 105 Alanylprolyladenosine residue as a donor substrate in the peptidyl transferase center of the ribosome: ball-and-stick skeletal model (without hydrogens). Atom designations are the same as in fig. 103. Atoms of the C-terminal prolyl residue are marked by the index i, those of the preceding alanyl residue by the index i − 1. (Kindly provided by Dr. V. I. Lim.)

the ribosomal peptidyl transferase center in a standard equivalent fashion, independent of their type. Using this principle as a guideline, some conclusions about the conformations of reacting substrates on the basis of purely stereochemical analysis may be attempted.

First, it should be remembered that the ribosome catalyzes transpeptidation with the proline residue as a substrate. In contrast to other amino acids, proline has a sterically limited angle of rotation around the C^{α}—N-bond, since this bond is involved in the ring structure. In the case of the proline residue in the donor substrate, this limitation will set a fixed angle, equal to about 60°, between the plane (N_i—C^{α}_i—C'_i) and the plane of the adjacent peptide group (N_i—C'_{i-1}—O_{i-1}) (fig. 105). In peptide chemistry, the angle given by the rotation around the C^{α}—N-bond is designated as φ; in this case its value is taken to be −60° since the plane of the peptide group is turned 60° counterclockwise when viewed from the C^{α}-atom. Since an amino acid residue should be positioned in the peptidyl transferase center in the standard way, angle φ should be adjusted to the same value by rotating it around the C^{α}—N-bond for each of the other 19 types of residues of the donor substrate (the C-terminal residue of the nascent peptide bound to tRNA and participating in transpeptidation is under consideration).

It follows from reaction chemistry that transpeptidation involves a nucleophilic attack on the carbon atom of the carbonyl group realized via the mechanism of S_n2 substitution.* Such an attack is known to proceed approximately perpendicularly to the plane of the ester group,

in this case $O^{3'}—C'(=O)—C^{\alpha}$ (fig. 106(a)). The attack should lead to an intermediate wherein the valence bonds of the attacked carbonyl carbon C_i' of the donor substrate are oriented tetrahedrally (tetrahedral intermediate) (fig. 106(b)).

On the other hand, it should be remembered that the attacking NH_2-group should be deprotonated, i.e. the unshared pair of electrons of the nitrogen atom should be free. Hence, the peptidyl transferase center should provide for the deprotonation of the aminoacyl residue of the acceptor substrate. Prior to peptide bond formation the nitrogen atom possesses three valency bonds which are directed toward the apexes of a tetrahedron, while the orbital of the unshared pair of electrons is directed toward the fourth apex. It follows from stereochemical analysis that, during the nucleophilic attack, the free valency of the attacking nitrogen atom should have a strictly defined direction: the plane formed by this direction and the $N_{i+1}—C^{\alpha}_{i+1}$-bond should be at an angle of about 120° to the plane ($N_{i+1}—C^{\alpha}_{i+1}—C'_{i+1}$) of the acceptor aminoacyl residue (fig. 106(a)); with any other orientations there can be no nucleophilic attack on the carbonyl group because of steric hindrances. After the peptide bond is formed, an angle φ of about $-60°$ is set in the newly added aminoacyl residue of the product (fig. 106(c)).

Taking all of the above into consideration, stereochemical analysis demonstrates that an effective nucleophilic attack in the ribosomal peptidyl transferase center can take place only if the angle between the

plane of the ester group $C_i'(=O)—O$ and the plane ($C_i'—C_i^{\alpha}—N_i$) defined by the rotation around the $C_i^{\alpha}—C_i'$-bond in the attacked aminoacyl residue of the donor substrate (peptidyl-tRNA) is about 60° (fig. 106(a)). After transpeptidation, it becomes an angle ψ (with an approximate value of $-60°$ (60° counterclockwise rotation around the $C^{\alpha}—C'$-bond when viewed from the C^{α}-atom) (fig. 106(c)).

* See, for example, P. Sykes (1981), *A Guide-Book to Mechanisms in Organic Chemistry*, 4th ed. (London and New York: Longman Publishers), pp. 200–239.

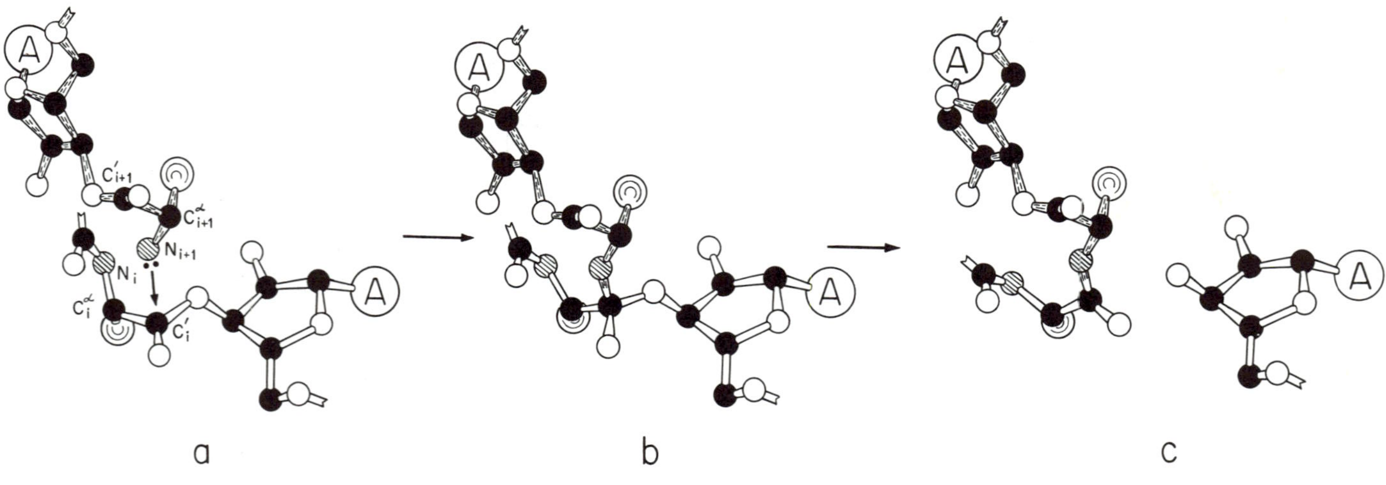

Figure 106 Stereochemistry of the transpeptidation reaction: ball-and-stick drawing (without hydrogens). Atom designations are the same as in fig. 105. Atoms of the acceptor aminoacyl residue are marked by the index i + 1. A is the adenine residue in both the donor and acceptor substrate. (a) Mutual positions of the donor (*lower*) and acceptor (*upper*) substrates in the peptidyl transferase center; the acceptor substrate nitrogen (N_{i+1}) attacks the donor carbonyl carbon (C'_i). (b) Tetrahedral intermediate resulting from the attack. (c) Products of the reaction: the elongated peptidyl adenosine residue (*left*) and the deacylated adenosine residue (*right*). (Kindly provided by Dr. V. I. Lim.)

Thus, the conformation of the donor aminoacyl residue in the ribosomal peptidyl transferase center should be similar to that of the aminoacyl residue in the α-helix ($\varphi = -50°$, $\psi = -60°$). Moreover, the acceptor aminoacyl residue attacks the donor residue in such an orientation that the transpeptidation reaction yields an elongated backbone $N_i—C_i^\alpha—C_i'—N_{i+1}—C_{i+1}^\alpha$ with parameters of the α-helical conformation.

Independent of this result of stereochemical analysis, the rule of the equivalent (universal) positioning of any aminoacyl residue in the ribosomal peptidyl tranferase center leads to the conclusion that the residue newly incorporated into the polypeptide chain always acquires a standard conformation of its backbone, i.e. standard values of torsion angles $C^\alpha—C'$ and $C^\alpha—N$. This implies that the peptidyl transferase center will generate a helical conformation of the synthesized peptide. It would not come as a surprise, therefore, if the peptide is built in a sterically and energetically favorable helical conformation, e.g. the α-helix. This initial conformation later rearranges into a unique three-dimensional structure. That the polypeptide chain folding *in vivo* does not, perhaps, begin from a random or extended state but rather is the result of a rearrangement of the helical structure[24] may provide unique opportunities for a quick and accurate search for the final conformations of the protein molecule.

References

1. M. Takanami (1964), "The effect of ribonuclease digests of aminoacyl-sRNA on a protein synthesis system," *Proc. Nat. Acad. Sci. U.S.A.* 52:1271–1276.
2. J. P. Waller, T. Erdös, F. Lemoine, S. Guttemann, and E. Sandrin (1966), "Inhibition of protein synthesis by aminoacyl-3′(2′)-adenosine," *Biochim. Biophys. Acta* 119:566–580.
3. R. E. Monro and K. A. Marcker (1967), "Ribosome-catalysed reaction of puromycin with a formylmethionine-containing oligonucleotide," *J. Mol. Biol.* 25:347–350.
4. R. R. Traut and R. E. Monro (1964), "The puromycin reaction and its relation to protein synthesis," *J. Mol. Biol.* 10:63–72.
5. I. Rychlik (1966), "Release of lysine peptides by puromycin from polylysyl-transfer ribonucleic acid in the presence of ribosomes," *Biochim. Biophys. Acta* 114:425–427.
6. T. Wagner, F. Cramer, and M. Sprinzl (1982), "Activity of the 2′ and 3′

isomers of aminoacyl transfer ribonucleic acid in the *in vitro* peptide elongation on *Escherichia coli* ribosomes," *Biochemistry* 21:1521–1529.

7. R. J. Harris and S. Pestka (1977), "Peptide bond formation," in *Molecular mechanisms of protein biosynthesis*, ed. H. Weissbach and S. Pestka, pp. 413–442 (New York: Academic Press).
8. A. A. Krayevsky and M. K. Kukhanova (1979), "The peptidyl-transferase center of ribosomes," Progress in nucleic acid research and molecular biology, ed. W. E. Cohn, vol. 23, pp. 1–51 (New York: Academic Press).
9. S. Fahnestock, H. Neumann, V. Shashoua, and A. Rich (1970), "Ribosome-catalysed ester formation," *Biochemistry* 9:2477–2483.
10. S. Fahnestock and A. Rich (1971), "Ribosome-catalysed polyester formation," *Science* 173:340–343.
11. J. Gooch and A. O. Hawtrey (1975), "Synthesis of thiol-containing analogues of puromycin and a study of their interaction with N-acetylphenylalanyl-transfer ribonucleic acid on ribosomes to form thioesters," *Biochem. J.* 149:209–220.
12. L. S. Victorova, L. S. Kotusov, A. V. Azhayev, A. A. Krayevsky, M. K. Kukhanova, and B. P. Gottikh (1976), "Synthesis of thioamide bond catalysed by *E. coli* ribosomes," *FEBS Letters* 68:215–218.
13. N. B. Tarussova, G. M. Jacovleva, L. S. Victorova, M. K. Kukhanova, and R. M. Khomutov (1981), "Synthesis of an unnatural P—N bond catalysed with *Escherichia coli* ribosomes," *FEBS Letters* 130:85–87.
14. C. T. Caskey, A. L. Beaudet, E. Scolnick, and M. Rosman (1971), "Hydrolysis of fMet-tRNA by peptidyl transferase," *Proc. Nat. Acad. Sci. U.S.A.* 68:3163–3167.
15. E. Scolnick, G. Milman, M. Rosman, and T. Caskey (1970), "Transesterification by peptidyl transferase," *Nature* 225:152–154.
16. A. A. Krayevsky, M. K. Kukhanova, and B. P. Gottikh (1975), "Peptidyl transferase centre of bacterial ribosomes: Substrate specificity and binding sites," *Nucleic Acids Res.* 2:2223–2236.
17. R. E. Monro (1967), "Catalysis of peptide bond formation by 50S ribosomal subunits from *Escherichia coli*," *J. Mol. Biol.* 26:147–151.
18. R. E. Monro, J. Černá, and K. A. Marcker (1968), "Ribosome-catalyzed peptidyl transfer: Substrate specificity at the P-site," *Proc. Nat. Acad. Sci. U.S.A.* 61:1042–1049.
19. R. E. Monro, T. Staehelin, M. L. Celma, and D. Vazquez (1969), "The peptidyl transferase activity of ribosomes," *Cold Spring Harbor Symp. Quant. Biol.* 34:357–366.
20. D. Vazquez, E. Battaner, R. Neth, G. Heller, and R. E. Monro (1969), "The function of 80S ribosomal subunits and effects of some antibiotics," *Cold Spring Harbor Symp. Quant. Biol.* 34:369–375.
21. S. Pestka (1977), "Inhibitors of protein synthesis," in *Molecular mechanisms*

of protein biosynthesis ed. H. Weissbach and S. Pestka, pp. 468–553 (New York: Academic Press).

22. M. Sundaralingam and S. K. Arora (1972), "Crystal structure of the aminoglycosyl antibiotic puromycin dihydrochloride pentahydrate: Models for the terminal 3′-aminoacyl-adenosine moieties of transfer RNAs and protein-nucleic acid interactions," *J. Mol. Biol.* 71:49–70.
23. H. P. M. deLeeuw, J. R. deJager, H. J. Koeners, J. H. vanBoom, and C. Altona (1977), "Puromycin and some of its analogues: Conformational properties in solution. A360-MHz nuclear-magnetic-resonance investigation," *Eur. J. Biochem.* 76:209–217.
24. V. I. Lim (1978), "Polypeptide chain folding through a highly helical intermediate as a general principle of globular protein structure formation," *FEBS Letters* 89:10–14.

Chapter 14

Elongation III: Translocation

14-1 Definition and experimental tests

As a result of transpeptidation, the newly elongated peptidyl-tRNA occupies the ribosomal A-site while deacylated tRNA occupies the P-site (fig. 107(left)). In other words, the reaction products are not released from the ribosome and occupy positions of the substrates.

Translocation is defined as an intraribosomal transition of the elongated peptidyl-tRNA from the position of the reaction product (from the A-site), where it is incapable of participating in transpeptidation as a peptidyl donor, to the position of the reaction substrate (to the P-site), where it is capable of reacting with the other (acceptor) substrate. This transition is directly coupled with the release of the second reaction product, deacylated tRNA, from the ribosomal P-site. Translocation is accompanied by the movement of the template polynucleotide relative to the ribosome at a distance of one codon and the appearance of competence to the binding of the second substrate, the aminoacyl-tRNA.

The change in the functional state of the ribosome as a result of

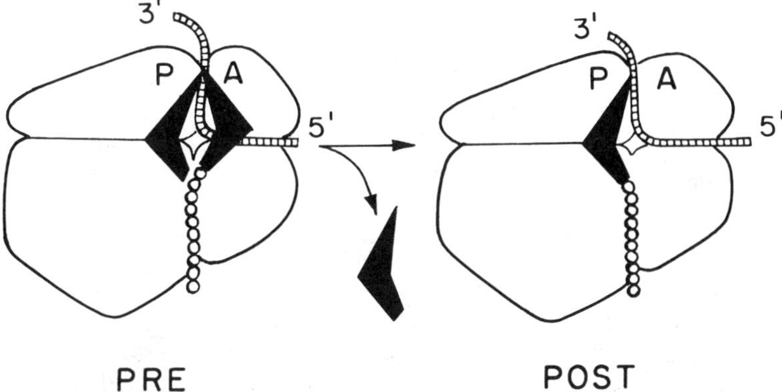

Figure 107 Schemes of the pretranslocation (*left*) and posttranslocation (*right*) states of the ribosome.

translocation is schematically presented in fig. 107. Summarily, peptidyl-tRNA moves to the P-site, deacylated tRNA is displaced from the P-site, the template moves at a distance of one codon, and the A-site with the new codon becomes vacant.

In accordance with the above, there are four ways of measuring translocation, based on the following criteria: (1) a transpeptidation reaction with a low-molecular-mass acceptor substrate, e.g. puromycin; (2) the binding of aminoacyl-tRNA; (3) the release of deacylated tRNA; and (4) the movement of the template polynucleotide.

1. The puromycin reaction is the simplest and most widely used measurement of translocation. Puromycin has been described above as a low-molecular-mass analog of aminoacyl-tRNA (chapter 11, fig. 88) serving as an acceptor substrate for the ribosomal peptidyl transferase center. The amino group of its aminoacyl moiety attacks the ester group of peptidyl-tRNA (or its analog), resulting in transpeptidation. The peptidyl puromycin formed is released from the ribosome. The peptidyl-tRNA of the posttranslocation state ribosome reacts with puromycin, whereas the peptidyl-tRNA of the pretranslocation state ribosome does not[1]. Therefore, if the peptidyl residue is labeled, the label released from the ribosome in response to puromycin addition can be used as a quantitative measure of the posttranslocation state in the population of ribosomes studied. Conversely, lack of competence to puromycin implies that the particles are in the pretranslocation state.

2. Binding of the aminoacyl-tRNA may also serve as a quantitative measure of translocation. If aminoacyl-tRNA is labeled, the codon-dependent binding of the label with the translating ribosome is impossible immediately following transpeptidation; the labeled aminoacyl-tRNA will bind only after translocation[2–4]. Consequently, an inability to bind aminoacyl-tRNA designates the pretranslocation state whereas competence to the aminoacyl-tRNA binding is an indication of the posttranslocation state.
3. The release of deacylated tRNA from the translating ribosome is coupled to translocation. Therefore, if the tRNA residue of peptidyl-tRNA retained by the ribosome is labeled, following transpeptidation the labeled deacylated tRNA will remain bound to the ribosome; only translocation will result in the release of the label into solution[5,6]. Thus, for the translating ribosome retention of the deacylated tRNA can be equated with the pretranslocation state, and the absence of deacylated tRNA testifies to the posttranslocation state. The release of deacylated tRNA from the ribosome in solution is a convenient quantitative test to trace the course of translocation (kinetics, rate, and amount) in the population of ribosomes[7].
4. The movement of the template polynucleotide as a test for translocation is a technically more difficult approach. Measurement of the movement may be indirect, i.e. when it is based on the appearance of competence to the binding of aminoacyl-tRNA specific to a codon following the codon that was previously positioned in the ribosome[8,9]. Measurement also can be direct, i.e. if the change of the template region screened (protected) by the ribosome is analyzed[10,11]. The direct test has demonstrated that the movement of the polynucleotide template relative to the ribosome by one nucleotide triplet accompanies the appearance of competence to puromycin and competence to bind aminoacyl-tRNA.

14-2 Participation of the elongation factor (EF-G or EF-2)

Translocation is catalyzed by a large protein, referred to as the elongation factor G (EF-G) in procaryotes[12], and as the elongation factor 2 (EF-2) in eucaryotes[2]. The molecular mass of EF-G is approximately 80,000 daltons. This protein is a single polypeptide chain with a length of 701 amino acid residues in the case of *Escherichia coli*[13], and is folded into several globular domains. The animal EF-2, somewhat larger than the procaryotic EF-G, has a molecular mass of about 95,000 daltons.

EF-G (or EF-2) interacts with GTP and the ribosome. This interaction induces GTPase activity and GTP is cleaved to GDP and orthophosphate. The interaction (complex formation) of EF-G and GTP with the pretranslocation ribosome results in a quick translocation, upon which EF-G, GDP, and orthophosphate are released from the ribosome.

The interaction of EF-G (or EF-2) with GTP is considerably weaker than the analogous interaction of EF-T_u (or EF-1) with GTP. The resultant EF-G·GTP complex is unstable and easily reversible (the EF-2·GTP complex is somewhat more stable). Nevertheless, this complex is being formed and its formation is a prerequisite for the interaction of EF-G with the ribosome. GTP is thought to induce a conformational change in the protein (EF-G or EF-2) which results in an affinity of the factor to the ribosome. GTP is highly specific in this respect and cannot be substituted for either by other nucleoside triphosphates or by any nucleoside mono- or diphosphates. GTP can, however, be replaced by the nonhydrolyzable GTP analogs, e.g. guanylyl methylene diphosphonate (GMP-PCP) or guanylyl imidodiphosphate (GMP-PNP) (see chapter 11, fig. 86).

The EF-G·GTP (or EF-2·GTP, in eucaryotic systems) complex can bind to the functioning (translating) ribosome as well as to the vacant ribosome, or even with the isolated 50S subunit. The binding site of EF-G on a 50S subunit appears to be located at the base of the L7/L12 stalk (see chapter 11, fig. 87); in the 70S ribosome it is found near the interface of the ribosomal subunits, in the region of the tRNA-binding sites. The binding of EF-G·GTP to the ribosome is prevented by the aminoacyl-tRNA bound in the ribosomal A-site, as well as by the presence of the other protein factor, EF-T_u (see chapter 11, section 11.2).

In all cases, the binding of the EF-G·GTP complex or the EF-2·GTP complex to the ribosome or its subunit results in cleavage (hydrolysis) of GTP. The ribosome seems to induce GTPase activity of EF-G; in other words, it is thought that the GTPase center is present on EF-G but is inactive in the absence of the ribosome (see chapter 11, section 11.3). If a ribosome is vacant or the large subunit is used instead of the whole ribosome, then the GTP is simply hydrolyzed, without being coupled with any events of elongation.

The EF-G-catalyzed hydrolysis of GTP on the ribosome results in EF-G being in a complex with GDP. However, GDP is incapable of supporting the EF-G affinity to the ribosome; therefore, the EF-G·GDP complex is released from the ribosome and subsequently dissociates.

Thus, the sequence of reactions assuming the participation of EF-G is as follows:

$$\text{EF-G} + \text{GTP} \rightleftharpoons \text{EF-G·GTP};$$
$$\text{EF-G·GTP} + \text{RS} \rightleftharpoons \text{RS·EF-G·GTP};$$
$$\text{RS·EF-G·GTP} + H_2O \rightarrow \text{RS·EF-G·GDP} + P_i;$$
$$\text{RS·EF-G·GDP} \rightarrow \text{RS} + \text{EF-G·GDP};$$
$$\text{EF-G·GDP} \rightleftharpoons \text{EF-G} + \text{GDP}.$$

If the vacant ribosome or its isolated 50S subunit participates in the reactions, the overall process is simply GTP hydrolysis proceeding according to the following overall equation:

$$\text{GTP} + H_2O \xrightarrow{\text{EF-G,RS}} \text{GDP} + P_i.$$

In other words, in this case EF-G, in combination with the ribosome, serves solely as GTPase. However, if these reactions involve the pretranslocation state ribosome, then in addition to GTP hydrolysis, participation of EF-G will result also in translocation.

The question then arises as to which of the sequential reactions involving EF-G is directly coupled to the translocation. For a long time translocation was thought to be an energy-consuming process and to be directly coupled with the GTP hydrolysis on the ribosome. Both of these assumptions turned out to be incorrect. The reasons why the first assumption—that an additional free-energy supply is required for translocation—is incorrect, will be given below. The assumption about the coupling of translocation to GTP hydrolysis has been rejected on the basis of direct experiments wherein a nonhydrolyzable GTP analog was used instead of GTP. When pretranslocation ribosomes interacted with EF-G and GMP-PCP or GMP-PNP, quick translocation was observed, as in the case of EF-G and GTP[14–16]. It followed that the attachment of EF-G·GTP to the ribosome was sufficient to promote (catalyze) translocation. There are two possible explanations here: (1) either the attachment (affinity) of the protein to the ribosome exerts some force, directly shifting the tRNA molecules in the ribosome, or (2) the kinetic barriers for spontaneous translocational movements become decreased when EF-G is attached to the pretranslocation ribosome.

Thus, translocation is catalyzed by EF-G with GTP; catalysis of the translocation results from the attachment of EF-G in a complex with GTP to the ribosome but is not a consequence of GTP cleavage.

14-3 Role of EF-G-mediated GTP hydrolysis

The use of a nonhydrolyzable GTP analog in EF-G-catalyzed translocation has demonstrated that, in this case, EF-G after translocation remains associated with the posttranslocation state ribosome. This relationship is the obvious result of the fact that the effector inducing EF-G affinity to the ribosome is not destroyed and therefore EF-G continues to retain the affinity and is held in the complex. However, the presence of EF-G on the ribosome precludes the presence of EF-T_u and aminoacyl-tRNA in the A-site. Therefore, as a result of the translocation effected by EF-G with a nonhydrolyzable GTP analog, peptidyl-tRNA acquires the capacity to react with the acceptor substrate (competence to puromycin), and deacylated tRNA is released from the ribosome; however, such posttranslocation state ribosomes cannot bind the next aminoacyl-tRNA and, hence, are unable to continue the elongation. In experiments conducted *in vitro*, EF-G together with the nonhydrolyzable GTP analog have been washed off such posttranslocation ribosomes, resulting in the capacity to bind aminoacyl-tRNA and to continue elongation[17,18]. This implies that during the normal process EF-G should become attached to the ribosome in order to induce translocation and then should leave the ribosome to allow the next step to occur. The role of GTP is that of an effector, which provides for the EF-G attachment to the ribosome; GTP hydrolysis implies that the effector is destroyed, this being necessary for the release of EF-G from the ribosome.

In this respect, the action of fusidic acid (fig. 108), an antibiotic

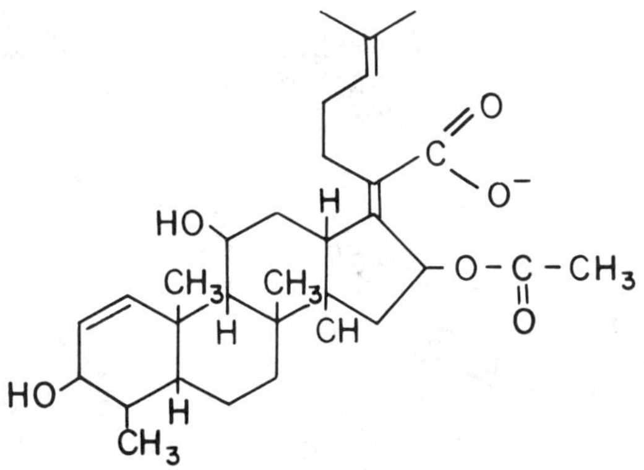

FUSIDIC ACID

Figure 108 Fusidic acid, an inhibitor affecting EF-G.

specifically affecting EF-G, proves to be interesting. EF-G in a complex with fusidic acid normally interacts with GTP and further with the ribosome, resulting in translocation, which in turn is followed by GTP cleavage to GDP and orthophosphate; fusidic acid, however, acts to increase the affinity of EF-G to the ribosome, and EF-G·GDP is not released after GTP hydrolysis[19]. As a consequence, in spite of translocation occurring, the next aminoacyl-tRNA cannot bind with the ribosomal A-site, and therefore elongation stops.

14-4 Sequence of events in EF-G-catalyzed translocation: Inhibitors

The sequence of events that take place during EF-G-promoted translocation is schematically given in fig. 109. The first event is the interaction of the pretranslocation state ribosome with EF-G·GTP. This event is inhibited by thiostrepton, a specific antibiotic binding to the 50S ribosomal subunit in the region of the base of the L7/L12 stalk, i.e. in approximately the same region where binding of EF-G has been demonstrated (see chapter 11, fig. 87). Protein L11 has been shown to be responsible directly for thiostrepton binding[20]. It is remarkable that some thiostrepton-resistant mutants and strains are devoid of protein L11 as a component of the 50S ribosomal subunit. It is likely that the molecule of bound thiostrepton blocks the site of the interaction between the ribosome and EF-G.

No stable (long-lived) complex between the pretranslocation state ribosome and EF-G·GTP or EF-G·GMP-PCP has been observed under experimental conditions. The interaction of the pretranslocation ribosome with EF-G·GTP or EF-G·GMP-PCP seems to induce immediate quick translocation. This is the second event in the scheme of fig. 109. The translocation event can be inhibited either by decreased temperature or by elevation of the magnesium ion concentration; e.g., at 4°C, or in 30 mM Mg^{2+} at a physiological temperature (30° to 37°C), translocation is virtually blocked[7]. However, even under these conditions a complex of the pretranslocation ribosome with EF-G·GTP or EF-G·GMP-PCP has not been isolated; in other words, the interaction appears to be quickly reversible. Some specific agents, e.g. antibiotic viomycin, also inhibit this event, i.e. translocation in the strict sense of the word[21].

The third event in the scheme of fig. 109 involves the hydrolysis of GTP and the release of orthophosphate. The posttranslocation state

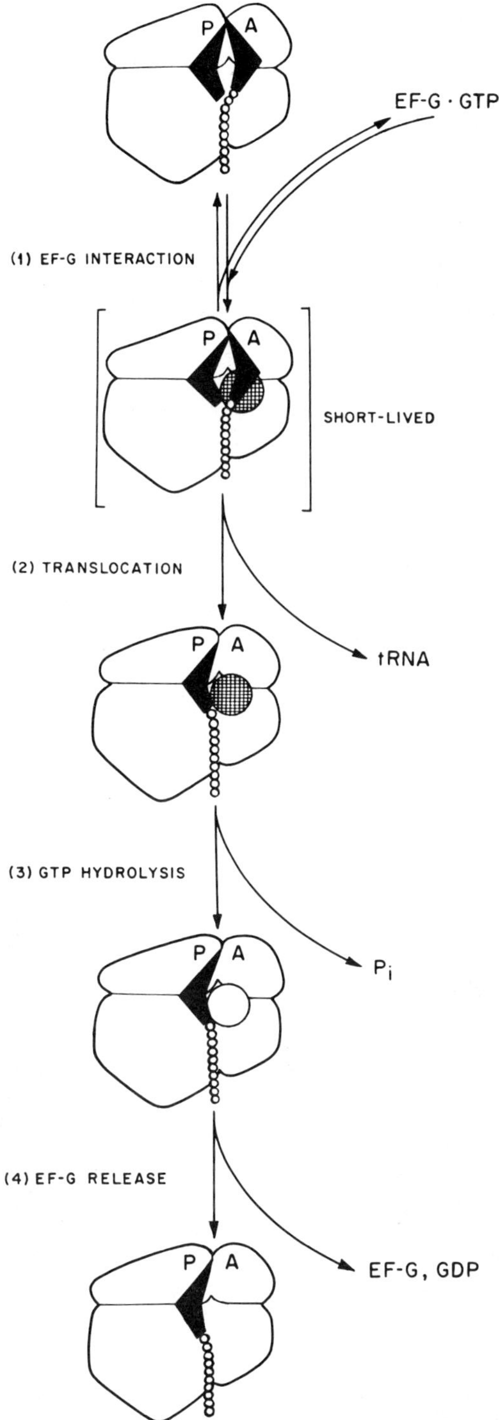

Figure 109 Sequence of events during translocation.

ribosome firmly retains EF-G until GTP hydrolysis occurs. Of course, if GTP is replaced by a nonhydrolyzable analog, e.g. GMP-PCP, this event does not take place and the posttranslocation ribosome will continue to be in a complex with EF-G·GMP-PCP.

After GTP hydrolysis has taken place, EF-G·GDP is found in a complex with the posttranslocation ribosome. This complex is quite unstable and dissociates, i.e. EF-G and GDP are released from the ribosome. This is the fourth event in the scheme. Fusidic acid specifically inhibits just this stage of the process by fixing the complex and preventing the release of EF-G·GDP from the ribosome[19].

On the whole, the sequence described by the scheme in fig. 109 is strikingly analogous to the sequence of events in EF-T_u-promoted binding of aminoacyl-tRNA (chapter 12, fig. 99): the initial transient interaction is followed by a fast (catalyzed) stage of the main event,

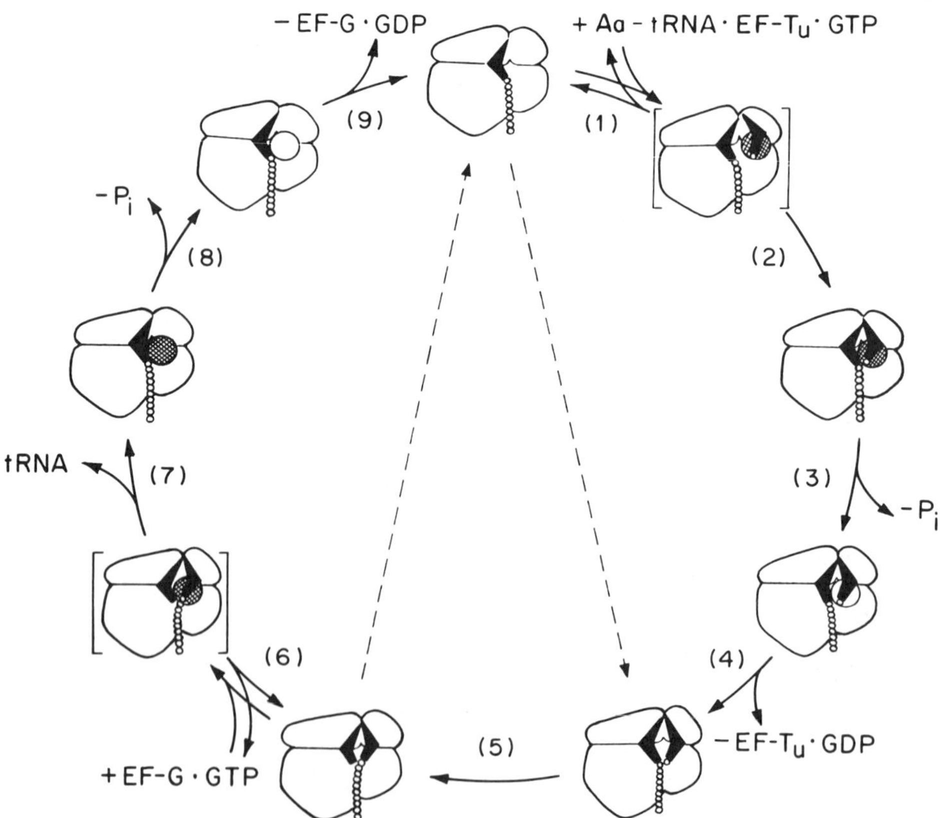

Figure 110 Complete sequence of events during the factor-catalyzed elongation cycle. Broken lines designate factor-free bypasses: nonenzymatic binding of aminoacyl-tRNA and nonenzymatic translocation.

which in turn is followed by GTP hydrolysis, and finally by the release of the protein factor together with GDP. Such symmetry of the elongation cycle is seen in fig. 110, where both factor-catalyzed processes are shown as sequences of consecutive events.

14-5 "Nonenzymatic" (factor-free) translocation

It has been established, using cell-free systems, that translocation can proceed in the absence of elongation factors and GTP[22,23]. This *"nonenzymatic" translocation* takes place far more slowly than the EF-G·GTP-catalyzed process. Nevertheless, it yields a normal post-translocation state of the ribosome, capable of continuing elongation. It can be concluded that translocation is a thermodynamically spontaneous event. The translocational mechanism appears to be an intrinsic property of the ribosome itself, and is not fully provided by the elongation factor.

Again the situation appears to be very similar to that of aminoacyl-tRNA binding: the processes may occur slowly in the factor-free mode, proceed spontaneously (downhill), and their mechanisms are provided by the ribosome; therefore the elongation factors catalyze only thermodynamically permissible and mechanistically ensured processes.

Factor-free (nonenzymatic) binding of aminoacyl-tRNA, ribosome-catalyzed transpeptidation, and factor-free (nonenzymatic) translocation constitute the factor-free elongation cycle. This cycle is designated by dashed shunting arrows in fig. 110. Repetition of this cycle results in a slow *factor-free elongation*. Cell-free systems have been used to perform factor-free translation of polyuridylic acid and of a number of synthetic heteropolynucleotides. It has thereby been shown that the polypeptide product corresponds completely to the coding sense of the template polynucleotide[24].

Whereas an increase in the Mg^{2+} concentration stimulates aminoacyl-tRNA binding, a decrease in Mg^{2+} stimulates translocation (fig. 111). At Mg^{2+} concentrations of about 3 mM, the rate of translocation approaches that taking place in the presence of EF-G with GTP[7] (but little aminoacyl-tRNA binding occurs). At Mg^{2+} concentrations of 30 mM, the translocation rate is close to zero[7] (but the binding of aminoacyl-tRNA is good). Thus, by alternating the Mg^{2+} concentration in the cell-free system, the action of the elongation factors might be simulated and the rate of the factor-free elongation cycle increased. It is probable that this approach will be used in the future.

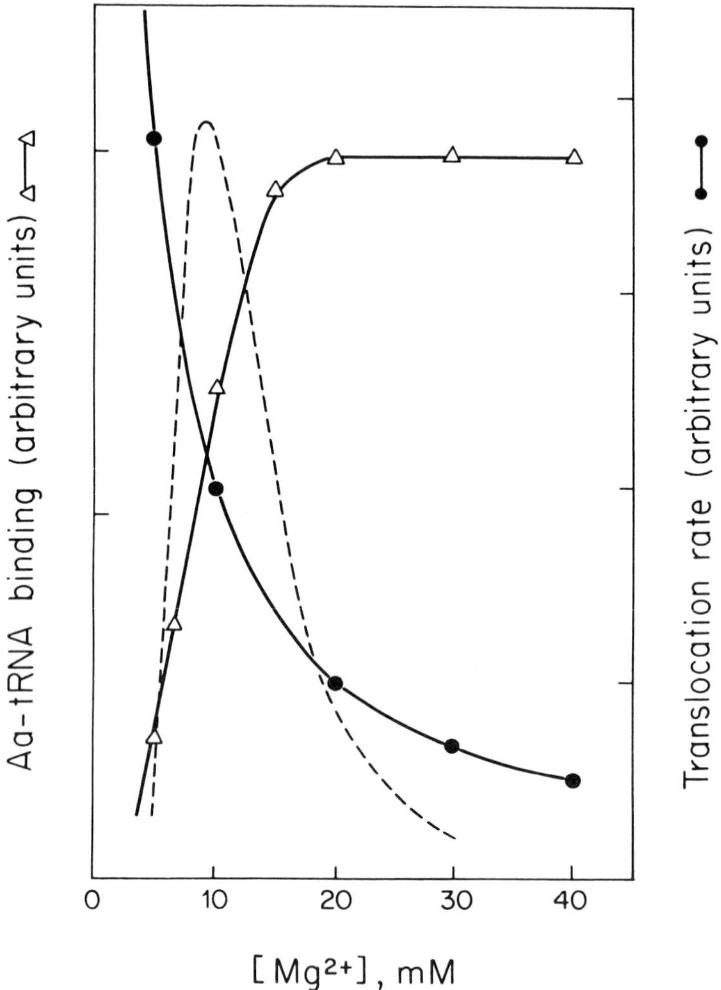

Figure 111 Dependence of the factor-free (nonenzymatic) binding of aminoacyl-tRNA to the ribosome and the factor-free (nonenzymatic) translocation on Mg^{2+} concentration. The lower the Mg^{2+} concentration, the worse the tRNA binding and the faster the translocation. The broken-line curve indicates the dependence of the factor-free translation rate on the Mg^{2+} concentration. (After N. V. Belitsina *et al.* (1975), *Dokl, Akad. Nauk S.S.S.R.* 224: 1205–1208.)

When comparing the slow factor-free translocation with the fast EF-G·GTP-catalyzed translocation, it should be taken into account that the elongation factor does not appear to decrease the heat energy of activation of the process; this suggests that catalysis in this case is primarily of an entropy type[25]. Analysis using inhibitors has demonstrated that the factor does not create a new reaction pathway that would proceed via intermediate stages to circumvent the high activation barrier, as is usually the case with an enthalpy-type catalyst. This follows from observations that various specific inhibitors of translocation, e.g. viomycin, spectinomycin, erythromycin, neomycin, kanamycin, gentamycin, and hygromycin B, affect the enzymatic as well as the nonenzymatic process, suggesting an identical translocational mechanism in both cases, involving identical targets. It is likely that the elongation factor catalyzes the process by creating better steric conditions in the ribosome for the same ribosome-intrinsic translocational pathway. A possible mechanism here is the simple fixation of one of the thermally fluctuating ribosomal substates which could be favorable for translocation (see below). EF-G as a large ligand of the ribosome could be capable of exerting such a fixation or orientation effect.

Thus, the existence of factor-free (nonenzymatic) translocation indicates that the translocational mechanism is intrinsic to the ribosome and is provided with energy without the involvement of GTP. Comparison of the properties of factor-free translocation and those of factor-dependent (enzymatic) translocation suggests that the catalytic contribution of EF-G is basically of an entropic nature.

14-6 Movement of the template during translocation

As has been noted, translocation involves shift of a template polynucleotide by one codon in the direction from the 5′-end to the 3′-end. During this shift and after its completion, pairing between the anticodon of peptidyl-tRNA and the template codon is retained; the codon-anticodon duplex appears to move as a whole, from the A-site to the P-site of the ribosome[26].

It is natural to wonder what plays the active part in the translocation: movement of the template or movement of the peptidyl-RNA. Several observations suggest that the shift of the template by one codon is driven by the translocational displacement of tRNA: through its anticodon tRNA pulls the codon of the template. An impressive demonstration of the lack of dependence of the translocation on the template polynucleotide was the discovery of ribosomal synthesis of a polypeptide from aminoacyl-tRNA in the absence of any template polynucleotide[27]; the elongation cycle, including EF-G·GTP-catalyzed translocation, was demonstrated in this case. The most direct evidence of the active (driving) role of tRNA and the passive (driven) role of mRNA in translocation was obtained by Riddle and Carbon[28]; in their experiment, mutant tRNA with a nucleotide quadruplet instead of a triplet as an anticodon suppressed the (+1) frame-shift mutation; in other words, it moved the mRNA in the ribosome, correspondingly, by four (and not by three) nucleotide residues.

Thus, the principal event in translocation is probably the movement of peptidyl-tRNA from the A-site to the P-site of the ribosome. The anticodon pulls the template codon associated with it, leading to a corresponding shift of the template relative to the ribosome by one triplet (normally). This results in the positioning in the A-site of the next (in the direction of the 3′-end) nucleotide triplet of the template, while the preceding (the 5′-side-adjacent) triplet, together with the anticodon of deacylated tRNA, leaves the P-site.

14-7 Energetics of translocation

The energy aspect of translocation was misunderstood for a long time due to various historical factors and due to traditional ways of thinking among biochemists. The participation of GTP in translocation was determined earlier than all other facts concerning this stage of the elongation cycle. Consideration of translocation as the process of mechanical displacement of large molecular masses and the observation of coupled cleavage of GTP into GDP and orthophosphate suggested an analogy to muscle contraction, which proceeds at the expense of energy of the ATP hydrolysis into ADP and orthophosphate. This analogy created a powerful psychological stimulus for inventing special problems of energy supply for translocation, which had to be solved at the expense of GTP cleavage. Most of the models of translocation proposed thus far assume that it is the energy of EF-G-mediated GTP hydrolysis that is used in one way or another for mechanical work involving the active movement or at least the active removal of ribosomal ligands (tRNA) from their binding sites; correspondingly, the function of contractile proteins is often ascribed to EF-G or to protein L7/L12. According to some models, the energy of GTP through EF-G is applied to peptidyl-tRNA occupying the A-site, and the developing force moves this tRNA together with its codon toward the P-site, displacing deacylated tRNA from the P-site. In other models, GTP energy is realized through the EF-G initially for the removal (pushing out) of the deacylated tRNA from the P-site; then the peptidyl-tRNA undergoes spontaneous transition from the A-site to the vacant P-site, to which it has a greater affinity.

As already mentioned, the GTP hydrolysis takes place *after* translocation. Hence, there is no apparent coupling between GTP hydrolysis and translocation. In addition, it has been shown that the hydrolysis involves the direct transfer of the phosphate residue from GTP to water without the formation of a phosphorylated intermediate that could be responsible for such coupling[29]. Thus, another mechanism should be assumed: translocation is coupled with the adsorption of EF-G on the ribosome whereas GTP hydrolysis is required for the desorption of EF-G. If work had been required to effect the transition of the ribosome from the pretranslocation to the posttranslocation state, it might be thought that the work is performed at the expense of the energy of the complex formation between the ribosome and GTP·EF-G, and the adsorption energy is then compensated by the energy of GTP hydrolysis. Such a mechanism would imply that the energy of GTP hydrolysis eventually is responsible for the work done, but not through direct coupling but by "lending," with the subsequent return.

In reality, it has been demonstrated that translocation can proceed spontaneously, without EF-G and GTP (nonenzymatic translocation). This implies that the process is thermodynamically permissible (downhill process) or, in other words, that the thermodynamic potential (free energy) of the pretranslocation state of the ribosome is higher than that of the posttranslocation state. There is no need to explain that in this situation energy expenditure for performing work (increasing potential) is not required. Thus, any thermodynamic contribution of EF-G with GTP in translocation should be rejected.

Nevertheless, in EF-G-catalyzed translocation, EF-G with GTP binds to the ribosome, the GTP is subsequently hydrolyzed, and so additional free energy is expended. But what is the purpose of this energy expension? It is apparent that energy generally can be expended either on some useful work against thermodynamic potential (uphill process), or on overcoming barriers in a spontaneous (downhill) process without accumulation of productive work. If the first of the two alternatives is excluded, it has to be recognized that the contribution of GTP is a purely kinetic one: at first, the interaction of GTP with EF-G provides for the attachment of EF-G·GTP to the ribosome and thereby decreases barriers in the course of translocation; thereafter, the GTP hydrolysis removes the barrier created by the EF-G itself for the subsequent stage of the elongation cycle[30]. Thus, GTP energy is expended solely to overcome barriers, and eventually it dissipates completely into heat. This process is called the catalysis of translocation. A peculiar feature of catalysis in this case is its energy dependence, which is similar to the catalysis of aminoacyl-tRNA binding with the participation of EF-T_u.

14-8 Molecular mechanisms of translocation

The molecular mechanisms of translocation have yet to be determined. Nevertheless, the evidence available on the structure of tRNA and the ribosome as well as on the properties of pretranslocation and posttranslocation complexes can be used to formulate a plausible stereochemical model of the interactions between the ribosome, two tRNAs, and mRNA, and of the changes in these relationships during translocation. Such a hypothetical model is presented below.

It has been noted in stereochemical consideration of the codon-anticodon interaction (chapter 12, section 12.1) that the anticodons of two tRNAs form double-helical structures of the A-form type with two adjacent mRNA codons (chapter 12, fig. 93). The 3′-ends of the two tRNA molecules are brought into close proximity, while their corners are somewhat apart, so that the planes of the two tRNA molecules are at an angle to each other (chapter 12, fig. 92). This situation continues after transpeptidation. Therefore the pretranslocation state ribosome contains a complex between deacylated tRNA (in the P-site) and peptidyl-tRNA (in the A-site) joined by a complementary hexaplet of mRNA. The corner of the peptidyl-tRNA is assumed to be positioned close to the heads of the 30S and 50S ribosomal subunits, while the corner of the deacylated tRNA molecule is assumed to be located in the region of the base of the L7/L12 stalk of the 50S subunit (chapter 11, fig. 85).

Translocation can be conceived as an operation of the *helical displacement* of the two tRNAs, including a clockwise turn (if one looks from their anticodons) and translation along the axis connecting the anticodons with the acceptor ends[31]. As a result, deacylated tRNA is displaced from the P-site and dissociates from the complex with its codon; peptidyl-tRNA is then in the P-site, and the corner of its tRNA residue is located at the base of the L7/L12 stalk; the A-site is now vacant. This is the posttranslocation state (fig. 112).

Which force is responsible for the movement of the peptidyl-tRNA from the A-site to the P-site during translocation? If the pathway of the displacement of the complex between two tRNAs and mRNA is determined by the construction of the ribosome, then the movement

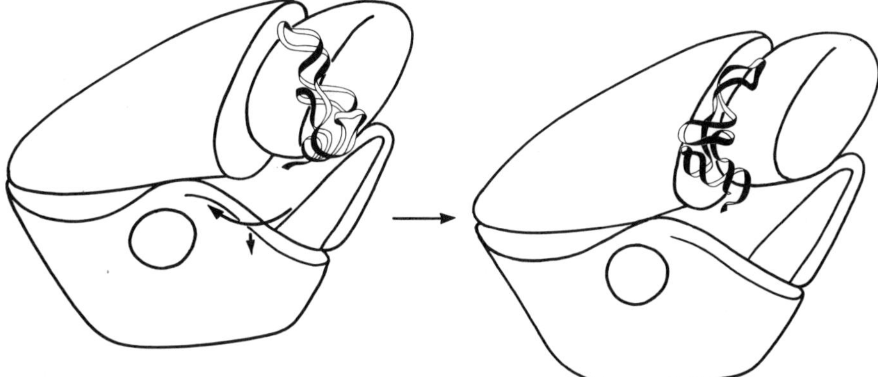

Figure 112 Schematic representation of the transition (helical displacement) of the tRNA molecule from the ribosomal A-site to the P-site during translocation. For the sake of simplicity, only one tRNA residue (that of peptidyl-tRNA) is shown in the pretranslocation state ribosome.

itself may be a consequence of thermal motion. Since the displacement is accompanied by dissociation of the deacylated tRNA, it should result in an entropy gain. At any rate, in the case of factor-free (nonenzymatic) translocation, there appears to be no other motive forces except thermal motion. It is likely that thermal motion similarly induces the displacement in the course of EF-G·GTP-catalyzed translocation, but the attachment of EF-G·GTP to the ribosome might create a specific structural environment wherein steric and energy barriers in the transition pathway are decreased (in particular, the affinity of the peptidyl-tRNA to the A-site may be decreased by EF-G). The ribosomal L7/L12 stalk appears to participate, to a certain extent, in the mechanism of translocation catalysis, perhaps contributing to creation of a favorable structural environment effected by EF-G with GTP.

14-9 Summary: The material and energy balance of the elongation cycle

The consecutive stages of the codon-directed binding of aminoacyl-tRNA, transpeptidation, and translocation create a cycle, resulting in: (1) the determination of the position of one amino acid residue in the polypeptide chain to be synthesized, (2) the formation of one peptide bond, (3) the deacylation of one molecule of aminoacyl-tRNA, (4) the hydrolysis of two molecules of GTP to GDP and orthophosphate, and (5) the shift (readout) of one nucleotide triplet of the template polynucleotide relative to the ribosome. Repetitions of this cycle create elongation; the number of cycles during elongation depends on the number of template codons (minus the initiation codon).

In regard to the material balance, one cycle involves the consumption of one molecule of aminoacyl-tRNA and two molecules of GTP (plus two water molecules) from solution. One molecule of deacylated tRNA, two molecules of GDP, and two molecules of orthophosphate are released into solution (chapter 11, fig. 79). Peptide elongation by one residue takes place concomitantly in the ribosome.

From this description the energy balance of the cycle can be summed up. The energy requirements of the cycle appear to be rather modest: they include determination of the position of the amino acid residue in the polypeptide chain ($\Delta G^{0\prime} \leq +2.5$ kcal/mole) and formation of the peptide bond ($\Delta G^{0\prime} \simeq +0.5$ kcal/mole). It is clear that these energy requirements of the complete cycle are amply covered by the free energy, which is liberated by deacylation of the aminoacyl-tRNA ($\Delta G^{0\prime} \simeq -7$ kcal/mole)[25,30]. Nevertheless, the hydrolysis of two molecules of GTP accompanies the cycle (chapter 11, fig. 79), resulting in the liberation of additional large amounts of free energy ($\Delta G^{0\prime} \simeq -15$ kcal/mole). Thus, elongation appears to be a wasteful process, noneconomical and with a low efficiency. The bulk of the free energy liberated during the cycle dissipates into heat.

However, if GTP is excluded even at one stage in the cycle (at the stage of aminoacyl-tRNA binding or at the stage of translocation), the process is greatly slowed and becomes fairly sensitive to unfavorable conditions, drugs, and other obstacles[25]. Therefore, a great excess of free energy is needed for the system in order to provide for high rates and high resistance of elongation. It is apparent that economy is not the main advantage providing for the survival of the system and of the corresponding organism in living nature.

References

1. R. R. Traut and R. E. Monro (1964), "The puromycin reaction and its relation to protein synthesis," *J. Mol. Biol.* 10:63–72.

2. L. Skogerson and K. Moldave (1968), "Evidence for the role of aminoacyl transferase II in peptidyl-transfer ribonucleic acid translocation," *J. Biol. Chem.* 243:5361–5367.

3. S. Pestka (1968), "Studies on the formation of transfer ribonucleic acid-ribosome complex. V. On the function of a soluble transfer factor in protein synthesis," *Proc. Nat. Acad. Sci. U.S.A.* 61:726–733.
4. A.-L. Haenni and J. Lucas–Lenard (1968), "Stepwise synthesis of a tripeptide," *Proc. Nat. Acad. Sci. U.S.A.* 61:1363–1369.
5. J. Lucas–Lenard and A.-L. Haenni (1969), "Release of transfer RNA during peptide chain elongation," *Proc. Nat. Acad. Sci. U.S.A.* 63:93–97.
6. A. Kaji, K. Igarashi, and H. Ishitsuka (1969), "Interaction of tRNA with ribosomes: Binding and release of tRNA," *Cold Spring Harbor Symp. Quant. Biol.* 34:167–177.
7. N. V. Belitsina and A. S. Spirin (1979), "Ribosomal translocation assayed by the matrix-bound poly(uridylic acid) column technique: Effects of different magnesium concentrations," *Eur. J. Biochem.* 94:315–320.
8. R. W. Erbe and P. Leder (1968), "Initiation and protein synthesis: Translation of di- and tri-codon messengers," *Biochem. Biophys. Res. Commun.* 31:798–803.
9. L. Skogerson, D. Roufa, and P. Leder (1971), "Characterization of the initial peptide of Q_β RNA polymerase and control of its synthesis," *Proc. Nat. Acad. Sci. U.S.A.* 68:276–279.
10. S. S. Thach and R. E. Thach (1971), "Translocation of messenger RNA and 'accommodation' of fMet-tRNA," *Proc. Nat. Acad. Sci. U.S.A.* 68:1791–1795.
11. S. L. Gupta, J. Waterson, M. L. Sopori, S. M. Weissman, and P. Lengyel (1971), "Movement of the ribosome along the messenger ribonucleic acid during protein synthesis," *Biochemistry* 10:4410–4421.
12. Y. Nishizuka and F. Lipmann (1966), "The interrelationship between GTPase and amino acid polymerization," *Arch. Biochem. Biophys.* 116:344–351.
13. Yu. A. Ovchinnikov, Yu. B. Alakhov, Yu. P. Bundulis, N. V. Dovgas, V. P. Kozlov, L. P. Motuz, and L. M. Vinokurov (1982), "The primary structure of elongation factor G from *Escherichia coli*," *FEBS Letters* 139:130–135.
14. N. Inoue–Yokosawa, C. Ishikawa, and Y. Kaziro (1974), "The role of guanosine triphosphate in translocation reaction catalyzed by elongation factor G," *J. Biol. Chem.* 249:4321–4323.
15. N. V. Belitsina, M. A. Glukhova, and A. S. Spirin (1975), "Translocation in ribosomes by attachment-detachment of elongation factor G without GTP cleavage: Evidence from a column-bound ribosome system," *FEBS Letters* 54:35–38.
16. J. Modolell, T. Girbés, and D. Vázquez (1975), "Ribosomal translocation promoted by guanylylimido diphosphate and guanylyl-methylene diphosphonate," *FEBS Letters* 60:109–113.
17. T. Girbés, D. Vázquez, and J. Modolell (1976), "Polypeptide-chain elongation promoted by guanyl-5′-yl imidodiphosphate," *Eur. J. Biochem.* 67:257–265.

18. N. V. Belitsina, M. A. Glukhova, and A. S. Spirin (1976), "Stepwise elongation factor G-promoted elongation of polypeptides on the ribosome without GTP cleavage," *J. Mol. Biol.* 108:609–613.

19. J. W. Bodley, F. J. Zieve, and L. Lin (1970), "Studies on translocation. IV. The hydrolysis of a single round of GTP in the presence of fusidic acid," *J. Biol. Chem.* 245:5662–5667.

20. J. Thompson, E. Cundliffe, and M. Stark (1979), "Binding of thiostrepton to a complex of 23S rRNA with ribosomal protein L11," *Eur. J. Biochem.* 98:261–265.

21. J. Modolell and D. Vázquez (1977), "The inhibition of ribosomal translocation by viomycin," *Eur. J. Biochem.* 81:491–497.

22. S. Pestka (1969), "Studies on the formation of transfer ribonucleic acid-ribosome complex. VI. Oligopeptide synthesis and translocation on ribosomes in the presence and absence of soluble transfer factors," *J. Biol. Chem.* 244:1533–1539.

23. L. P. Gavrilova, O. E. Kostiashkina, V. E. Koteliansky, N. M. Rutkevitch, and A. S. Spirin (1976), "Factor-free ('non-enzymic') and factor-dependent systems of translation of polyuridylic acid by *Escherichia coli* ribosomes," *J. Mol. Biol.* 101:537–552.

24. N. M. Rutkevitch and L. P. Gavrilova (1982), "Factor-free and one-factor-promoted poly(U,C)-dependent synthesis of polypeptides in cell-free systems from *Escherichia coli*," *FEBS Letters* 143:115–118.

25. A. S. Spirin (1978), "Energetics of the ribosome," Progress in nucleic acid research and molecular biology, ed. W. E. Cohn, vol. 21, pp. 39–62 (New York: Academic Press).

26. A. J. M. Matzke, A. Barta, and E. Kuechler (1980), "Mechanism of translocation: Relative arrangement of tRNA and mRNA on the ribosome," *Proc. Nat. Acad. Sci. U.S.A.* 77:5110–5114.

27. N. V. Belitsina, G. Z. Tnalina, and A. S. Spirin (1981), "Template-free ribosomal synthesis of polylysine from lysyl-tRNA," *FEBS Letters* 131:289–292.

28. D. L. Riddle and J. Carbon (1973), "Frameshift suppression: A nucleotide addition in the anticodon of a glycine transfer RNA," *Nature New Biol.* 242:234.

29. M. R. Webb and J. F. Eccleston (1981), "The stereochemical course of the ribosome-dependent GTPase reaction of elongation factor G from *Escherichia coli*," *J. Biol. Chem.* 256:7734–7737.

30. A. B. Chetverin and A. S. Spirin (1982), "Bioenergetics and protein synthesis," *Biochim. Biophys. Acta* 683:153–179.

31. A. Rich (1974), "How transfer RNA may move inside the ribosome," in *Ribosomes*, ed. M. Nomura, A. Tissières, and P. Lengyel, pp. 871–884 (Cold Spring Harbor, N.Y.: Cold Spring Harbor Laboratory).

Further reading

Brot, N. (1977). Translocation. In *Molecular mechanisms of protein biosynthesis* (H. Weissbach and S. Pestka, eds.), pp. 375–411. New York: Academic Press.

Cundliffe, E. (1980). Antibiotics and prokaryotic ribosomes: Action, interaction, and resistance. In *Ribosomes: Structure, function, and genetics* (G. Chambliss, G. R. Craven, J. Davies, K. Davis, L. Kahan, and M. Nomura, eds.), pp. 555–581. Baltimore: University Park Press.

Hardesty, B.; Culp, W.; and McKeehan, W. (1969). The sequence of reactions leading to the synthesis of a peptide bond on reticulocyte ribosomes. *Cold Spring Harbor Symp. Quant. Biol.* 34:331–345.

Johnson, A. E.; Adkins, H. J.; Matthews, E. A.; and Cantor, C. R. (1982). Distance moved by transfer RNA during translocation from the A site to the P site on the ribosome. *J. Mol. Biol.* 156:113–140.

Kaziro, Y. (1978). The role of guanosine 5′-triphosphate in polypeptide chain elongation. *Biochim. Biophys. Acta* 505:95–127.

Kaziro, Y.; Inoue, N.; Kuriki, Y.; Mizumoto, K.; Tanaka, M.; and Kawakita, M. (1969). Purification and properties of factor G. *Cold Spring Harbor Symp. Quant. Biol.* 34:385–393.

Leder, P. (1973). The elongation reactions in protein synthesis. *Adv. Protein Chem.* 27:213–242.

Leder, P.; Bernardi, A.; Livingston, D.; Loyd, B.; Roufa, D.; and Skogerson, L. (1969). Protein biosynthesis: Studies using synthetic and viral mRNAs *Cold Spring Harbor Symp. Quant. Biol.* 34:411–417.

Lucas–Lenard, J.; Tao, P.; and Haenni, A.–L. (1969). Further studies on bacterial polypeptide elongation. *Cold Spring Harbor Symp. Quant. Biol.* 34:455–462.

Moldave, K.; Galasinski, W.; Rao, P.; and Siler, J. (1969). Studies on the peptidyl tRNA translocase from rat liver. *Cold Spring Harbor Symp. Quant. Biol.* 34:347–356.

Parmeggiani, A., and Gottschalk, E. M. (1969). Isolation and some properties of the amino acid polymerization factors from *Escherichia coli*. *Cold Spring Harbor Symp. Quant. Biol.* 34:377–384.

Pestka, S. (1969). Translocation, aminoacyl-oligonucleotides, and antibiotic action. *Cold Spring Harbor Symp. Quant. Biol.* 34:395–410.

Skoultchi, A.; Ono, Y.; Waterson, J.; and Lengyel, P. (1969). Peptide chain elongation. *Cold Spring Harbor Symp. Quant. Biol.* 34:437–454.

Spirin, A. S. (1969). A model of the functioning ribosome: Locking and unlocking of the ribosome subparticles. *Cold Spring Harbor Symp. Quant. Biol.* 34: 197–207.

Tocchini–Valentini, G. P.; Felicetti, L.; and Rinaldi, G. M. (1969). Mutants of *Escherichia coli* blocked in protein synthesis: Mutants with an altered G factor. *Cold Spring Harbor Symp. Quant. Biol.* 34:463–468.

Weissbach, H. (1980). Soluble factors in protein synthesis. In *Ribosomes: Structure, function, and genetics* (G. Chambliss, G. R. Craven, J. Davies, K. Davis, L. Kahan, and M. Nomura, eds.), pp. 377–411. Baltimore: University Park Press.

Chapter 15

Elongation IV: Rate Modulation

15-1 Discontinuities in elongation

It follows from the rate of reading-out the mRNA by ribosomes (chapter 4, section 4.4), that the time of each elongation cycle, including aminoacyl-tRNA binding, transpeptidation, and translocation, is about 0.06–0.1 sec (at 37°C). A frequency of about 10–15 cycles per second, however, can be considered only as an average value for the entire elongation. There is evidence that the elongation rate may be not constant during translation of an mRNA chain.

First of all, it was observed that during synthesis of the globin chains in rabbit reticulocytes the size distribution of nascent peptides was not continuous but rather displayed discrete size classes, thus suggesting some discontinuities in the process of translation (elongation)[1]. Similar results were obtained with the synthesis of bacteriophage MS2 coat protein in MS2-infected *Escherichia coli* cells[2]. Discontinuous translation was also observed in intact cells and in cell-free systems for a number of other proteins, e.g., silk fibroin[3], vitellogenin and serum albumin[4], tobacco mosaic virus proteins[5], encephalomyocarditis virus proteins[6], bacterial colicins[7], etc. All these observations have led to the conclusion that the ribosomes move along the mRNA chain during elongation with a nonuniform rate: from time to time, at a few specific sites of mRNA, they can slow down their advance or stop temporarily. In other words, more or less long pauses may occur during elongation, at least in some cases.

Three mechanisms, not necessarily excluding each other, can be proposed to explain these translational discontinuities (pauses)[6].

(1) The first explanation is that the ribosome can be retarded at the codons corresponding to minor (present in small amounts) isoacceptor tRNA species (or, less likely, to tRNA species possessing low efficiency of codon recognition and ribosome binding)[3,5]. Analyses of codon usage in different mRNAs demonstrate that the mRNAs coding for abundant cellular proteins selectively use synonymous codons corresponding to isoacceptor tRNA species which are present in the cell in relatively large amounts[8,9]. The synonymous codons corresponding to minor tRNA species are used rarely, if at all, in these mRNAs (e.g., in *Escherichia coli*, the arginine codons CGA and CGG, as well as AGA and AGG; the isoleucine codon AUA; the leucine codon CUA; the serine codons UCA and UCG; the proline codon CCC; the glycine codons GGA and GGG). It is possible that, when the translating ribosome encounters a rare codon within mRNA, it needs some time to wait for minor tRNA coming from the surrounding medium. In other words, low concentrations of some tRNA species may be a rate-limiting factor resulting in translational pauses at the corresponding codons.

The codons corresponding to minor tRNAs can be called *modulating codons*, as they supposedly have a modulating role, i.e., they regulate the rate of translation[10,11]. The more modulating codons are contained within an mRNA, the less expression of this mRNA is expected. At the same time, the cell can selectively change the effectivity of expression (translation) of a weakly expressed mRNA through *adaptation* of isoacceptor tRNA concentrations to codon frequencies in the particular mRNA. For instance, it is known that during the massive synthesis of fibroin in silk glands the intracellular pool of tRNA isoacceptors undergoes very significant changes that optimize the availability of tRNA species required for fibroin mRNA translation; in particular, the glycine, alanine, and serine isoacceptor tRNA species become predominant, in accordance with the predominance of the corresponding glycine, alanine and serine codons in the fibroin mRNA[12–14].

(2) The speed of ribosome movement along the mRNA chain may be uneven also because of different stability of secondary and tertiary structures at different regions of mRNA[2,6,7]. In particular, in order to have a stable double-stranded region of mRNA unwound and open for translation, the ribosome requires more waiting time than during

translation of other, less structured regions of mRNA. In such cases the discontinuities in elongation will arise, and the sites of the translational pauses will correspond to the entry of ribosomes at relatively stable double helices. This hypothesis is consistent with predictions for secondary structures of mRNAs encoding some discontinuously synthesized proteins[7].

In connection with the hypothesis under consideration, an observation made recently by Svitkin and Agol[6] is of special interest. They have demonstrated that during translation of encephalomyocarditis virus RNA a marked translational barrier causing a significant delay in the time of expression of the subsequent coding sequence exists at a specific site of the RNA. It is remarkable that this barrier can be overcome by addition of elongation factor 2 (EF-2), in excess over the catalytic amounts of EF-2 required for the maximal rates of elongation on the different RNA regions. Hence, the eucaryotic EF-2 possesses some regulatory functions in elongation, in addition to the catalysis of translocation. It is possible that this function is related to the RNA-binding capability of EF-2 (absent in the case of procaryotic EF-G)[15]; it is tempting to believe that the RNA-binding capability of EF-2 may serve for unwinding or destabilization of some structural barriers in mRNA, such as stable double-stranded helices or special tertiary interactions.

(3) Some regulatory proteins or small ribonucleoproteins may exist which interact with the translating ribosome and selectively stop or impede elongation at defined points. One example of such specific repressors of elongation in eucaryotes is known: this is a 7S RNA-containing ribonucleoprotein particle which recognizes a special N-terminal hydrophobic sequence of nascent polypeptide on the translating ribosome, attaches itself to the ribosomes, and stops further elongation, until the ribosome interacts with the endoplasmic reticulum membrane (see chapter 19, section 19.2). It cannot be excluded that similar mechanisms are used for modulating the elongation rate at some other phases of protein synthesis, e.g., at some steps of protein folding or protein assembly on the translating ribosome.

15-2 Selective regulation of elongation rate on different mRNAs

In eucaryotes differential changes of elongation rates on different mRNAs seem to be possible during cell development, as well as in response to hormone treatment or environmental influences. One of the examples is the different elongation rates on two populations of mRNAs after heat shock in *Drosophila* cells: elongation on the abundant preheat-shock species becomes slowed down, while it is fast on the heat shock mRNA species[16].

Another example of regulated differential changes in the elongation rates has been provided by Ilan and associates[17]: injection of estrogen into chickens results in the induction of vitellogenin synthesis in the liver with the initial elongation rate being equal to about 9 residues per second; meanwhile, the elongation rate of total liver proteins decreases from 7 to 4.5 residues per second; 2 days after the injection, and thereafter, the elongation rate of all the liver proteins becomes equal to 2 to 3 residues per second.

The mechanisms of the selective regulation of elongation rates on different mRNAs in eucaryotes are unknown. Some of the mechanisms discussed above (section 15.1) may be involved here also; in particular, possible changes in the isoacceptor tRNA population, as well as changes in concentrations of elongation factors and aminoacyl-tRNA synthetases, could contribute to the selective regulation of elongation. Of course, the direct participation of hypothetical repressors of elongation, e.g., proteins capable of binding to coding regions of mRNA, cannot be excluded either.

15-3 Total regulation of elongation rate

Slowing-down of elongation upon virus infections

Virus infection of the eucaryotic cell often leads to a decline in the rate of elongation of all polypeptides to be synthesized on host ribosomes. This phenomenon of total slowing-down of elongation is especially typical for picornavirus infections; it has been reported for poliovirus-infected HeLa cells[18], mengovirus-infected mouse Ehrlich ascites tumor cells[19], and encephalomyocarditis virus-infected L cells[20]. It has been concluded that picornaviral infection induces a decrease in the efficiency of ribosomes in elongation. There is an indication that the decrease in elongation activity is caused by some alterations in the pH 5 fraction[19] which contains aminoacyl-tRNA synthetases, tRNAs, and elongation factors.

Inhibition of elongation has been also reported for vaccinia virus-infected cells, and this effect can be reproduced *in vitro*, in a cell-free system, by addition of vaccinia virus cores[21]. This observation suggests the direct effect of a virus component on the elongation machinery participants.

Total repression of elongation by bacterial and plant toxins

Several protein toxins of bacterial and plant origin have turned out to be powerful inhibitors of the eucaryotic protein-synthesizing systems. It is the elongation phase of translation that is blocked by these toxins. All of these toxins possess the catalytic (enzymatic) mechanism of action. Their action seems to be aimed primarily at the stage of translocation in the elongation cycle of the eucaryotic ribosome. Of these toxins, the diphtheria toxin has been studied in the greatest detail[22–24].

Diphtheria toxin, a protein with a molecular mass of about 60,000 daltons, is secreted by *Corynebacterium diphtheriae* cells carrying a lysogenic bacteriophage β; the protein is encoded by one of the phage genes and not by the bacterial genome. The molecule of this protein is a covalently continuous polypeptide chain arranged in at least two relatively independent globular domains (A and B); the domains are additionally connected by a disulfide bridge:

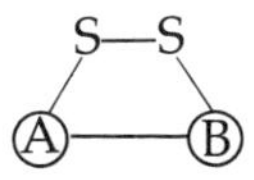

S—S
Ⓐ——Ⓑ

The C-terminal domain B, with a molecular mass of about 39,000 daltons, has a lectinlike action: it is capable of specific binding with an unidentified surface receptor of animal cells. The binding of the protein to the cell surface results in the following series of events. The protein enters the cytoplasmic membrane by a mechanism whose nature remains unclear and, after its entry, the interdomain peptide bond is proteolytically cleaved, concomitantly with the disulfide bridge reduction. As a consequence, two fragments—A and B—are formed from the original protein. The N-terminal fragment A, with a molecular mass of 21,150 daltons, then enters the cytoplasm. It is this fragment that serves as an inhibitor of protein synthesis in the cell. The fragment is a highly specific enzyme performing ADP-ribosylation of just one amino acid residue in EF-2. Such a modification impairs the normal functioning of EF-2. Since fragment A possesses catalytic action, one toxin molecule is sufficient to modify all the EF-2 molecules and therefore to kill the cell.

It should be pointed out that the original molecules of diphtheria toxin do not have such an inhibitory effect on protein synthesis; the initial toxin is a zymogen, which is converted into a catalytically active protein (A-fragment) only after its cleavage. On the other hand, fragment A by itself does not possess cytotoxic action since it cannot penetrate the intact cell.

Nicotinamide adenine dinucleotide (NAD) serves as the donor of the ADP-ribose residue, which is enzymatically transferred to EF-2:

$$\text{EF-2} + \text{NAD} \xrightleftharpoons{\text{Fragment A}} \text{ADP-ribosyl-EF-2} + \text{nicotinamide}.$$

The reaction is reversible and, *in vitro*, in conditions of nicotinamide excess, the protein can be de-ADP-ribosylated, the intact EF-2 thereby being recovered.

The aminoacyl residue onto which the ADP-ribose is transferred is a unique histidine derivative, the so-called diphtamide:

IX

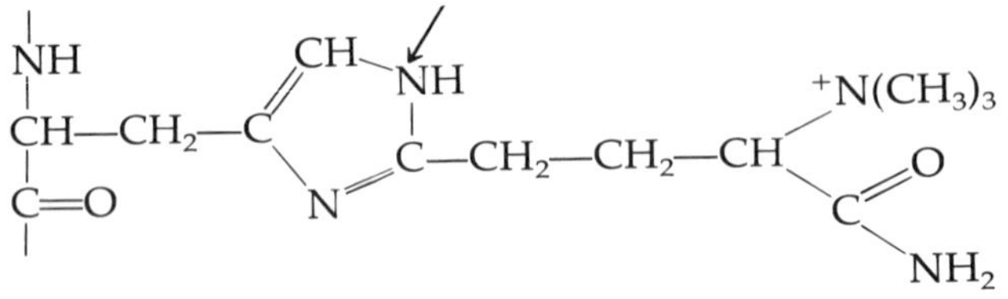

NH — CH — CH_2 — C (imidazole ring: CH, NH ↙, C, N) — C — CH_2 — CH_2 — CH($^+N(CH_3)_3$) — C(=O)NH_2

C=O

The ADP-ribosyl residue is transferred to the nitrogen of the imidazole ring designated by an arrow.

Although many details about the conversion of diphtheria toxin into the enzymatically active A-fragment and the enzymatic action of the A-fragment on EF-2 are known, the mechanism of the resulting protein synthesis inactivation continues to be a riddle. Surprisingly, ADP-ribosylation does not lead, as might be expected, to inactivation of the known partial functions of the factor. After ADP-ribosylation, the EF-2 continues to be capable of interacting with GTP as an effector, and the ADPR-EF-2·GTP complex binds with the ribosome, just as the original EF-2·GTP does. The binding occurs at the same ribosomal site. Furthermore, this binding may result in the translocation of the peptidyl-tRNA from the A-site to the P-site and may induce the release of deacylated tRNA. The interaction of the modified factor plus GTP with the ribosome is accompanied by GTPase activity. It has been pointed out in a number of studies, however, that, in the case of ADP-ribosylated EF-2, the GTPase activity may be lowered and, under these conditions, the promotion of translocation requires not catalytic but stoichiometric amounts of the factor with respect to ribosomes. It has been suggested that inhibition of the protein synthesis due to ADP-ribosylation of EF-2 is the result either of the decreased affinity of a modified EF-2 to the factor-binding site of the ribosome, or of the decreased rate of EF-2 reutilization (recycling), particularly because of a low dissociation rate of the ribosome·ADPR-EF-2·GDP complex. It remains to be determined, however, why ADP-ribosylation of EF-2 results in complete arrest of the protein synthesis while the partial functions of EF-2 are affected only to some extent, if at all.

There is one property of EF-2 that is not related directly to promoting translocation, and which is lost after ADP-ribosylation. This property is a nonspecific RNA-binding capacity, which is characteristic of the eucaryotic elongation factors (just like eucaryotic aminoacyl-tRNA synthetetases; see chapter 3, section 3.3) and distinguishes them from their procaryotic counterparts[15]. Due to this property, a considerable amount of EF-2, just as of the EF-1 of the eucaryotic cell, is compartmentalized around the polyribosomes being in a loose and reversible association with their RNA (perhaps with both mRNA and ribosomal RNA). ADP-ribosylation of EF-2 results in the disappearance of the nonspecific RNA-binding capacity of the factor and to its complete decompartmentalization from polyribosomal structures[25]. It may well be that this effect resulting in a drastic reduction of the local concentration of the factor in the vicinity of the polyribosomes is responsible at least in part for the inhibition of protein synthesis caused by the diphtheria toxin A-fragment.

Exotoxin A of *Pseudomonas aeruginoza*[24] has an action mechanism similar to that of diphtheria toxin. This protein, with a molecular mass of 71,500 daltons, also interacts with the surface of the eucaryotic cell through its lectin domain. After the protein enters the membrane, it is cleaved to yield fragments A and B, with molecular masses of 27,000 daltons and 45,000 daltons, respectively, and then fragment A passes into the cytoplasm. Fragment A is an enzyme transferring the ADP-ribose residue from NAD to the same diphtamide residue of EF-2 that was discussed above. This results in inhibition of protein synthesis. There is, however, no immunological cross-reactivity between *Pseudomonas aeruginoza* toxin and diphtheria toxin; in addition, receptors for these two toxins on the cell membrane are different.

The toxin produced by *Shigella dysenteriae* is also a powerful inhibitor of protein synthesis in the cells of a number of vertebrate animals. This protein toxin consists of one polypeptide A-chain, with a molecular mass of 30,500 daltons, and six (or seven) relatively short B-chains, with a molecular mass of about 5000 daltons each (A_1B_6). The B-moiety of the protein appears to be responsible for the interaction of the toxin with the cytoplasmic membrane receptor of the animal cell and for the subsequent entry of the toxin into the membrane.

Proteolytic cleavage of the toxin A chain in the membrane yields two fragments, A_1 (molecular mass 27,500 daltons) and A_2 (molecular mass 3000 daltons); the A_1-fragment passes into the cytoplasm and inhibits the protein synthesis. The inhibition results from the enzymatic activity of fragment A_1. (It should be pointed out that enzymatic activity is not displayed prior to cleavage.) The enzymatic action in this case is targeted to the 60S subunit of the eucaryotic 80S ribosome. Although the true nature of the enzymatic modification is unclear, the process requires neither NAD nor other cofactors. Ribosomes with modified 60S subunits are capable of performing transpeptidation; in other words, the peptidyl transferase center is not impaired. It is assumed that the inhibition of protein synthesis by the A_1-fragment of the toxin is due to some impairment in the function of aminoacyl-tRNA binding or translocation; it is possible that the factor-binding site on the 60S ribosomal subunit is somehow damaged by the toxin.

A number of powerful toxins that inhibit protein synthesis in target cells are found among plant lectins specifically interacting with the D-galactose residues of the glycoproteins present in the cell membrane of animal cells[26,27]. These toxins include ricin from castor beans, *Ricinus communis*; abrin from *Abrus precatorius*; modeccin from *Modecca digitata*; and viscumin from the mistletoe, *Viscum album*. These proteins show a striking similarity to the bacterial toxins, as regards their molecular-functional organization.

Ricin is a two-subunit protein (glycoprotein) with a molecular mass of 62,000 daltons. The B-subunit (molecular mass 31,400 daltons) is a lectin in the strict sense of the word and is capable of binding with galactose residues on the external surface of the animal cell membrane. The A-subunit (molecular mass 30,000 daltons) has unidentified enzymatic activity and is responsible for the inhibition of protein synthesis in the cytoplasm. The two subunits are linked by a disulfide bridge. The attachment of the toxin molecule to the membrane is followed by entry of the molecule into the membrane, disulfide bridge reduction, and release of the liberated A-subunit into the cytoplasm. The enzymatic activity of the A-subunit does not appear to require either NAD or any other cofactors, its action being assumed to involve some hydrolytic activity. The A-subunit affects 60S ribosomal particles. The nature of the modification of these particles remains to be determined; however, there is indirect evidence that some protein component of 60S subunits is impaired by the enzymatic subunit of the toxin. It may well be that the modification of the 60S subunit affects the function of its interaction with the elongation factors. It has been reported that the main result of toxin action is inhibition of EF-2-catalyzed translocation[28].

Other plant toxins are organized in a similar way and have similar action, although chemically they are different proteins. The so-called pokeweed antiviral protein (PAP) from *Phytalacca americana* is particularly interesting. Its molecular mass is 27,000 daltons and it is an analog of the A-subunit of ricin. Correspondingly, it does not possess lectin activity and cannot interact with the cell membrane. Therefore it does not affect intact cells but strongly inhibits protein synthesis *in vitro* in eucaryotic cell-free systems. PAP is an enzyme modifying the 60S ribosomal subunit (probably one of the ribosomal proteins). The main result of this modification is the inhibition of EF-2-catalyzed translocation[28], although it is also possible that the functional interaction of the modified ribosome with both elongation factors, EF-2 and EF-1, can be impaired.

Thus, various protein toxins of bacterial and plant origin make use of the same principle of cytotoxic action based on a two-subunit or two-domain structure: one subunit (or fragment) interacts with the membrane and is responsible for the transmembrane transport, while the other is released into the cell and exhibits an enzymatic activity there, resulting in the inhibitory modification of some components of the protein-synthesizing (or some other) system. This principle observed in living nature may be exploited to deliver any enzyme protein inside the cell, if such a protein is artificially conjugated or crosslinked with a suitable membrane-interacting protein[29,30].

Even the first experiments yielded promising results. Using a simple procedure of disulfide exchange, it was possible to conjugate the enzymatic A-fragment of diphtheria toxin or the A-subunit of ricin with a nontoxic plant lectin (e.g. with concanavalin A or lectin of *Wistaria floribunda*) and to obtain a cytotoxic effect; it is clear that the lectin moiety of the chimeric protein was responsible for the delivery of the inhibitory component into the cell. However, just as in the case of the original toxins, the effect was not tissue specific. High tissue specificity of the chimeric toxin can be obtained if the A-fragment of diphtheria toxin or the A-subunit of ricin is conjugated with a peptide hormone interacting with the specific receptor of the cell membrane of a given type (e.g. with the chorionic gonadotropin, or epidermal growth factor, or insulin). In such cases, the enzymatic component is delivered into the cell, inhibits protein synthesis, and kills the cell. Furthermore, the membrane-interacting component may be an antibody (or its Fab-fragment) against some surface antigen, which is specific for a membrane of only one cell type. Then, by conjugating the diphtheria toxin A-fragment or ricin A-subunit with such an antibody, an extremely tissue-specific chimeric toxin can be obtained which will selectively kill only specific target cells. When an antibody against the surface antigen of a tumor cell serves as the membrane-binding moiety of the chimeric toxin, such a toxin should selectively kill tumor cells without affecting other cell types[31].

Regulation of the elongation rate through endogenous modifications of elongation factors

Studies of the protein toxins exerting enzymatic activities demonstrate that they have very specific targets. The best studied case is that of the diphtheria toxin. The target of the diphtheria toxin is a unique amino acid residue, diphthamide, in elongation factor 2 (see above). This residue in EF-2 is surprisingly conservative in evolution, being universal among eucaryotes including animals, plants, and fungi and occurring even in archaebacteria. At the same time, diphthamide seems to be not vital for EF-2 functioning in translocation: viable toxin-resistant mutants of cultured animal cells have been obtained which lost this modification of histidine in EF-2 without diminishing their protein-synthesizing activity[32]. It is unlikely that diphthamide in EF-2 has originated and been conserved in evolution of eucaryotes just for the purpose of being the target for toxins. It may be rather thought that diphthamide serves as a specific target for some endogenous (intracellular) regulatory effects at the level of EF-2, and some bacteria just use the same target for intoxication of eucaryotic cells.

Indeed, endogenous ADP-ribosyltransferase which specifically modifies the diphthamide residue in EF-2 has been recently discovered in normal mammalian cells[33–35]. The enzyme seems to be associated with polyribosomes[35], i.e., present in the same cellular compartment that comprises also elongation factors (in the case of the eucaryotic cell)[36]. The function of the endogenous ADP-ribosylation of EF-2 may consist in affecting an activity of EF-2 different from the catalysis of translocation. It is known that ADP-ribosylation of EF-2 by diphtheria toxin results in the loss of the nonspecific RNA-binding capability of EF-2 and, consequently, the removal of an excess EF-2 from polyribosomes (decompartmentation)[25]. Endogenous ADP-ribosylation might regulate the degree of compartmentation of EF-2 on polyribosomes of the eucaryotic cell. The loose association of EF-2 with polyribosomes may have several aims; first, it maintains an increased local concentration of EF-2 near the sites of its functioning, which is important in the large volume of the eucaryotic cell[36]; second, an excess EF-2 reversibly bound to mRNA may have a special function in overcoming the translational barriers, e.g., stable secondary structure[6]. Thus, the elongation rate could be modulated by regulation of compartmentation of EF-2 on polyribosomes through the mechanism of endogenous ADP-ribosylation.

It is possible that an analogous regulation of the degree of compartmentation on polyribosomes exists also in the case of the other elongation factor, EF-1. It has been recently found that the polyribosomal fraction of the eucaryotic cell contains a latent phosphokinase which under certain conditions can be activated and phosphorylates specifically the α-subunit of EF-1; as a result, EF-1 loses its non-specific affinity for high molecular mass RNA and leaves polyribosomes[37]. It cannot be excluded that the phosphorylation of EF-1α may affect the rate of elongation and serve for regulation of the translational process in the cell.

References

1. A. Protzel and A. J. Morris (1974), "Gel chromatographic analysis of nascent globin chains. Evident of nonuniform size distribution," *J. Biol. Chem.* 249:4600.
2. W. G. Chaney and A. J. Morris (1979), "Nonuniform size distribution of nascent peptides. The effect of messenger RNA structure upon the rate of translation," *Arch. Biochem. Biophys.* 194:283–291.
3. P. M. Lizardi, V. Mahdavi, D. Shields, and G. Candelas (1979), "Discontinuous translation of silk fibroin in a reticulocyte cell-free system and in intact silk gland cells," *Proc. Nat. Acad. Sci. U.S.A.* 76:6211–6215.
4. B. Wieringa, J. van der Zwaag–Gerritsen, J. Mulder, G. Ab, and M. Gruber (1981), "Translation *in vivo* and *in vitro* of mRNAs coding for vitellogenin, serum albumin and very-low-density lipoprotein II from chicken liver. A difference in translational efficiency," *Eur. J. Biochem.* 114:635–641.
5. A. K. Abraham and A. Pihl (1980), "Variable rate of polypeptide chain elongation *in vitro*. Effect of spermidine", *Eur. J. Biochem.* 106:257–262.
6. Yu. V. Svitkin and V. I. Agol (1983), "Translational barrier in central region of encephalomyocarditis virus genome. Modulation by elongation factor 2 (eEF-2)," *Eur. J. Biochem.* 133:145–154.
7. S. Varenne, M. Knibiehler, D. Cavard, J. Morlon, and C. Lazdunski (1982), "Variable rate of polypeptide chain elongation for colicins A, E2, and E3," *J. Mol. Biol.* 159:57–70.
8. T. Ikemura (1981), "Correlation between the abundance of *Escherichia coli* transfer RNAs and the occurrence of the respective codons in its protein genes," *J. Mol. Biol.* 146:1–21.
9. M. Gouy and C. Gautier (1982), "Codon usage in bacteria: correlation with gene expressivity," *Nucleic Acids Res.* 10:7055–7074.
10. W. Fiers, R. Contreras, F. Duerinck, G. Haegeman, D. Iserentant, J. Merregaert, W. Min Jou, F. Molemans, A. Raeymaekers, A. Van den

Berghe, G. Volckaert, and M. Ysebaert (1976), "Complete nucleotide sequence of bacteriphage MS2 RNA: primary and secondary structure of the replicase gene," *Nature* 260:500–507.

11. H. Grosjean and W. Fiers (1982), "Preferential codon usage in prokaryotic genes: the optimal codon-anticodon interaction energy and the selective codon usage in efficiently expressed genes," *Gene* 18:199–209.
12. J.-P. Garel (1976), "Quantitative adaptation of isoacceptor tRNAs to mRNA codons of alanine, glycine, and serine," *Nature* 260:805–806.
13. J.-P. Garel, R. L. Garber, and M. A. Q. Siddiqui (1977), "Transfer RNA in posterior silk gland of *Bombyx mori*: polyacrylamide gel mapping of mature transfer RNA, identification and partial structural characterization of major isoacceptor species," *Biochemistry* 16:3618–3624.
14. G. Chavaney, A. Chevallier, M. Fournier, and J. P. Garel (1979), "Adaptation of iso-tRNA concentrations to mRNA codon frequency in the eukaryotic cell," *Biochimie* 61:71–78.
15. S. P. Domogatsky, T. N. Vlasik, T. A. Seryakova, L. P. Ovchinnikov, and A. S. Spirin (1978), "Difference in RNA-binding ability between eukaryotic and prokaryotic elongation factors of translation," *FEBS Letters* 96:207–210.
16. D. G. Ballinger and M. L. Pardue (1983), "The control of protein synthesis during heat shock in Drosophila cells involves altered polypeptide elongation rates," *Cell* 33:103–114.
17. L. Gehrke, R. E. Bast, and J. Ilan (1981), "An analysis of rates of polypeptide chain elongation in avian liver explants following *in vivo* estrogen treatment. II. Determination of the specific rates of elongation of serum albumin and vitellogenin nascent chains," *J. Biol. Chem.* 256:2522–2530.
18. D. F. Summers, J. V. Maizel, and J. E. Darnell (1967), "Decrease in size and synthetic activity of poliovirus polysomes late in the infectious cycle," *Virology* 31:427–435.
19. P. B. Hackett, E. Egberts, and P. Traub (1978), "Translation of ascites and mengovirus RNA in fractionated cell free systems from uninfected and mengovirus-infected Ehrlich-ascites-tumor cells," *Eur. J. Biochem.* 83:341–352.
20. T. V. Ramabhadran and R. E. Thach (1981), "Translational elongation rate changes in encephalomyocarditis virus-infected and interferon-treated cells," *J. Virol.* 39:573–583.
21. P. W. L. Tas and O. H. W. Martini (1983), Inhibition of the elongation step of protein synthesis by vaccinia virus," *FEBS Letters* 153:427–430.
22. R. J. Collier (1975), "Diphtheria toxin: mode of action and structure," *Bact. Rev.* 39:54–85.
23. A. M. Pappenheimer (1977), "Diphtheria toxin," *Ann. Rev. Biochem.* 46:69–94.

24. S. van Heyningen (1980), "ADP-ribosylation by bacterial toxins," The enzymology of post-translational modification of proteins, ed. R. B. Freedman and H. C. Hawkins, vol. 1, pp. 387–422 (London and New York: Academic Press).

25. A. S. Sitikov, E. K. Davydova, J. A. Bezlepkina, L. P. Ovchinnikov, and A. S. Spirin (1984), "Eukaryotic elongation factor EF-2 loses its nonspecific affinity for RNA and leaves polyribosomes as a result of ADP-ribosylation," *FEBS Letters* 176:406–410.

26. S. Olsnes and A. Phil (1976), "Abrin, ricin, and their associated agglutinins," in *Receptors and recognition. Series B: The specificity and action of animal, bacterial, and plant toxins*, ed. P. Cuatrecasas, pp. 130–173 (London: Chapman and Hall).

27. S. Olsnes and A. Pihl (1982), "Toxic lectins and related proteins," in *Molecular actions of toxins and viruses*, ed. P. Cohen and S. van Heyningen, pp. 51–105 (Amsterdam: Elsevier, North-Holland).

28. S. L. Gessner and J. D. Irvin (1980), "Inhibition of elongation factor 2-dependent translocation by the pokeweed antiviral protein and ricin," *J. Biol. Chem.* 255:3251–3253.

29. S. Olsnes (1981), "Directing toxins to cancer cells," *Nature* 290:84.

30. S. Olsnes and A. Pihl (1982), "Chimeric toxins, *Pharmac. Ther.* 15:355–381.

31. K. A. Krolick, C. Villemez, P. Isakson, J. W. Uhr, and E. S. Vitetta (1980), "Selective killing of normal or neoplastic B cells by antibodies coupled to the A chain of ricin," *Proc. Nat. Acad. Sci. U.S.A.* 77:5419–5423.

32. J. M. Moehring, T. J. Moehring, and D. E. Danley (1980), "Posttranslational modification of elongation factor 2 in diphtheria-toxin-resistant mutants of CHO-K1 cells," *Proc. Nat. Acad. Sci. U.S.A.* 77:1010–1014.

33. H. Lee and W. J. Iglewski (1984), "Cellular ADP-ribosyl-transferase with the same mechanism of action as diphtheria toxin and *Pseudomonas* toxin A," *Proc. Nat. Acad. Sci. U.S.A.* 81:2703–2707.

34. W. J. Iglewski, H. Lee, and P. Muller (1984), "ADP-ribosyl-transferase from beef liver which ADP-ribosylates elongation factor-2," *FEBS Letters* 173:113–118.

35. A. S. Sitikov, E. K. Davydova, and L. P. Ovchinnikov (1984), "Endogenous ADP-ribosylation of elongation factor 2 in polyribosome fraction of rabbit reticulocytes," *FEBS Letters* 176:261–263.

36. A. S. Spirin and L. P. Ovchinnikov (1984), "Compartmentation of proteins of the translation machinery on eukaryotic polyribosomes," in *Progress in bioorganic chemistry and molecular biology*, ed. Yu. A. Ovchinnikov, pp. 71–82 (Amsterdam and London: Elsevier Science Publishers).

37. E. K. Davydova, A. S. Sitikov, and L. P. Ovchinnikov (1984), "Phosphorylation of elongation factor 1 in polyribosome fraction of rabbit reticulocytes," *FEBS Letters* 176:401–405.

Chapter 16

Initiation of Translation and Its Regulation in Procaryotes

16-1 Significance of initiation

The initiation of protein biosynthesis does not simply imply the beginning of elongation. First of all, since the start of the mRNA-coding sequence does not coincide with the 5′-end of the polynucleotide chain but is removed from the 5′-end, sometimes by a significant distance, a precise recognition of the first codon located internally in the chain is required. Such a recognition not only defines the beginning of the polypeptide chain to be synthesized but also sets a frame for the subsequent triplet-by-triplet readout of mRNA. It follows that this process is vital for the entire amino acid sequence of the polypeptide. Briefly, it is the initiation that determines the fixed point on the template polynucleotide, from which triplet-by-triplet, comma-free translation begins (see chapter 2).

Furthermore, initiation constitutes the principal step at which protein synthesis is controlled at the translational level. Regulation at the translational level in procaryotes may be discussed entirely in terms of either permitting or preventing the initiation of the mRNA-coding sequence readout by ribosomes (see below). Selective or preferential translation of certain mRNA species or mRNA cistrons and translational inactivation of other mRNA species or cistrons are achieved precisely in this way. In addition, differential rates of initiation with different mRNAs (or different cistrons) determine the ratio of production of corresponding proteins.

In accordance with the above, a special complex mechanism for initiation of translation provides both for the precise recognition of the coding sequence starting point, implying the phasing of elongation, and for the action of positive or negative regulatory signals.

16-2 Initiation codons, initiator tRNA, and protein initiation factors

Initiation codons. Translation of natural mRNA or its individual cistrons in the case of polycistronic mRNAs, usually begins at the AUG triplet or, far less frequently, at the GUG or UUG triplet[1–7]. These triplets can serve as the initiation codons of the template. In some rarer cases AUU and AUA triplets may also serve as initiation codons. The initiation codon provides an initial point of triplet readout along the template toward its 3′-end.

Initiator tRNA. Initiation codons are recognized by a special initiator tRNA[8]. The anticodon of this tRNA is CAU and it is capable of pairing with the initiation codon AUG as well as with GUG, UUG, AUU, or AUA in some cases. It is apparent that unusual noncanonical pairing can occur here either at the first position of the codon (G·U or U·U), or, in rare cases, at the third position of the codon (U·C or A·C); in all cases, however, this pairing differs from Crick wobbling at the third position of codons during elongation (see chapter 12, section 12.1). This may be explained by the fact that codon-anticodon interaction during initiation takes place in the ribosomal P-site, but not in the A-site as in the course of elongation[9] (see below).

The structure of the initiator tRNA is organized similarly to that of the usual (elongator) tRNAs, although there are certain differences[10–12]. Thus, the 5′-terminal nucleotide residue of the initiator tRNA is not paired with the opposite nucleotide residue of the 3′-terminal region (fig. 113). It appears to provide greater flexibility for the acceptor end of the initiator tRNA; X-ray analysis of the initiator tRNA crystals has demonstrated that the 3′-end can curl back toward the 5′-end and

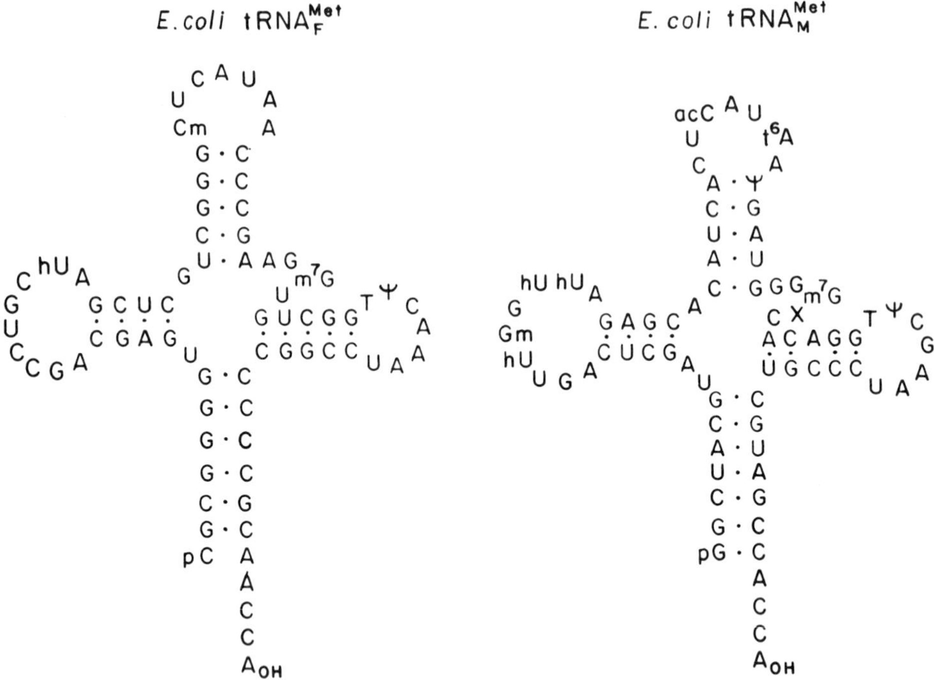

Figure 113 Nucleotide sequence and the secondary structure of the procaryotic initiator tRNA (tRNA$_F^{Met}$, *left*)[10] compared with the procaryotic methionine tRNA participating in elongation (tRNA$_M^{Met}$, *right*)[11].

does not continue the helical organization of the acceptor stem as in the case of the usual tRNA species. Moreover, the nucleotide residues of the dihydrouridylic loop of the initiator tRNA (positions 16 and 17) are more tightly packed together and with the core of tRNA, compared to other tRNA species where they appear to be loosely accommodated near the corner of the L-shaped molecule. An interesting difference has also been found in the anticodon loop: although the stacking structure of the anticodon and the two purine residues adjacent to it from the 3′-side is very similar in the initiator tRNA and other tRNA species, in the case of the initiator tRNA the invariant U in position 33 adjacent to the anticodon from the 5′-side is turned outside (whereas it is turned inside the loop and faces the phosphate of the third anticodon residue in usual tRNAs). For this reason the structure of the anticodon in the initiator tRNA looks somewhat distorted compared to the anticodons of other tRNAs.

The initiator tRNA has an affinity to the usual methionyl-tRNA synthetase and, correspondingly, is capable of accepting methionine. Therefore, two classes of tRNA are acylated by methionine in the cell:

the usual $tRNA^{Met}$ recognizing the methionine codon AUG in the course of elongation; and the initiator tRNA which can recognize the AUG triplet, as well as the GUG and UUG (and sometimes AUU and AUA) triplets, during initiation. In contrast to Met-$tRNA^{Met}$, the initiator Met-tRNA serves as a substrate for a special formyl transferase which transfers the formyl group from the formyl tetrahydrofolate to the amino group of the methionine residue, yielding formylmethionyl-tRNA[13]:

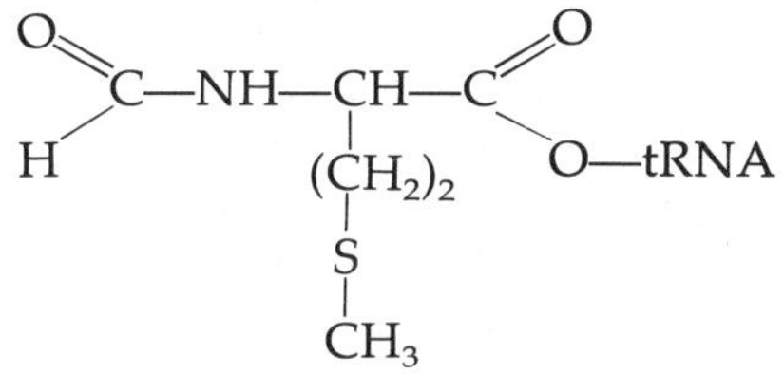

$$\mathrm{H{-}C({=}O){-}NH{-}CH[(CH_2)_2{-}S{-}CH_3]{-}C({=}O){-}O{-}tRNA}$$

Therefore, the initiator tRNA, in contrast to $tRNA^{Met}$, is usually designated by the letter *F*: $tRNA_F^{Met}$. In its aminoacylated and formylated state (F-Met-$tRNA_F^{Met}$) it plays a part in initiation. Correspondingly, formyl methionine is always the first amino acid residue of any polypeptide chain to be synthesized by the procaryotic ribosome[14,15]. During subsequent elongation the formyl residue is cleaved off by formylase; the methionine residue is often, although not always, cleaved from the growing polypeptide chain by a special aminopeptidase[16,17].

Protein initiation factors. Three proteins are required for initiation in procaryotic systems[18–22]; these are referred to as initiation factors IF-1, IF-2, and IF-3 (see reviews by Revel (1972, 1977), and Grunberg-Manago and Gros (1977)).

IF-1 is a small basic protein with a molecular mass of about 9000 daltons. This factor has not been found in some bacterial species.

In contrast, IF-2 is a large acidic protein possessing an SH-group important for the function. This protein is the principal initiation factor. It has been isolated in two forms differing somewhat in molecular mass; one of these, IF-2a, has a molecular mass of about 100,000 daltons whereas the other, IF-2b, has a molecular mass of approximately 90,000 daltons. Both forms appear to function equally in initiation. IF-2 has an affinity to GTP and forms an unstable complex with it. IF-2 and GTP can interact with the F-Met-tRNA and with the 30S ribosomal subunit. GTP in this process can be substituted for by its nonhydrolyzable analogs.

IF-3 is a slightly basic protein with a molecular mass of about 21,000 to 23,000 daltons. Two functionally equivalent forms have been described: one has a polypeptide chain 181 aminoacyl residues long, and the other lacks 6 amino acids at the N-terminus.

16-3 State of ribosomes before initiation

Free nontranslating 70S ribosomes under ionic and temperature conditions favorable for initiation are in dynamic equilibrium with their 30S and 50S subunits (see chapter 10, section 10.1):

$$70S \rightleftarrows 30S + 50S.$$

It has been demonstrated that the 30S ribosomal subunit, rather than the 70S ribosome, participates in the initial associations with mRNA and the subsequent initiation of translation[23,24]. The 30S ribosomal subunit is taken from the equilibrium mixture by mRNA and then becomes involved in the first initiation stages. The 50S subunit attaches at later initiation stages to the 30S-mRNA initiation complex[25]. The initiation factors, e.g. IF-3 and apparently IF-1, contribute to the shift of equilibrium toward the dissociation of free nontranslating ribosomes into ribosomal subunits[26] (see below).

In the case of the polycistronic template, the ribosome after termination at the preceding cistron may pass to the nontranslating state but still remain in association with mRNA for a time. Then, it can either be released from mRNA into solution or pass directly to the initiation of translation at the next cistron without dissociating from mRNA. The probability of a given ribosome reinitiating at the next cistron seems to be determined by the length and structure of the intercistronic region: the chances of initiation without dissociation are greater with shorter, less structured intercistronic regions. In this case, only the 30S ribosomal subunits seem to remain associated with the template, while the 50S subunits after termination come to reversible equilibrium with the pool of free 50S subunits.

16-4 Association of the ribosome with the template polynucleotide

A number of structural requirements should be met for the complex between the free nontranslating ribosomal particle (30S subunit) and mRNA in the region of the initiation codon to be formed. First, this region of mRNA should be sufficiently exposed to the interaction; in any case, the initiation codon and adjacent sections should not be hidden in the stable secondary or tertiary structure of the template polynucleotide. Furthermore, some special conformation in the ribosome-binding region of the template may positively contribute to its recognition by the ribosomal particle.

The most apparent universal structural feature of the ribosome-binding regions of procaryotic mRNA is the polypurine nucleotide sequence (the so-called Shine–Dalgarno sequence)[27], located 3 to 10 nucleotides away from the initiation codon toward the 5′-end. This sequence preceding the initiation codon is complementary, to a greater or lesser extent, to the pyrimidine-rich 3′-terminal region of the ribosomal 16S RNA (. . . $GAUCACCUCCUUA_{OH}$ in *E. coli*). Between three and nine residues (usually four or five) may be complementary. In most cases it is the CCUCCU sequence of the 3′-terminal region of the 16S RNA (in *E. coli*) that is complementary (although usually only in part) to the polypurine preinitiation sequence of mRNA[28]. Several examples are given in fig. 114. It is generally assumed that this complementary pairing of the 3′-terminal region of 16S RNA with the preinitiation polypurine block of mRNA participates directly in the association between the free ribosomal 30S subunit and the template polynucleotide.

It should be pointed out, however, that the initiation of translation by free ribosomes is also possible at least to some extent in the case of templates lacking the Shine–Dalgarno sequence. For example, mRNA coding for the cI-repressor of the phage λ does not have such a sequence before the initiation codon; even so, initiation is observed, although it is not very effective[29]. Bacterial ribosomes may perform initiation with certain heterologous (eucaryotic) RNA devoid of the Shine–Dalgarno sequence, e.g. with RNA 4 of the brome mosiac virus; initiation requires all three bacterial initiation factors and takes place at the normal initiation codon[30]. It may be concluded that the Shine–Dalgarno sequence contributes greatly to the effective association of ribosomal particles with the template, but probably is not an absolute requirement.

If a ribosome reads a polycistronic mRNA and does not dissociate from the template after termination of a given cistron, then reinitiation of translation at the next cistron does not appear to require the Shine–Dalgarno sequence.

16-5 Sequence of events during initiation[31–38]

The dissociation of nontranslating 70S ribosomes into 30S and 50S subunits precedes the initiation of translation in all cases. It has

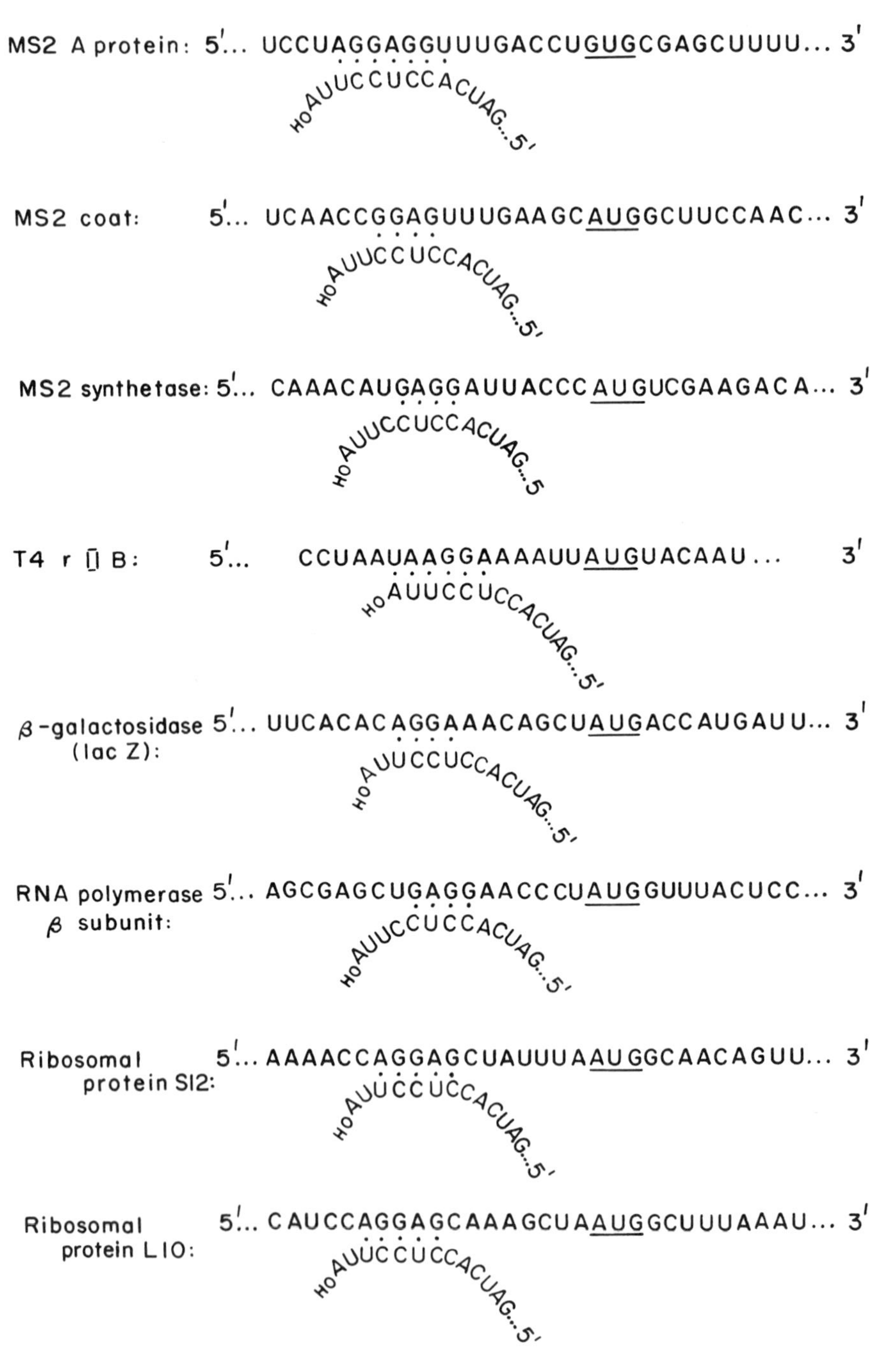

Figure 114 Nucleotide sequence preceding the initiation codon (AUG or GUG) in a number of mRNAs coding for phage and bacterial proteins and the pairing of their polypurine clusters (Shine–Dalgarno sequences) with the 3′-terminal polypyrimidine cluster of ribosomal 16S RNA.

already been mentioned that under physiological conditions, nontranslating ribosomes are in reversible equilibrium with their subunits. IF-3 has a strong affinity to the individual 30S subunit and, when bound to it, the subunit is prevented from reassociating with the 50S subunit; in this way the 30S subunit is removed from the equilibrium and prepared for initiation:

$$\begin{array}{l} 70S \rightleftarrows 30S + 50S \\ \qquad\quad\;\; + \\ \qquad\quad\; IF\text{-}3 \\ \qquad\quad\;\; \downarrow\uparrow \\ \qquad\quad 30S{\cdot}IF\text{-}3 \end{array}$$

Two other initiation factors, IF-1 and IF-2, also possess a degree of affinity to the 30S subunit and may bind to it:

$$30S{\cdot}IF\text{-}3 + IF\text{-}1 + IF\text{-}2 \rightleftarrows 30S{\cdot}IF\text{-}3{\cdot}IF\text{-}2{\cdot}IF\text{-}1.$$

The 30S ribosomal subunit with a either IF-3 or with all bound initiation factors can associate with the initiation region of mRNA. This region may be located near the 5′-end or far from it; in the case of polycistronic mRNA there may be several such regions along the mRNA chain. What is important is that this region be exposed to interaction, possess suitable conformation, and contain the preinitiation polypurine Shine–Dalgarno sequence as well as the initiation triplet AUG (or GUG, UUG, etc.). In the absence of initiation factors, the 30S ribosomal subunit itself is also capable of recognizing the initiation region of mRNA and binding to it. IF-3 appears to promote association between the 30S subunit and mRNA, perhaps by augmenting the complementary interaction of the 3′-end of the ribosomal RNA with the Shine-Dalgarno sequence or by contributing an additional interaction between the ribosomal subunit and the template.

The next stage of initiation is marked by the beginning of IF-2 action. This factor may either be already associated with the 30S ribosomal subunit or be in solution. It interacts with GTP and the initiator F-Met-$tRNA_F$, resulting in the cooperative formation of an initiation 30S complex and the release of IF-3:

$$\begin{aligned} &30S{\cdot}mRNA{\cdot}IF\text{-}3 + IF\text{-}2 + GTP + F\text{-}Met\text{-}tRNA_F \rightleftarrows \\ &\qquad\qquad 30S{\cdot}mRNA{\cdot}F\text{-}Met\text{-}tRNA_F{\cdot}IF\text{-}2{\cdot}GTP + IF\text{-}3. \end{aligned}$$

Here, just as in the case of elongation factors, GTP plays the part of effector, increasing the affinity of IF-2 to the ribosome (in this case to the 30S ribosomal subunit) and to aminoacylated tRNA (F-Met-$tRNA_F$). It appears that, initially, GTP interacts with IF-2 and then the complex IF-2·GTP binds F-Met-tRNA; F-Met-tRNA recognizes the initiation codon on the 30S subunit, and IF-2·GTP complexed both with the F-Met-tRNA and with the ribosomal particle provides for an additional interaction. The formation of this initiation 30S complex is not accompanied by GTP cleavage; as already indicated, GTP in this process may be substituted for by one of its noncleavable analogs, e.g. guanylyl methylene diphosphonate (GMP-PCP) or guanylyl imidodiphosphate (GMP-PNP).

There is, however, an alternative pathway of the initiation 30S complex formation. It has been demonstrated that the 30S ribosomal subunit, even without mRNA, possesses an intrinsic affinity to the initiator F-Met-tRNA, as well as to IF-2. This provides conditions suitable for an interaction between the F-Met-tRNA·IF-2·GTP complex and the 30S subunit in the absence of the template:

$$30S{\cdot}IF\text{-}3 + IF\text{-}2 + GTP + F\text{-}Met\text{-}tRNA_F \rightleftarrows 30S{\cdot}F\text{-}Met\text{-}tRNA_F{\cdot}IF\text{-}2{\cdot}GTP \quad (+IF\text{-}3?).$$

After this process has taken place the 30S subunit, together with F-Met-tRNA, specifically binds the initiation region and the initiation codon of mRNA. In this case the anticodon of F-Met-tRNA may contribute to the recognition of the initiation triplet on mRNA. Although, at present, most investigations of cell-free translation systems focus on the first of the two pathways, both pathways may coexist in the cell, and it may be that the second pathway is, in fact, preferable.

The two pathways of 30S initiation complex formation are presented schematically in fig. 115.

The part played by the third (and smallest) initiation factor, IF-1, in the course of initiation 30S complex formation has not yet been fully clarified. On one hand, there is evidence that this factor contributes to a higher dissociation rate of nontranslating 70S ribosomes into subunits. On the other hand, experiments have shown that it stabilizes the binding of the two other initiation factors, IF-3 and IF-2, with the 30S ribosomal subunit, being also bound itself in their presence (this is cooperative binding of the three initiation factors). IF-1 is present in the final initiation 30S complex, which contains mRNA and F-Met-tRNA, and appears to stabilize the complex, while IF-3 is released as a consequence of F-Met-tRNA binding.

The initiation 30S complex now has a strong affinity to the free 50S ribosomal subunit, and they undergo association, yielding the initiation 70S complex. As a result, F-Met-tRNA occupies the P-site of the ribosome. The IF-2 associated with F-Met-tRNA and the 30S ribosomal subunit interacts with the 50S subunit of the complex, directly contacting the region at the base of the L7/L12 stalk. This interaction induces GTPase activity and GTP is hydrolyzed to yield GDP and

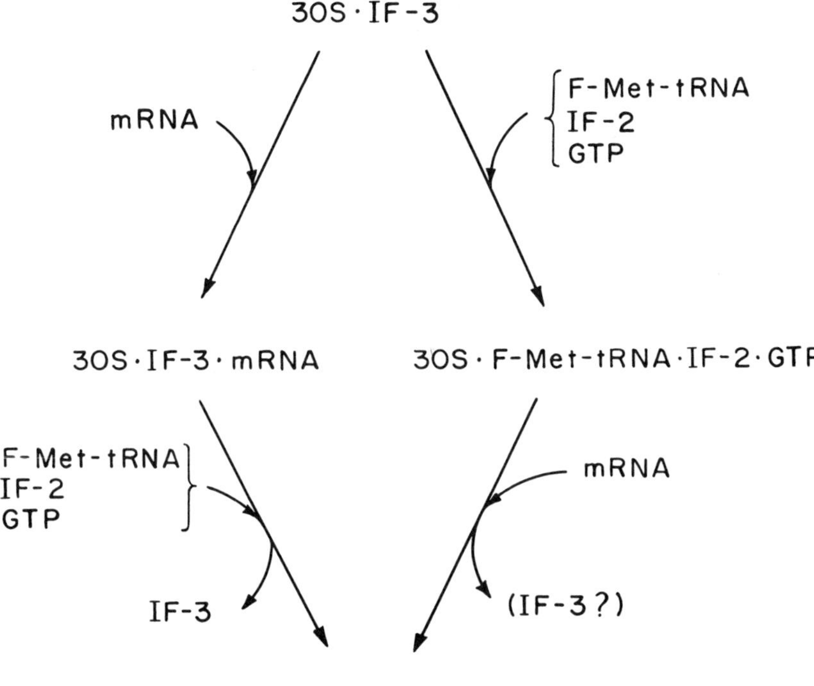

Figure 115 Two possible pathways of 30S initiation complex formation. *Left,* Initial association of the 30S ribosomal subunit with mRNA followed by the binding of the initiator tRNA. *Right,* Initial binding of the initiator tRNA with the 30S subunit followed by association with mRNA. Both pathways lead to the same final complex.

inorganic phosphate. As a consequence, the affinity of IF-2 to F-Met-tRNA and the ribosome is reduced drastically; IF-2 and GDP are released into solution. There is evidence that IF-1 promotes the release of IF-2 after GTP hydrolysis. The 70S ribosome remains in association with mRNA and with the initiator F-Met-tRNA at the P-site where the initiation codon is bound to the anticodon of the initiator tRNA. Thus:

$$\begin{aligned}&\text{30S·mRNA·F-Met-tRNA·IF-2·GTP·IF-1} + \text{50S} \rightleftarrows\\&\qquad \text{70S·mRNA·F-Met-tRNA·IF-2·GTP·IF-1} \rightarrow\\&\qquad \text{70S·mRNA·F-Met-tRNA} + \text{IF-2} + \text{GDP} + P_i + \text{IF-1}.\end{aligned}$$

(An alternative interpretation exists that IF-1 leaves the initiation complex after joining of the 50S subunit and before hydrolysis of EF-2-associated GTP). The final initiation 70S complex can now accept

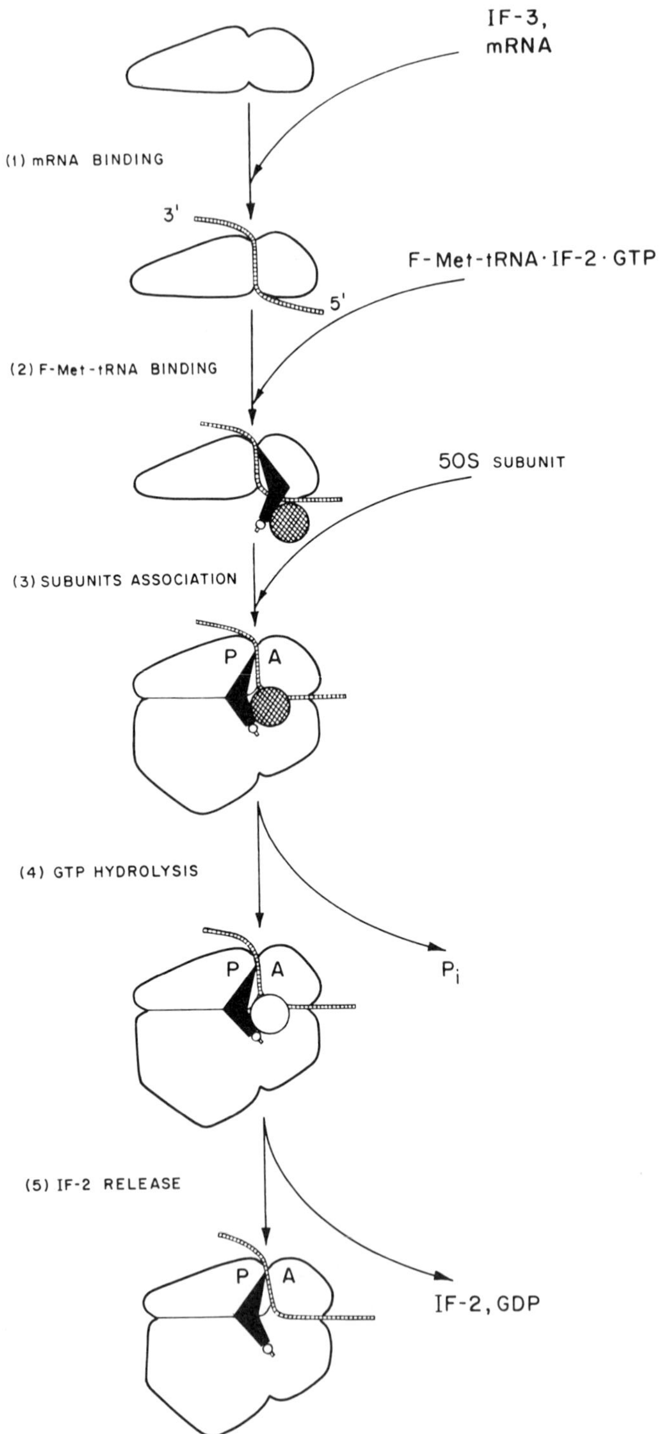

Figure 116 Sequence of events during initiation of translation in procaryotes.

the aminoacyl-tRNA (with EF-T_u and GTP) into the vacant A-site and begin elongation.

The sequence of events during the formation of the initiation complex is shown schematically in fig. 116. If a nonhydrolyzable GTP analog is used instead of GTP, then association with a 50S subunit and the formation of the intermediate initiation 70S complex take place normally; this includes the correct positioning of F-Met-tRNA in the P-site. However, IF-2 is not subsequently released and thus prevents the binding of the next aminoacyl-tRNA molecule at the A-site; as a result, the beginning of elongation is blocked. At the same time, under experimental conditions the IF-2 together with the nonhydrolyzable GTP analog may be physically washed off the 70S complex; the resultant complex does not then differ from the complex formed via the step of GTP cleavage: F-Met-tRNA and the initiation codon are positioned at the P-site, the A-site is vacant, and elongation can begin[39–41]. Hence, as in the case of elongation, GTP hydrolysis in the course of initiation is not coupled to any of the "active" events of the process, e.g. the positioning of the initiator tRNA and the corresponding initiation codon in the P-site, or the association and mutual adjustment of the two ribosomal subunits. GTP hydrolysis appears necessary only for the release of the protein factor IF-2, which has already carried out its role and now interferes with the next stage (that of binding the next aminoacyl-tRNA molecule). The similarity between IF-2 and EF-G is particularly striking: both proteins have a similar molecular mass; neither forms stable complexes with GTP and tRNA; both operate with N-blocked aminoacyl-tRNA derivatives; and, finally, both participate in the positioning of such tRNA derivatives in the ribosomal P-site. It may well be that these proteins are homologous and that IF-2 could have evolved from EF-G in the course of the evolutionary specialization of the initiation stage of translation.

16-6 Initiation in the absence of some initiation components

*The hyena is a beast of prey, she hunts at night, only at full moon, and if there is no moon, she does all the same.**

As already noted, the association between the 30S ribosomal subunit and mRNA requires the participation of the Shine–Dalgarno

* Taken from folklore of the Caucasus, as told by Professor Vladimir Alexandrov.

polypurine sequence, located on the template before the initiation codon. Association leading to initiation can, however, take place even in the absence of the Shine–Dalgarno sequence, although it is then less effective.

Initiation also requires IF-1, at least in the *E. coli* systems with natural mRNA. The requirement of IF-1, however, is not absolutely strict and initiation may occur in the absence of IF-1 (it should be mentioned that in certain species of bacteria, e.g. *Bacillus stearothermophilus* and *Caulobacter crescentus*, neither IF-1 nor its analogs have been found[42,43]).

It is also known that when certain synthetic template polynucleotides (e.g. poly(UG)) are used, the initiation involving the initiator F-Met-tRNA, initiation codon GUG, IF-2, and GTP may take place in the absence of IF-3[44].

The requirement of IF-2 and GTP for initiation is the most stringent. It should be pointed out, however, that the 30S ribosomal subunit has an intrinsic preferential affinity to $tRNA^F$ as compared to other tRNA species. At a high magnesium ion concentration, provided the AUG codon is present, the initiator tRNA may bind selectively to the ribosomal subunit in the absence of IF-2 and GTP; as a result of such a factor-free binding, F-Met-tRNA becomes attached to the ribosomal P-site and forms the normal initiation complex which is prepared for subsequent elongation[45,46].

Although AUG is a common initiation codon in procaryotic systems, GUG and UUG are also used, although far less frequently; furthermore, in some cases *in vivo*, initiation involving the same F-Met-$tRNA_F$ with the CAU anticodon takes place at several other triplets (AUU or AUA) as well. There are examples where elongation is interrupted as a result of a mutation leading to a termination codon, but the ribosome does not dissociate from the template and reinitiates translation using the nearest GUG, UUG, or CUG triplet[7]. In addition, in some mutated mRNA the initiation triplet AUG preceded by the Shine–Dalgarno sequence is changed into AUA[47,48] or ACG[49]; although the effectiveness of initiation on such mRNAs is reduced drastically, a certain level of initiation with the normal initiator F-Met-$tRNA_F$ and all the initiation factors persists. Under specific artificial conditions in cell-free translation systems, poly(U) may be used as a template for initiation involving F-Met-$tRNA_F$ and the initiation factors[50].

Finally, initiation of translation *in vitro* is possible even in the absence of the normal initiator F-Met-$tRNA_F$. Fairly satisfactory initiation involving the initiation factors may be obtained *in vitro* using certain N-blocked aminoacyl-tRNA species, e.g. N-acetylphenylalanyl-tRNA (N-Ac-Phe-$tRNA^{Phe}$), in the system with poly(U) as a template[51]. Blocking the amino group of aminoacyl-tRNA appears necessary for the interaction with IF-2 and for a better fit into the ribosomal P-site, as well as for good reactivity with the next aminoacyl-tRNA molecule occupying the A-site. However, even this requirement is not vital: in the absence of initiation factors, any aminoacyl-tRNA can get into the P-site of a vacant ribosome in correspondence with the template codon positioned there; thereafter, another aminoacyl-tRNA molecule can enter the A-site and react with the first to yield the first peptide bond. This process is, of course, much slower and far less accurate than normal initiation, but in the absence of components of the initiation machinery, it can take place; it simply requires higher concentrations of Mg^{2+} to provide sufficient affinity between ribosomes and the ligands, e.g. tRNA and mRNA[52]. The translation of poly(U) using only Phe-tRNA as a substrate in the simplest cell-free systems begins precisely in this way.

16-7 Regulation of initiation: Control of protein synthesis at the translational level

It is generally accepted that in procaryotes protein synthesis is controlled mainly at the level of transcription. Indeed, metabolic instability of mRNA in procaryotic cells, involving its rapid synthesis and rapid degradation, provides for a fast change of templates depending on environmental conditions and cell requirements. At the same time, however, the existence of polycistronic templates in procaryotes often demands differential control of the individual cistron activity in order to provide for quantitatively different or temporary uncoupled production of proteins coded by a given polynucleotide. Moreover, in a number of cases the accumulation of excessive amounts of the product of translation may be used to shut down the translation of corresponding mRNA; in this way, a very fine tuning between the level of protein production and the extent of cell requirement in this protein can be achieved. In all known cases, it is the stage of initiation that provides the point of regulation at the translational level in procaryotes.

There are several ways of controlling initiation. First, a quantitatively different level of protein production on different mRNAs, or on different cistrons of a given polycistronic mRNA, may be due to the initiation regions of templates having different "strengths": some initiation sequences may associate with ribosomal particles with high affinity and yield initiation complexes with high rates, while others do this less effectively. Specifically, the strength of the mRNA initiation region may depend on the length of the Shine–Dalgarno sequence and the degree of its complementarity to the 3′-end of the 16S ribosomal RNA[53]. The differences in the primary structures of initiation regions will set a fixed difference in the productivity of different mRNAs or mRNA cistrons in translation.

Second, initiation can be negatively controlled because of the three-dimensional structure of the initiation region of mRNA: folding into a stable secondary and tertiary structure may block initiation[54,55]. In such a case, only the unfolding of the structure, either by ribosomes reading the preceding cistron in a polycistronic mRNA (see below), or by a special agent destabilizing the three-dimensional structure of a corresponding mRNA region, opens the way for initiation.

Third, very selective negative control of initiation can be accomplished by special proteins that bind specifically to initiation regions of mRNA and in this way block ("repress") the association with the ribosome (see below). This is a true repression of translation. The function of such translational repressors are often played by proteins whose main function in the cell is quite different.

Two systems where translation is regulated at the initiation stage will be considered as examples.

Regulation of translation of MS2 phage RNA[56]

The RNA of the MS2 bacteriophage contains three cistrons separated by nontranslated sequences and one cistron which overlaps two others (see chapter 2, section 2.4, and the scheme in fig. 6). The A-cistron is the closest to the 5′-end of this polycistronic mRNA (1182 nucleotide residues in length, including the termination codon); it codes for the A-protein, or "maturation protein," which has a length of 393 amino acid residues. Furthermore in the 3′-ward direction comes the C-cistron (393 nucleotide residues in length, including the termination codon UAA); this cistron codes for the phage coat protein which is 129 amino acid residues long. Nearest to the 3′-end is the S-cistron (1638 nucleotide residues long, including the termination codon UAG); it codes for a subunit of the RNA replicase, its length being equal to 544 amino acid residues. The L-cistron (228 nucleotide residues long, including its termination codon UAA) codes for a small lysis protein which is 75 amino acid residues long; it overlaps, out of frame, the end of the C-cistron, the nontranslated spacer sequence, and the beginning of the S-cistron. (It should be mentioned that in the course of the synthesis of the coat protein and the RNA replicase subunit, the N-terminal methionines are cleaved off, and therefore the number of amino acid residues in the proteins is less by one residue than the number of sense template codons).

The three nonoverlapping cistrons are preceded by polypurine Shine–Dalgarno sequences, which are shown in fig. 114. Based on the degree of their complementarity to the 3′-terminal sequence of the 16S RNA, one may assume that the "recognition strength" of the initiation region by the ribosome, and consequently the initiation rate, are far greater for the A-cistron than for the C- and S-cistrons. Indeed, using isolated ribosome-binding RNA fragments of the related phage R17, it has been demonstrated that the 30-nucleotide-long fragment containing the initiation codon of the A-cistron binds to the 30S ribosomal subunit far better than similar fragments containing the initiation codons of either the C- or S-cistron[55]. This relationship, however, is not observed when intact MS2 or R17 RNA is used: the A-cistron as well as the S-cistron is incapable of binding to the ribosome because its initiation region is masked in the secondary and tertiary structure of the RNA. Only the C-cistron appears to be exposed for immediate interaction with free ribosomal particles and thus can be involved in the initiation of translation, independent of the translation of other cistrons.

In accordance with the above, the translation of intact MS2 RNA in cell-free systems, and perhaps *in vivo* as well, begins with the initiation of the coat protein synthesis. After the translation of the C-cistron has begun, ribosomes move along its sequence in the direction of the S-cistron and unfold the RNA secondary or tertiary structure in the course of their progression. This results in the opening of the S-cistron initiation region[54]. Hence, even before the first ribosome has completed translation of the C-cistron and, thus, before the first coat protein molecule has been synthesized, the initiation region of the S-cistron becomes accessible for initiation and, correspondingly, synthesis of the RNA replicase subunit is initiated.

The completion of C-cistron translation by the first ribosomes results in the appearance of free coat protein molecules in the system. As translation proceeds, this protein accumulates; eventually it will be used in the self-assembly of mature phage particles. The C-protein, however, in addition to its role in phage particle assembly, has been found to possess a strong specific affinity to the region of MS2 RNA between the C- and S-cistrons, including the initiation codon of the S-cistron[57]. The protein binds to this region and represses the initiation of S-cistron translation. The repression seems to result from the labile secondary structure (shown in chapter 2, fig. 11) being stabilized by the phage coat protein, and thus the initiation codon of the S-cistron becoming inacessible[58]. Hence, shortly after the translation of the S-cistron has been allowed by the translation of the preceding cistron, the S-cistron translation is repressed due to an accumulation of the protein product coded by the preceding cistron. Under these conditions, the ribosomes that have already begun to translate the S-cistron continue, and eventually complete, the synthesis of an appropriate number of RNA replicase subunit molecules. This number is sufficient to form the active molecules of the RNA replicase which begin to replicate MS2 RNA. At the same time, the repression of the further synthesis of this protein prevents an unnecessary overproduction of the enzyme. In this way the phage coat protein, which plays the part of S-cistron repressor, performs the regulatory function in translation.

In order to form the active RNA replicase molecule, the product of the phage S-cistron must associate with the three host cell proteins, which also perform other functions in a normal cell. Two of these proteins are the elongation factors EF-T_u and EF-T_s, and the third is the ribosomal protein S1[59]. In other words, the complete active RNA replicase is a protein with a quaternary structure consisting of four different subunits, only one of which is coded by phage RNA. The RNA replicase is an RNA-dependent RNA polymerase using the original chain ("+" chain) of the MS2 RNA to form the complementary chain ("−" chain) and then, using it as a template, to produce numerous copies of the original + chain. The A-cistron cannot be translated until MS2 RNA replication is started. Its initiation region appears to be masked by the secondary and tertiary structure of the intact RNA. It may be exposed for *in vitro* translation and thus can be initiated as a result of some artificial treatments, e.g. partial nuclease or heat-induced degradation of the intact polynucleotide chain, or mild treatment with formaldehyde, which disrupts base pairing. At the early period of + chain replication, however, when the chain is still growing, the three-dimensional structure of the 5′-terminal section containing the initiation codon of the A-cistron has not yet been fully formed; it is this period that seems to be used for initiating translation of the A-cistron under normal *in vivo* conditions[60,61]. Since the mature virus particle contains only one molecule of A-protein per 180 molecules of coat protein, the relatively brief period, during which the initiation of A-cistron translation is possible, appears to be sufficient for the required production of the A-protein. Thereafter, the elongated MS2 RNA folds in such a way that the initiation region of the A-cistron becomes involved into some three-dimensional structure which makes the initiation region inaccessible to free ribosomes.

In the case of polycistronic MS2 RNA, the termination of translation of each of the cistrons is accompanied by the release of ribosomes from the template. Correspondingly, each cistron begins to be translated as a result of independent initiation by free ribosomes from the medium. The situation with the L-cistron, however, appears to be different. For some structural reasons, no effective association between free ribosomes and the initiation region of the L-cistron in MS2 RNA can be obtained. It is assumed that the initiation of L-cistron translation is the result of a spontaneous +1 frame shift occurring by chance from time to time during C-cistron translation: if the frame shift happens, ribosomes will terminate at the UAA triplet which is located three nucleotides away from the initiation codon of the L-cistron toward the 5′-end; then, having no time to dissociate from the template, they will reinitiate at the nearest AUG triplet[62] (chapter 2, fig. 7). It is not surprising, then, that the productivity of the L-cistron should be low, which is indeed the case.

Regulation of the synthesis of ribosomal proteins

The bacterial cell is known to avoid overproduction of the ribosomal proteins. Generally speaking, the ribosomal proteins are synthesized in amounts required just for ribosomal assembly, in accordance with the amount of ribosomal RNA formed; under normal conditions the cell contains no significant excess of free ribosomal proteins[63,64]. Surprisingly, coordinated levels of production of nearly all ribosomal proteins in equimolar amounts are achieved even though their genes are not organized as a single regulated block, but are represented by approximately 16 independent operons, which are scattered throughout the cell genome[65]. It happens that the coordinated and virtually stoichiometric production of the ribosomal proteins and the prevention of their overproduction are maintained by a controlling mechanism which provides the repression of translation by protein excess (translational feedback control)[66].

A large proportion of the genes coding for ribosomal proteins (31 out of 52) are present in two main clusters on the *E. coli* chromosome. One of these clusters is located in the *str-spc* region at the 72nd min, and the other in the *rif* region at the 89th min. The *str-spc* region contains four operons coding for 27 ribosomal proteins, EF-T_u, and EF-G, as well as for the α-subunit of RNA polymerase. The *rif* region contains two operons coding for four ribosomal proteins, as well as for the β- and β'-subunits of RNA polymerase. Each operon produces a polycistronic mRNA. The cistrons and their order in these polycistronic mRNAs are shown schematically in fig. 117.

Studies conducted by Nomura and associates have demonstrated

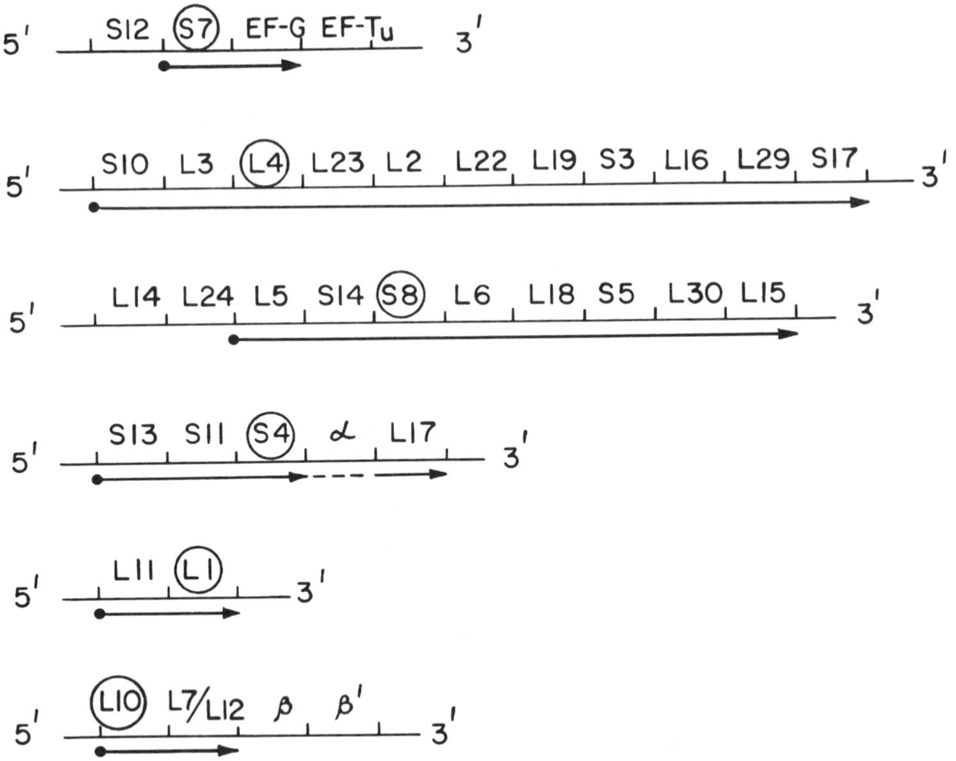

5′ S12 (S7) EF-G EF-Tu 3′

5′ S10 L3 (L4) L23 L2 L22 L19 S3 L16 L29 S17 3′

5′ L14 L24 L5 S14 (S8) L6 L18 S5 L30 L15 3′

5′ S13 S11 (S4) α L17 3′

5′ L11 (L1) 3′

5′ (L10) L7/L12 β β' 3′

Figure 117 Scheme of the sequential arrangement of ribosomal protein cistrons along polycistronic mRNA chains[66,67]. α, β, and β' are the cistrons of the corresponding subunits of RNA polymerase. The origin of the arrow under the sequence designates the point of action of a repressor protein; the arrow extends to the cistron subject to the control. The circled ribosomal proteins serve as repressors; they bind to the regions of the polycistronic mRNAs corresponding to the origins of the arrows.

that in the case of each polycistronic mRNA, one of the translation products, a ribosomal protein, serves as a repressor of the translation of a corresponding mRNA (these products are circled in fig. 117). This effect has been demonstrated both in experiments *in vivo* and in cell-free systems. Experiments *in vivo* have shown that the synthesis of the ribosomal proteins coded by the corresponding mRNA is inhibited when the overproduction of one of the proteins circled in fig. 117 is induced[67]. Superproduction has been achieved by using a plasmid containing *lac*-operator and *lac*-promotor fused to the gene of the corresponding ribosomal protein; synthesis of this protein could then be induced by isopropyl thiogalactoside, an inducer of *lac*-operon. Induction of proteins S7, L4, S8, S4, L1, or L10 result in the inhibited synthesis of only those ribosomal proteins that are coded by the polycistronic mRNA possessing the cistron of the corresponding protein. Experiments *in vitro* have brought even more direct results: adding one of the above proteins, e.g. S7, L4, S8, S4, L1, or L10, to the cell-free translation system leads to a selective inhibition of synthesis of only that set of proteins which is coded by the mRNA containing the cistron corresponding to the added protein[68]. The addition of other ribosomal proteins does not result in any inhibition of translation.

Synthesis of some of the proteins coded by the listed polycistronic mRNAs, however, is not inhibited when the repressory ribosomal proteins are added. Protein S12, for example, continues to be synthesized after the protein S7 is added to the cell-free system or in the *in vivo* version of the experiment with the selective induction of protein S7[69]. Similarly, the synthesis of proteins L14 and L24 does not stop in response to the addition or overproduction of protein S8[70]. It is noteworthy that the cistrons of the proteins not controlled by proteins S7 or S8 are found to be proximal to the 5′-end of the polycistronic mRNA.

All of these observations may be best explained by assuming that the repressor protein binds specifically to the initiation region of one cistron and blocks the translation of all cistrons located in the direction of the 3′-end. Protein S8, for example, binds with the origin of the cistron of protein L5 and, as a result, the translation of all subsequent downstream-located (but not upstream-located) cistrons is repressed. The implication is that in these cases free ribosomes cannot initiate the translation of each cistron independently. The so-called *sequential translation* appears to occur instead: ribosomes that have terminated the translation of the preceding cistron do not dissociate from the template but pass directly to reinitiation at the next cistron. Such sequential translation of cistrons provides for the equimolar production of ribosomal proteins coded by a given polycistronic mRNA. The lack of the dissociation of ribosomes (or at least of their 30S subunits) from mRNA after termination and the absence of independent initiation on internal cistrons by free ribosomes constitute the most likely explanation of the observations made[71].

There are some exceptions, however, in the same polycistronic mRNAs. For example, it has been shown that the translation of mRNA cistrons for EF-T_u and for RNA polymerase subunits $\beta + \beta'$ and α is not repressed by proteins S7, L10, and S4, respectively (fig. 117). Therefore, it appears that the initiation regions of these mRNA cistrons are capable of binding the free ribosomes, providing for independent initiation of translation[66]. On the other hand, it is known that in the case of the synthesis of proteins L10 and L7/L12, the production of protein L7/L12 is four times greater than the production of protein L10, in accordance with their stoichiometry in the ribosome; it may be assumed that the independent initiation of translation of the L7/L12 cistron takes place here and the initiation rate for this cistron is far greater than that for the L10 cistron. At the same time, as has already been indicated, protein L10 represses the translation of both proteins L10 and L7/L12. From this, one has to admit that, as with the S-cistron of MS2 RNA, the initiation of translation of the L7/L12 cistron cannot occur unless the preceding cistron is read; translation of the L10 cistron seems to make accessible the initiation region of the L7/L12 cistron for interaction with free ribosomes and thus for independent initiation[72].

Identifying the attachment sites of the repressor ribosomal proteins on polycistronic mRNAs is particularly interesting. The experiments have been performed using deletion mutants which produce mRNAs devoid of certain regions. It has been demonstrated that if the origin of the structural gene for protein S13 and the preceding nucleotide sequence is deleted, protein S4 is unable to repress translation of the corresponding polycistronic mRNA[73] (fourth line in fig. 117). In contrast, protein S7 can repress its own synthesis if the origin of its polycistronic mRNA (first line in fig. 117), including the S12 cistron, is absent[73]. Protein L1 has been shown to exert repressory action upon its bicistronic mRNA (fifth line in fig. 117) only in the presence of the 5′-terminal sequence, preceding the L11 cistron[71]. It follows that the attachment sites of the repressor proteins should be located at the origins of the cistrons from which the repression of the sequential translation begins in polycistronic mRNA. Correspondingly, for protein S7 the site should lie somewhere between the cistrons of proteins S12 and S7 or at the origin of the S7 cistron; for protein S4 the attachment site should be located before or at the beginning of the S13 cistron; and for protein L1 this site should be before the L11 cistron. Continuing this line of argument, the site of the repressor action for protein L4 should be located prior to or at the beginning of the S10 cistron; for protein L10, prior to or at the beginning of its own cistron; and for protein S8, between the cistrons of proteins L24 and L5 or at the beginning of the L5 cistron.

It is known that proteins S4, S7, S8, L1, and L4 play an important part in ribosomal structure and self-assembly: they are core proteins that bind tightly to specific sites on the ribosomal RNA. The attachment sites of these proteins on ribosomal RNA have been identified (see chapter 9). Nomura and associates assumed that these ribosomal proteins, playing the role of repressors, bind to mRNA through the same RNA-binding centers that participate in the binding to ribosomal RNA. Then, the structures of the RNA regions binding a

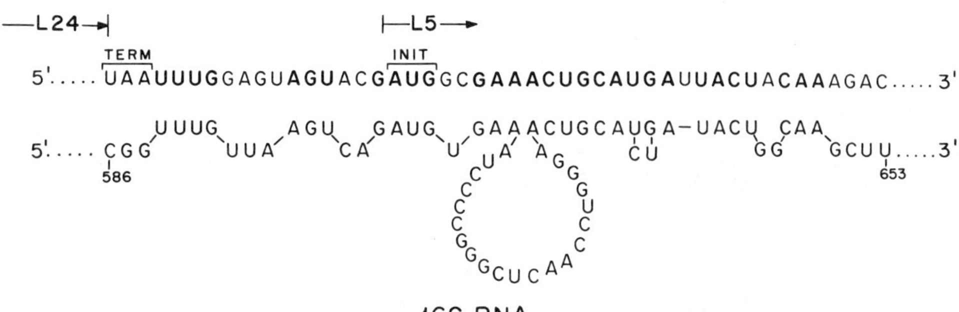

Figure 118 Comparison of the nucleotide sequences of the polycistronic mRNA region serving as a site of the repressor action of protein S8 (*top*) and the 16S ribosomal RNA region recognized by protein S8 in the course of ribosomal assembly (*bottom*)[74]. Identical sequences are in boldface in the upper chain.

given ribosomal protein should be similar in ribosomal RNA and in mRNA. Knowing the region of the binding site of a given repressor on polycistronic mRNA, Nomura and associates compared its primary structure and predicted secondary structure with the corresponding primary and secondary structures of the protein-binding region of ribosomal RNA. The result, although expected, was striking: the structures were found to be similar[73,74].

Figure 118 presents a comparison of the primary structures of the intercistronic L24–L5 region and the origin of the L5 cistron assumed to be the binding site of the repressory protein S8 (top) and the region of the ribosomal 16S RNA that binds protein S8 (bottom). High homology, including seven identical sections ranging from 3 to 7–9 residues in length, is apparent. Even greater homology follows from the comparison of predicted secondary structures of the two regions recognized by protein S8[74] (see fig. 119).

Homology of the primary and predicted secondary structures has also been shown between the mRNA sequence containing the end of the S12 protein cistron, the 100-nucleotide-long intercistronic spacer, and the beginning of the S7 protein cistron, on the one hand, and the 16S ribosomal RNA sequence located near the 3′-end and binding specifically protein S7, on the other[73]. Similarly, there is structural homology between the beginning of the S13 protein cistron and the preceding (5′-proximal) 50-nucleotide-long sequence in mRNA, on the one hand, and, on the other, the protein S4-binding region of 16S ribosomal RNA, comprising nucleotides 490 to 550[73]. In the S10 cistron

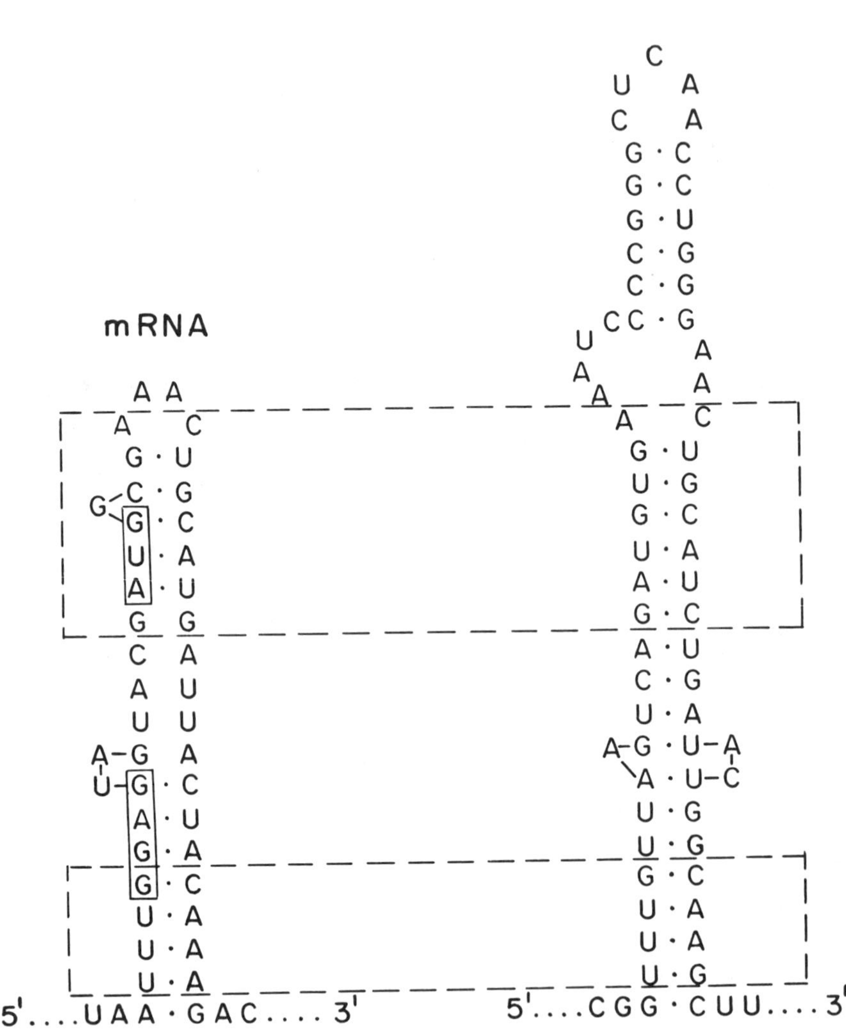

Figure 119 Comparison of the predicted secondary structure of the polycistronic mRNA region recognized by protein S8 during the repression of translation (*left*; initiation codon and Shine–Dalgarno sequence are boxed) and that of the 16S ribosomal RNA region recognized by protein S8 during ribosomal assembly (*right*)[74]. Homologous helices are enclosed by the broken lines.

and the adjacent 5′-proximal section of the corresponding polycistronic mRNA, a nonanucleotide sequence and a number of shorter sequences can be found which are identical to, and arranged in the same order along the chain as, those in the protein L4-binding region of the 23S ribosomal RNA.

In all of these cases, the model of repressory action of the corresponding ribosomal protein is analogous to the repression of S-cistron by the phage MS2 or R17 coat protein (see above): the preinitiation mRNA region and the initiation codon are involved in a rather labile secondary and tertiary structure which does not, by itself, prevent initiation; however, when the specific repressor protein recognizes this structure and binds to it, the structure becomes stable and makes the preinitiation sequence and the initiation codon inaccessible for interactions with the ribosome and the initiator tRNA. Indeed, the regions of mRNA structurally homologous to the protein-binding regions of ribosomal RNA generally include the preinitiation sequence as well as the initiation codon (see fig. 119).

The case of polycistronic mRNA coding for the ribosomal proteins L10 and L7/L12 (sixth line in fig. 117) appears to differ from the cases considered above. It has been demonstrated, in direct experiments, that protein L10 or, even better, its complex with protein L7/L12 (pentameric complex $[L7/L12]_4{\cdot}L10$, see chapter 8, section 8.3), binds specifically to a region of the 5′-terminal leader sequence located at least several dozen nucleotide residues upstream from the initiation codon of the first cistron, and this binding represses translation[75]. A region in the leader noncoding sequence of mRNA shows considerable homology with the region of ribosomal 23S RNA containing the presumable specific binding site for the $L10{\cdot}(L7/L12)_4$ complex. Thus, protein L10, or its complex with protein L7/L12, recognizes a certain structure in mRNA distant from the initiation codon. Nevertheless, their interaction blocks the initiation of translation. It has been found that initiation also can be blocked by numerous point mutations as well as by deletions in this distant region of the leader sequence. It is likely that the appropriate secondary and tertiary structures at this region of the leader sequence provide a positive contribution to the initiation of translation of this mRNA, so that perturbations of the structure, including those induced by the binding of protein L10 or of the complex $(L7/L12)_4{\cdot}L10$, result in blocked initiation of translation.

The hypothesis that a repressory ribosomal protein employs a common active center for binding to ribosomal RNA in the course of ribosome self-assembly, and for binding to mRNA in the course of translational repression, has been confirmed by another series of facts. It has been demonstrated that ribosomal RNA added to the translation system removes the repression exerted by a corresponding ribosomal protein. Thus, in experiments *in vitro* the repressory action of protein L1 upon the synthesis of proteins L1 and L11, as well as the inhibition of the synthesis of proteins L10 and L7/L12 by the L10·$(L7/L12)_4$ complex, can be prevented specifically by adding ribosomal 23S, but not 16S RNA[71,72].

Proceeding from these observations, Nomura and his associates proposed an elegant model for the coordinated regulation of synthesis of ribosomal proteins[66,73]. This model is based on the idea that there is competition between ribosomal RNA and mRNA for binding to the core ribosomal proteins. Such proteins as S4, S7, S8, L1, L4, as well as the protein complex L10·$(L7/L12)_4$, possess strong affinity to specific sites on ribosomal RNA. Therefore, after they have been synthesized, they become involved immediately in the assembly of ribosomal subunits through their direct binding to the 16S and 23S RNA. Intrinsic high affinity to ribosomal RNA and the cooperativity of ribosomal assembly involving other ribosomal proteins result in the sequestration of the newly formed ribosomal proteins in the course of the particle assembly. Under these conditions, mRNA molecules are unable to compete for the binding of these proteins and, therefore, do not associate with them; so they can be translated normally. However, when the number of ribosomal proteins increases compared to the amount of available ribosomal RNA, a free pool of such proteins is formed. This leads to the binding of the corresponding key proteins to their mRNAs, and the result is inhibited initiation and repressed translation. The strict sequential translation of polycistronic mRNA coding for a set of ribosomal proteins enables just one repressor protein and the site of its attachment for each mRNA to be sufficient for the coordinated control of translation of the whole set of proteins coded by a given mRNA. This simple mechanism provides a direct regulatory relationship between the assembly of ribosomes and the synthesis of ribosomal proteins.

References

1. B. F. C. Clark and K. A. Marcker (1966), "The role of N-formyl-methionyl-sRNA in protein biosynthesis," *J. Mol. Biol.* 17:394–406.

2. T. A. Sundararajan and R. E. Thach (1966), "Role of the formyl-methionine codon AUG in phasing translation of synthetic messenger RNA," *J. Mol. Biol.* 19:74–90.

3. R. E. Thach, K. F. Dewey, J. C. Brown, and P. Doty (1966), "Formylmethionine codon AUG as an initiator of polypeptide synthesis," *Science* 153:416–418.

4. D. A. Kellog, B. P. Doctor, J. E. Loebel, and M. W. Nirenberg (1966), "RNA codons and protein synthesis. IX. Synonym codon recognition by multiple species of valine-, alanine-, and methionine-sRNA," *Proc. Nat. Acad. Sci. U.S.A.* 55:912–919.

5. H. G. Khorana, H. Büchi, H. Ghosh, N. Gupta, T. M. Jacob, H. Kössel, R. Morgan, S. A. Narang, E. Ohtsuka, and R. D. Wells (1966), "Polynucleotide synthesis and the genetic code," *Cold Spring Harbor Symp. Quant. Biol.* 31:39–49.

6. A. J. Wahba, M. Salas, and W. M. Stanley (1966), "Studies on the translation of the genetic message. II. Translation of oligonucleotide messengers of specified base sequence," *Cold Spring Harbor Symp. Quant. Biol.* 31:103–111.

7. J. G. Files, K. Weber, and J. H. Miller (1974), "Translational reinitiation: Reinitiation of *lac* repressor fragments at three internal sites early in the *lac* i gene of *Escherichia coli*," *Proc. Nat. Acad. Sci. U.S.A.* 71:667–670.

8. K. Marcker and F. Sanger (1964), "N-formyl-methionyl-sRNA," *J. Mol. Biol.* 8:835–840.

9. M. S. Bretscher (1966), "Polypeptide chain initiation and the characterization of ribosomal binding sites in *E. coli*," *Cold Spring Harbor Symp. Quant. Biol.* 31:289–295.

10. S. K. Dube, K. A. Marcker, B. F. C. Clark, and S. Cory (1968), "Nucleotide sequence of N-formyl-methionyl-transfer RNA," *Nature* 218:232–233.

11. S. Cory, K. A. Marcker, S. K. Dube, and B. F. C. Clark (1968), "Primary structure of a methionine transfer RNA from *Escherichia coli*," *Nature* 220:1039–1040.

12. N. H. Woo, B. A. Roe, and A. Rich (1980), "Three-dimensional structure of *Escherichia coli* initiator $tRNA_f^{Met}$," *Nature* 286:346–351.

13. K. A. Marcker (1965), "The formation of N-formyl-methionyl-sRNA," *J. Mol. Biol.* 14:63–70.

14. J. M. Adams and M. R. Capecchi (1966), "N-formyl-methionyl-sRNA as the initiator of protein synthesis," *Proc. Nat. Acad. Sci. U.S.A.* 55:147–155.

15. R. E. Webster, D. L. Engelhardt, and N. D. Zinder (1966), "*In vitro* protein synthesis: Chain initiation," *Proc. Nat. Acad. Sci. U.S.A.* 55:155–161.

16. M. R. Capecchi (1966), "Initiation of *E. coli* proteins," *Proc. Nat. Acad. Sci. U.S.A.* 55:1517–1524.

17. J. M. Adams (1968), "On the release of the formyl group from nascent protein," *J. Mol. Biol.* 33:571–589.

18. G. Brawerman and J. M. Eisenstadt (1966), "A factor from *E. coli* concerned with the stimulation of cell-free polypeptide synthesis by exogenous

ribonucleic acid. II. Characteristics of the reaction promoted by the stimulation factor," *Biochemistry* 5:2784–2789.

19. M. Revel and F. Gros (1966), "A factor from *E. coli* required for the translation of natural mRNA," *Biochem. Biophys. Res. Commun.* 25:124–132.
20. M. Revel, G. Brawerman, J. C. Lelong, and F. Gros (1968), "Function of three protein factors and of ribosomal subunits in the initiation of protein synthesis in *E. coli*," *Nature* 219:1016–1021.
21. W. M. Stanley, M. Salas, A. J. Wahba, and S. Ochoa (1966), "Translation of the genetic message: Factors involved in the initiation of protein synthesis," *Proc. Nat. Acad. Sci. U.S.A.* 56:290–295.
22. K. Iwasaki, S. Sabol, A. Wahba, and S. Ochoa (1968), "Translation of the genetic message. VII. Role of initiation factors in formation of the chain initiation complex with *E. coli* ribosomes," *Arch. Biochem. Biophys.* 125:542–547.
23. M. Nomura and C. V. Lowry (1967), "Phage f2 RNA directed binding of formyl methionyl tRNA to ribosomes and the role of 30S ribosomal subunits in initiation of protein synthesis," *Proc. Nat. Acad. Sci. U.S.A.* 58:946–953.
24. C. Guthrie and M. Nomura (1968), "Initiation of protein synthesis: A critical test for the 30S subunit model," *Nature* 219:232–235.
25. M. Nomura, C. V. Lowry, and C. Guthrie (1967), "The initiation of protein synthesis: Joining of the 50S ribosomal subunit to the initiation complex," *Proc. Nat. Acad. Sci. U.S.A.* 58:1487–1493.
26. A. R. Subramanian and B. D. Davis (1971), "Activity of initiation factor F_3 in dissociating *E. coli* ribosomes," *Nature New Biol.* 228:1254–1268.
27. J. Shine and L. Dalgarno (1974), "The 3′-terminal sequence of *Escherichia coli* 16S ribosomal RNA: Complementarity to nonsence triplets and ribosome binding sites," *Proc. Nat. Acad. Sci. U.S.A.* 71:1342–1346.
28. J. A. Steitz (1980), "RNA·RNA interactions during polypeptide chain initiation," in *Ribosomes: Structure, function, and genetics*, ed. G. Chambliss, G. R. Craven, J. Davies, K. Davis, L. Kahan, and M. Nomura, pp. 479–495 (Baltimore: University Park Press).
29. M. Ptashne, K. Backman, M. Z. Humayun, A. Jeffrey, R. Maurer, B. Meyer, and T. Sauer (1976), "Autoregulation and function of a repressor in bacteriophage lambda," *Science* 194:156–161.
30. N. V. Belitsina and G. Zh. Tnalina (1981), "Translation of RNA 4 from plant brome mosaic virus in an *Escherichia coli* cell-free system with pure translation factors," *Mol. Biol.* (*U.S.S.R.*) 15:1234–1244.
31. M. Revel, M. Herzberg, and H. Greenshpan (1969), "Initiator protein dependent binding of messenger RNA to the ribosome," *Cold Spring Harbor Symp. Quant. Biol.* 34:261–275.
32. R. E. Thach, J. W. B. Hershey, D. Kolakofsky, K. F. Dewey, and E.

Remold–O'Donnell (1969), "Purification and properties of initiation factors F_1 and F_2," *Cold Spring Harbor Symp. Quant. Biol.* 34:277–284.

33. A. J. Wahba, Y.-B. Chae, K. Iwasaki, R. Mazumder, M. J. Miller, S. Sabol, and M. A. G. Sillero (1969), "Initiation of protein synthesis in *Escherichia coli*. I. Purification and properties of the initiation factors," *Cold Spring Harbor Symp. Quant. Biol.* 34:285–290.
34. A. J. Wahba, K. Iwasaki, M. J. Miller, S. Sabol, M. A. G. Sillero, and C. Vasquez (1969), "Initiation of protein synthesis in *Escherichia coli*. II. Role of the initiation factors in polypeptide synthesis," *Cold Spring Harbor Symp. Quant. Biol.* 34:291–299.
35. U. Maitra and J. Dubnoff (1969), "Protein factors involved in polypeptide chain initiation in *E. coli*," *Cold Spring Harbor Symp. Quant. Biol.* 34:301–306.
36. G. Brawerman (1969), "Role of initiation factors in the translation of messenger RNA," *Cold Spring Harbor Symp. Quant. Biol.* 34:307–312.
37. G. Jay and R. Kaempfer (1974), "Sequence of events in initiation of translation: A role for initiator transfer RNA in the recognition of messenger RNA," *Proc. Nat. Acad. Sci. U.S.A.* 71:3199–3203.
38. G. Jay and R. Kaempfer (1975), "Initiation of protein synthesis: Binding of messenger RNA," *J. Biol. Chem.* 250:5742–5748.
39. R. Benne and H. O. Voorma (1972), "Entry site of formylmethionyl-tRNA," *FEBS Letters* 20:347–351.
40. J. S. Dubnoff, A. H. Lockwood, and U. Maitra (1972), "Studies on the role of guanosine triphosphate in polypeptide chain initiation in *Escherichia coli*," *J. Biol. Chem.* 247:2884–2894.
41. C. Coutsogeorgopoulos, R. Fico, and J. T. Miller (1972), "On the function of guanosine triphosphate in the formation of N-acetyl-phenylalanyl puromycin," *Biochem. Biophys. Res. Commun.* 47:1056–1062.
42. A. C. Kay, M. Gaffe, and M. Grunberg-Manago (1975), "Stimulation by ATP of protein initiation in a prokaryotic organism, *B. stearothermophilus*," *FEBS Letters* 58:112–118.
43. A. C. Kay, M. Graffe, and M. Grunberg-Manago (1976), "Purification and properties of two initiation factors from *Bacillus stearothermophilus*," *Biochimie* 58:183–199.
44. D. P. Suttle and J. M. Ravel (1974), "The effects of initiation factor 3 on the formation of 30S initiation complex with synthetic and natural messengers," *Biochem. Biophys. Res. Commun.* 57:386–393.
45. M. S. Bretscher and K. A. Marcker (1966), "Polypeptidyl-ribonucleic acid and aminoacyl-ribonucleic acid binding sites on ribosomes," *Nature* 221:380–384.
46. P. S. Rudland and S. K. Dube (1969), "Specific interaction of an initiator tRNA fragment with 30S ribosomal subunits," *J. Mol. Biol.* 43:273–280.

47. T. Taniguchi and C. Weissmann (1978), "Site-directed mutations in the initiator region of the bacteriophage Q_β coat cistron and their effect on ribosome binding," *J. Mol. Biol.* 118:533–565.

48. D. Belin, J. Hedgpeth, G. B. Selzer, and P. H. Epstein (1979), "Temperature-sensitive mutation in the initiation codon of the rIIB gene of bacteriophage T4," *Proc. Nat. Acad. Sci. U.S.A.* 76:700–704.

49. J. J. Dunn, E. Buzash-Pollert, and F. W. Studier (1978), "Mutations of bacteriophage T7 that affect initiation of synthesis of the gene 0.3 protein," *Proc. Nat. Acad. Sci. U.S.A.* 75:2741–2745.

50. K. van der Laken, H. Bakkar-Steeneveld, and P. van Knippenberg (1979), "Polyuridylic acid-dependent binding of fMet-tRNA to *Escherichia coli* ribosomes and incorporation of formyl-methionine into polyphenylalanine," *FEBS Letters* 100:230–234.

51. J. Lucas-Lenard and F. Lipmann (1967), "Initiation of polyphenylalanine synthesis by N-acetylphenylalanyl-sRNA," *Proc. Nat. Acad. Sci. U.S.A.* 57:1050–1057.

52. T. Nakamoto and D. Kolakofsky (1966), "A possible mechanism for initiation of protein synthesis," *Proc. Nat. Acad. Sci. U.S.A.* 55:606–613.

53. J. A. Steitz, K. U. Sprague, D. A. Steege, R. C. Yuan, M. Laughrea, P. B. Moore, and A. J. Wahba (1977), "RNA·RNA and protein·RNA interactions during the initiation of protein synthesis," in *Nucleic acid-protein recognition*, ed. H. J. Vogel, pp. 491–508 (New York: Academic Press).

54. H. F. Lodish and H. D. Robertson (1969), "Regulation of in vivo translation of bacteriophage f2 RNA," *Cold Spring Harbor Symp. Quant. Biol.* 34:655–673.

55. J. A. Steitz (1973), "Discriminatory ribosome rebinding of isolated regions of protein synthesis initiation from the ribonucleic acid of bacteriophage R17," *Proc. Nat. Acad. Sci. U.S.A.* 70:2605–2609.

56. C. Weissmann, M. A. Billeter, H. M. Goodman, J. Hindley, and H. Weber (1973), "Structure and function of phage RNA," *Ann. Rev. Biochem.* 42:303–328.

57. A. Bernardi and P. F. Spahr (1972), "Nucleotide sequence at the binding site for coat protein on RNA of bacteriophage R17," *Proc. Nat. Acad. Sci. U.S.A.* 69:3033–3037.

58. J. Gralla, J. A. Steitz, and D. M. Crothers (1974), "Direct physical evidence for secondary structure in an isolated fragment of R17 bacteriophage mRNA," *Nature* 248:204–208.

59. R. Kamen, M. Kondo, W. Romer, and C. Weissmann (1972), "Reconstitution of Q_β replicase lacking subunit α with protein-synthesis interference factor i," *Eur. J. Biochem.* 31:44–51.

60. H. D. Robertson and H. F. Lodish (1970), "Messenger characteristics of nascent bacteriophage RNA," *Proc. Nat. Acad. Sci. U.S.A.* 67:710–716.

61. D. Kolakofsky and C. Weissmann (1971), "Possible mechanism for

translation of viral RNA from polysome to replication complex," *Nature New Biol.* 231:42–46.

62. R. A. Kastelein, E. Remaut, W. Fiers, and J. van Duin (1982), "Lysis gene expression of RNA phage MS2 depends on a frameshift during translation of the overlapping coat protein gene," *Nature* 295:35–41.
63. N. O. Kjeldgaard and K. Gausing (1974), "Regulation of biosynthesis of ribosomes," in *Ribosomes*, ed. M. Nomura, A. Tissières, and P. Lengyel, pp. 369–392 (Cold Spring Harbor, N.Y.: Cold Spring Harbor Laboratory).
64. K. Gausing (1980), "Regulation of ribosome biosynthesis in *E. coli*," in *Ribosomes: Structure, function, and genetics*, ed. G. Chambliss, G. R. Craven, J. Davies, K. Davis, L. Kahan, and M. Nomura, pp. 693–718 (Baltimore: University Park Press).
65. L. Lindahl and J. Zengel (1982), "Expression of ribosomal genes in bacteria," Advances in genetics, ed. E. W. Caspari, vol. 21, pp. 53–111 (New York: Academic Press).
66. M. Nomura, S. Jinks-Robertson, and A. Miura (1982), "Regulation of ribosome biosynthesis in *Escherichia coli*," in *Interaction of translational and transcriptional controls in the regulation of gene expression*, ed. M. Grunberg-Manago and B. Safer, pp. 91–104 (New York: Elsevier).
67. D. Dean and M. Nomura (1980), "Feedback regulation of ribosomal protein gene expression in *Escherichia coli*," *Proc. Nat. Acad. Sci. U.S.A.* 77:3590–3594.
68. J. L. Yates and M. Nomura (1980), "*E. coli* ribosomal protein L4 is a feedback regulatory protein," *Cell* 21:517–522.
69. D. Dean, J. L. Yates, and M. Nomura (1981), "Identification of ribosomal protein S7 as a repressor of translation within the *str* operon of *E. coli*," *Cell* 24:413–419.
70. D. Dean, J. L. Yates, and M. Nomura (1981), "*Escherichia coli* ribosomal protein S8 feedback regulates part of *spc* operon," *Nature* 289:89–91.
71. J. L. Yates and M. Nomura (1981), "Feedback regulation of ribosomal protein synthesis in *E. coli*: Localization of the mRNA target sites for repressor action of ribosomal protein L1," *Cell* 24:243–249.
72. J. L. Yates, D. Dean, W. A. Strycharz, and M. Nomura (1981), "*E. coli* ribosomal protein L10 inhibits translation of L10 and L7/L12 mRNAs by acting at a single site," *Nature* 294:190–192.
73. M. Nomura, J. L. Yates, D. Dean, and L. E. Post (1980), "Feedback regulation of ribosomal protein gene expression in *Escherichia coli*: Structural homology of ribosomal RNA and ribosomal protein mRNA," *Proc. Nat. Acad. Sci. U.S.A.* 77:7084–7088.
74. P. O. Olins and M. Nomura (1981). "Translational regulation by ribosomal protein S8 in *Escherichia coli*: Structural homology between rRNA binding site and feedback target on mRNA," *Nucleic Acids Res.* 9:1757–1764.

75. M. Johnsen, T. Christensen, P. P. Dennis, and N. P. Fiil (1982), "Autogenous control: Ribosomal protein L10–L12 complex binds to the leader sequence of its mRNA," *EMBO Journal* 1:999–1004.

Further reading

Grunberg-Manago, M. (1980). Initiation of protein synthesis as seen in 1979. In *Ribosomes: Structure, function, and genetics* (G. Chambliss, G. R. Craven, J. Davies, K. Davis, L. Kahan, and M. Nomura, eds.), pp. 445–477. Baltimore: University Park Press.

Grunberg-Manago, M., and Gros, F. (1977). Initiation mechanisms of protein synthesis. Progress in nucleic acid research and molecular biology (W. E. Cohn, ed.), vol. 20, pp. 209–284. New York: Academic Press.

Revel, M. (1972). Polypeptide chain initiation: The role of ribosomal protein factors and ribosomal subunits. In *The mechanism of protein synthesis and its regulation* (L. Bosch, ed.), pp. 87–131. Amsterdam and London: North-Holland.

——— (1977). Initiation of messenger RNA translation into protein and some aspects of its regulation. In *Molecular mechanisms of protein biosynthesis* (H. Weissbach and S. Pestka, eds.), pp. 245–321. New York: Academic Press.

Rudland, P. S., and Clark, B. F. C. (1972). Polypeptide chain initiation and the role of a methionine tRNA. In *The mechanism of protein synthesis and its regulation* (L. Bosch, ed.), pp. 55–86. Amsterdam and London: North-Holland.

Scherer, G. F. E.; Walkinshaw, M. D.; Arnott, S.; and Morré, D. J. (1980). "The ribosome binding sites recognized by *E. coli* ribosomes have regions with signal character in both the leader and protein coding segments," *Nucleic Acids Res.* 8:3895–3907.

Weissbach, H. (1980). Soluble factors in protein synthesis. In *Ribosomes: Structure, function, and genetics* (G. Chambliss, G. R. Craven, J. Davies, K. Davis, L. Kahan, and M. Nomura, eds.), pp. 377–411. Baltimore: University Park Press.

Chapter 17

Initiation of Translation and Its Regulation in Eucaryotes

17-1 Characteristics of eucaryotic mRNA

In contrast to procaryotic mRNA, eucaryotic mRNA is complexed with proteins and exists as messenger ribonucleoproteins (mRNP, or informosomes). Furthermore, it is metabolically stable; there is no ongoing fast degradation of mRNA accompanied by its intensive resynthesis. Eucaryotic mRNA is monocistronic. As a rule, it possesses a specifically modified (capped) 5′-end. All this is associated with a number of distinctive features regarding the initiation of translation and its regulation in eucaryotes.

First, a considerable fraction of mRNA in eucaryotic cells does not take part in translation at a given time[1,2]; rather, it exists in the form of free cytoplasmic mRNPs (free informosomes)[3]. These free *nontranslatable* particles have a protein-to-RNA weight ratio of about 3:1 and, correspondingly, they possess a relatively low buoyant density, about 1.4 g/cm^3 in CsCl (whereas the buoyant density of eucaryotic ribosomes is about 1.55–1.59 g/cm^3 in CsCl). The particles contain a specific set of proteins whose function is, as yet, unclear. It is assumed that at least some of these proteins play the part of translation repressors, preventing ribosomes from attaching to mRNA and initiating translation[4]. It should be stressed, however, that the existence of such repressor proteins in eucaryotes has not been demonstrated directly. The repressor function is also ascribed to certain cytoplasmic low-molecular-mass RNAs found complexed with free mRNP[5].

In any case, mRNA in eucaryotes may exist a long time in a nontranslatable, stored (masked) form. The stored mRNAs of unfertilized animal eggs and plant seeds provide impressive examples of this[1,2]. Fertilization of oocytes as well as germination of seeds results in a general *and* a selective activation of preexisting mRNA translation. In processes of embryonic development and cell differentiation of eucaryotic organisms, the synthesis of mRNA and the accumulation of mRNA in the cytoplasm, probably in the form of mRNP, may take place long before this RNA is used in translation. For example, dividing myoblasts are characterized by the synthesis and accumulation of nontranslatable myosin mRNA in the form of mRNP, and only the subsequent fusion of cells and the transition to the differentiated state of myotubes induce the translation of the accumulated myosin mRNA[6]. Translation of presynthesized mRNA is activated by mechanisms whose nature has still to be determined. In any case, the transition of mRNA into a translatable form is accompanied by a change in the spectrum of the proteins complexed with it. (The mRNA may be isolated from eucaryotic polyribosomes in the form of mRNP either, but the protein composition of such *translatable* mRNPs is different from that of free mRNP[7].)

The metabolic stability (long lifetime) of eucaryotic mRNA makes control at the level of translation particularly important in the general pattern of protein synthesis regulation. Specifically, along with the signals for template activation, i.e. for the initiation of translation, the signals for inactivating templates, i.e. for the arrest of translation, become necessary. Hormonal regulation of translation provides an example of both the switching on and the shutting down of translation of certain mRNA species. It has been shown that injecting female sex hormones (estrogens) into chicks induces cytodifferentiation of the oviducts, accumulation of the ovalbumin mRNA, and synthesis of the ovalbumin. In the oviducts of chicks withdrawn from estrogen treatment, the syntheses of the mRNA and ovalbumin stop; it can, however, be demonstrated that the presynthesized ovalbumin mRNA is partly preserved in the cytoplasm in a nontranslatable or low-translatable form. Readministration of the hormone (estrogen or progesterone) immediately reactivates translation by stimulating the initiation[8].

Synthesis of δ-crystallin has been shown to decrease at a certain stage of chick embryo eye lens differentiation. At the same time, the amount of δ-crystallin mRNA in lens cells is maintained at approximately the same level[9].

Heat shock triggers the synthesis of a few special proteins, while translation of most of the preexisted cellular mRNAs becomes reduced. Cell recovery at a normal temperature is accompanied by reactivation of total mRNA translation and cessation of heat-shock protein synthesis. In the case of frog (*Xenopus laevis*) oocytes, the heat-shock response and the recovery from heat treatment have been shown to be controlled exclusively at the translational level[10]: the heat-shock mRNA in frog oocytes is synthesized during oogenesis and stored in a masked form at a normal temperature, but can be reversibly activated for translation by a high temperature. Again, the masking and demasking mechanisms remain to be determined.

It is believed that the need for a more developed control system at the translational level in eucaryotes, as compared with procaryotes, has resulted in the evolutionary emergence of a more complex initiation mechanism operating with a far greater number of initiation proteins (see below). Instead of the complementary RNA-RNA interaction involving the preinitiation Shine–Dalgarno sequence of procaryotic mRNAs (while the 5′-end of the template is not necessary for initiation), eucaryotic ribosomes with the participation of a special cap-binding protein complex recognize eucaryotic mRNAs at the capped 5′-end (see below). Perhaps, this mechanism provides more opportunities for controlling initiation. It may be thought that eucaryotic mRNAs are monocistronic because the 5′-end of the template polynucleotide in eucaryotes is required for binding certain proteins and initiating translation. It is, as a rule, impossible for eucaryotic ribosomes to recognize the internal cistrons of polycistronic mRNAs.

17-2 Initiation codon, initiator tRNA, and protein initiation factors

Initiation codon. The AUG codon seems to be the only initiation codon in naturally occurring eucaryotic mRNAs[11]. In many instances, this is the first AUG triplet from the 5′-end of the template polynucleotide[12], although this is not necessarily the case for all mRNAs.

Initiator tRNA. Special $tRNA_F$ acylated by methionyl-tRNA synthetase serves as the initiator tRNA recognizing the AUG codon[13]. The situation with eucaryotes in this respect is similar to that of procaryotes. However, in contrast to procaryotic initiator tRNA, the methionyl-$tRNA_F$ of eucaryotes does not undergo formylation[15]. Hence, the only difference between the initiator methionyl-$tRNA_F$ and the usual (elongator) methionyl-$tRNA^{Met}$ lies in certain structural features of the $tRNA_F$ residue itself. These features are what make $tRNA_F$ capable of interacting with the initiation factors and the vacant ribosomal subunit, and incapable of participating in elongation.

The primary and secondary structures (in the cloverleaf form) of these two eucaryotic tRNA species[16,17] are presented in fig. 120. It can be seen that the differences are not that great. The most salient feature of initiator tRNA in eucaryotes is the presence of the GAUC sequence (or GAψC in higher plants and echinoderms) rather than the universal GTψC sequence in the elongator tRNAs in positions 51–54. In addition, in most eucaryotes the dihydrouridylic loop of the initiator tRNA does not contain dihydrouridine. In contrast to the initiator tRNAs of procaryotes and mitochondria, the 5′-terminal nucleoside

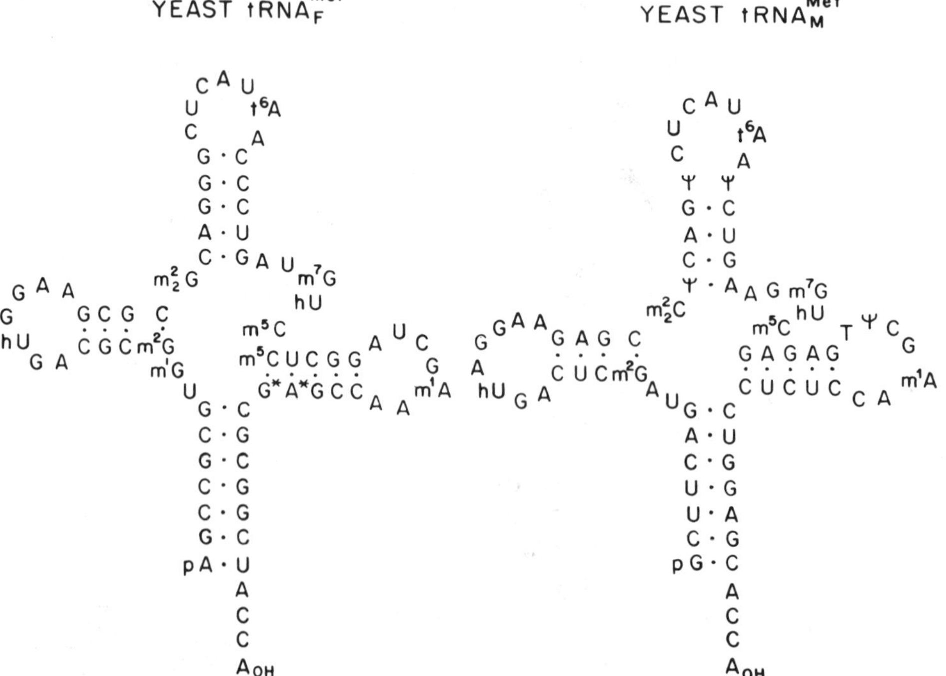

Figure 120 Nucleotide sequence and secondary structures of the eucaryotic initiator tRNA ($tRNA_F^{Met}$, *left*)[16] compared with the eucaryotic methionine tRNA participating in elongation ($tRNA_M^{Met}$, *right*)[17].

residue of eucaryotic initiator tRNAs is paired with the nucleoside of the 3′-terminal region, just as in all elongator tRNAs. It is not known, however, which structural features are responsible for the functional specificity of eucaryotic initiator tRNA.

Initiation factors. At present, about a dozen protein factors are known which are required for the eucaryotic ribosomes to initiate translation of the eucaryotic mRNA. These factors are referred to as eIF-1, eIF-2, eIF-2A, eIF-2B, eIF-3, eIF-4A, eIF-4B, eIF-4C, eIF-4D, eIF-4E, eIF-4F, eIF-5, etc. (the "e" refers to "eucaryotic," and "IF" to "initiation factor"). The list of initiation factors and several of their physical and functional properties are given in table 2. Strictly speaking, nearly all of these initiation factors were isolated in a more or less individual state from mammalian reticulocytes and tested in the globin mRNA translation system[18–20]. Evidence for other eucaryotic systems is more fragmentary and applies mainly to such principal initiation factors as eIF-2, eIF-3, and eIF-5.

Initiation strictly requires eIF-2, eIF-3, and eIF-5, whereas eIF-1, eIF-2B, eIF-4A, eIF-4B, eIF-4C, eIF-4E, and eIF-4F show more or less stimulatory action. The part played by eIF-2A and eIF-4D in initiation and the contribution of these factors to the overall process are still unclear, and there are doubts as to whether they are truly necessary for initiation. Protein eIF-2A *in vitro* stimulates the binding of the initiator tRNA to the 40S ribosomal subunit in the absence of GTP. Protein eIF-4D stimulates the reaction between the initiator methionyl-$tRNA_F$ and puromycin in the initiation 80S complex.

All initiation factors, with the exception of eIF-4A and eIF-4D, appear to be RNA-binding proteins: they possess nonspecific affinity to all high-molecular-mass RNAs and are capable of forming more or less stable complexes with such RNAs[21–24].

The eIF-2 is a large protein. It interacts with GTP and Met-$tRNA_F$. Functionally, it is analogous to procaryotic IF-2. However, eIF-2 consists of three nonidentical subunits, with molecular masses of about 32,000 daltons (α-subunit), 35,000 daltons (β-subunit), and 55,000 daltons (γ-subunit)[25]. (When molecular mass of the β-subunit is estimated from gel electrophoresis in the presence of sodium dodecyl sulfate the value of about 50,000 daltons is obtained, probably as a result of its anomalous behavior due to the high content of basic amino acid residues).

The eIF-3 is an even larger, more complex protein. It has a molecular mass of at least 500,000 daltons, and its sedimentation coefficient is approximately 15S. This protein consists of about a dozen subunits with molecular masses ranging from 30,000 to 150,000 daltons. It has a high affinity to the 40S ribosomal subunit and is often regarded as a functional analog of the procaryotic IF-3. There is some

Table 2 Eucaryotic initiation factors: physical and functional properties

Designation	*Synonyms*	*Molecular mass* $\times 10^{-3}$ *(daltons)*	*Subunits and their molecular masses* $\times 10^{-3}$ *(daltons)*	*Affinity to 40S subunit*	*Affinity to mRNA*	*Stimulation of Met-tRNA binding to 40S subunits*	*Stimulation of mRNA binding to 40S subunits*
eIF-1		12		±	+	+	+
eIF-2		*ca.* 120	32(α), 35(β), 55(γ)	+?	+	+	+
eIF-2A		65	—	+?	?	+	–
eIF-2B	antiHCl, eRF, ESP, GEF, RF, SRF	*ca.* 300	30, 40, 55, 65, and 80	–	?	+	–
eIF-3		*ca.* 600	about 10 subunits, from 30 to 150	++	+	+	+
eIF-4A		50	—	–	–	–	+
eIF-4B		80	—	–	+	–	+
eIF-4C		17	—	+	+	+	+
eIF-4D		16	—	–	–	–	–
eIF-4E	CBP-I	24	—	–	+	–	+
eIF-4F (includes eIF-4A, eIF-4E, and a 220,000-dalton protein)	CBP-II	300	24, 50, 220	–	+	–	+
eIF-5		160	—	–	+	–	–
eIF-6		25	—	–	?	–	–

evidence, however, that the functions of the eIF-3 are far more complicated; in fact, the protein may be tissue specific and play a part in the selective recognition (discrimination) of mRNAs.

The eIF-5 is also a large protein but it consists of a single polypeptide chain with a molecular mass of about 160,000 daltons. Apparently, procaryotic systems do not possess any analog of eIF-5. This factor is indispensable for the binding of the 60S ribosomal subunit to the initiation 40S complex, and for the coupled cleavage of eIF-2-bound GTP into GDP and orthophosphate. In other words, it promotes the formation of 80S particles at the final stage of initiation.

Nearly all of the remaining initiation factors, including eIF-1, eIF-4A, eIF-4B, eIF-4C, eIF-4D, and eIF-4E, consist of one polypeptide chain each. The eIF-1, eIF-4C, and eIF-4D are small proteins with molecular masses ranging from 15,000 to 20,000 daltons; the first two exhibit stimulatory action at the mRNA-binding stage. The eIF-4A and eIF-4B are larger proteins with molecular masses of about 50,000 and 80,000 daltons, respectively; their participation in the binding of mRNA to the 40S ribosomal subunit appears more vital.

The eIF-4E is an important protein; it frequently is referred to as a cap-binding protein, or CBP-I, and has a molecular mass of 24,000 daltons[26]. It binds specifically to the modified (capped) 5′-end of mRNA and stimulates the association between 40S ribosomal subunits and templates. There is evidence that this protein exists and functions as a subunit of the larger cap-binding, multiprotein 8-10S complex CBP-II, which is also referred to as eIF-4F. The complex factor eIF-4F includes eIF-4A, eIF-4E and a subunit (one polypeptide chain) with a molecular mass of 220,000 daltons. The eIF-4F (CBP-II complex) functions together with free eIF-4A, as well as with eIF-4B.

In contrast to the situation in procaryotic systems, the stage of mRNA binding in the course of initiation in eucaryotes includes the cleavage of ATP to ADP and orthophosphate. The ATP hydrolysis is coupled with the eIF-4F (CBP-II), eIF-4A, and eIF-4B taking part in mRNA binding[27,28]. It is likely that the cap-binding proteins perform ATP-dependent unwinding and destabilization of secondary structure at the 5′-terminal region of eucaryotic mRNA, thereby facilitating the association of this region with ribosomal particles[29].

The large protein eIF-2B, which consists of five different subunits, has auxiliary significance[30–34]. It forms a complex with eIF-2 in which the affinity of eIF-2 for GDP is decreased, and the affinity for GTP is increased, so that the effective exchange of eIF-2-bound GDP for free GTP is provided. (Otherwise, at physiological concentrations of GTP and GDP, more than 90% of the eIF-2 molecules would be bound with GDP, thus existing in the inactive state). The resultant eIF-2B·eIF-2·GTP complex directly binds initiator Met-tRNA$_F$, and then the

Met-tRNA$_F$·eIF-2·GTP complex is transferred from eIF-2B onto the initiating 40S ribosomal particle. Hence, eIF-2B catalyzes the reuse of eIF-2 after its release (in the form eIF-2·GDP) from the ribosomes finishing initiation (see fig. 123).

17-3 State of ribosomes before initiation

As is the case with procaryotes, the nontranslating eucaryotic ribosomes must pass into a dissociated state prior to the initiation of translation[35,36]. The eucaryotic 80S ribosomes, however, are rather stable, and it must be assumed that their eventual dissociation into subunits, after termination of translation, is achieved only by the action of protein factors. In any case, the cytoplasmic extracts of eucaryotic cells contain the "native" 40S and 60S subunits, which differ from the 40S and 60S subunits derived from 80S ribosomes by dissociation in a low magnesium ion concentration. In contrast to the "derived" subunits, the native subunits are unable to associate into 80S ribosomes in moderate Mg^{2+} concentrations. The native subunits, the 40S subunit in particular, show a far greater protein content than the derived subunits. Correspondingly, the buoyant density of the derived 40S subunit in CsCl is about 1.52 g/cm^3, whereas for the native 40S subunit, it varies from 1.49 g/cm^3 down to 1.40 g/cm^3[36,37]. A number of initiation factors have been found in native 40S subunits in varying amounts (the native subunits often serve as the source for preparing initiation factors)[38,39]. All native 40S subunits contain eIF-3 in the equimolar amount. Native 60S subunits may have the elongation factors attached.

Thus, eucaryotic cell cytoplasm and cytoplasmic extracts always contain a pool of free native 40S and 60S subunits which are incapable of associating and which contain bound initiation factors, as well as a number of other proteins. The native 40S subunits with bound eIF-3 and certain other initiation factors begin the initiation process[24].

17-4 Formation of the complex between the 40S ribosomal subunit and the initiator tRNA

The eIF-2, in the presence of GTP, recognizes specifically the initiator methionyl-tRNA and forms a ternary complex with it[40,41]:

$$\text{eIF-2} + \text{GTP} + \text{Met-tRNA}_F \rightleftarrows \text{Met-tRNA}_F\cdot\text{eIF-2}\cdot\text{GTP}.$$

The ternary complex has a strong affinity to the native 40S subunit, resulting in the formation of the initiation 43S complex[41,42]:

$$40S + Met\text{-}tRNA_F{\cdot}eIF\text{-}2{\cdot}GTP \rightleftarrows 40S{\cdot}Met\text{-}tRNA_F{\cdot}eIF\text{-}2{\cdot}GTP.$$

As has been mentioned eIF-2B, interacting with eIF-2, stimulates the formation of the ternary complex Met-tRNA$_F$·eIF-2·GTP and its entry into the initiation 43S complex (see section 17.7, fig. 123).

The presence of other initiation factors on the 40S subunit, particularly eIF-3 and eIF-4C, and possibly eIF-1, stabilizes the initiation 43S complex[24]. In the process of complex formation, GTP may be substituted for by its noncleavable analog.

Thus, in eucaryotic systems the binding of the small ribosomal subunit with the initiator tRNA appears to precede the association with mRNA. This is the first step in the initiation of translation.

17-5 Association of the ribosomal 40S subunit with mRNA[24,42]

The complex between the native 40S subunit and the initiator methionyl-tRNA, initiation factors, and GTP then associates with mRNA. The eIF-3 is indispensable at this stage; however, its action mechanism is still unclear. The assumption is that, when bound with the native 40S subunit, eIF-3 takes part in forming the mRNA recognition site. It is also thought to facilitate the unfolding of the secondary structure of the template polynucleotide in the course of initiation. There is evidence that eIF-3 plays a role in discriminating between different mRNAs. The presence of eIF-2 on the 40S subunit also contributes to the mRNA binding and, probably, to mRNA discrimination (this protein is known to possess a strong RNA-binding capability even in the isolated state). Other factors, e.g. eIF-1, eIF-4A, eIF-4B, eIF-4C, and cap-binding proteins (eIF-4E and eIF-4F), also play a part in the association between the 40S ribosomal subunit and mRNA. At this stage, ATP is hydrolyzed to ADP and orthophosphate. This results in the formation of the initiation 48S complex:

$$\begin{aligned} \text{XI}\quad & 40S{\cdot}eIF\text{-}3{\cdot}eIF\text{-}4C{\cdot}Met\text{-}tRNA{\cdot}eIF\text{-}2{\cdot}GTP \\ & + eIF\text{-}1 + mRNA{\cdot}CBP + ATP + H_2O \xrightarrow{eIF\text{-}4A,\ eIF\text{-}4B} \\ & 40S{\cdot}mRNA{\cdot}Met\text{-}tRNA{\cdot}eIF\text{-}2{\cdot}GTP{\cdot}eIF\text{-}3{\cdot}eIF\text{-}4C{\cdot}eIF\text{-}1 \\ & + CBP + ADP + P_i \end{aligned}$$

In most cases, an important prerequisite for the initial binding of mRNA is the presence of the cap structure at its 5′-end (see chapter 2, fig. 5), and of the cap-binding proteins complexed with the cap.

Indeed, most of the eucaryotic mRNAs are capped, and when the cap is removed the rate of mRNA translation is reduced drastically. It has been demonstrated that the cap, in the presence of the cap-binding proteins, increases the affinity of mRNA to the ribosomal particles. The addition of chemical analogs of caps, e.g. m^7GMP, m^7GDP, or m^7GpppN^m (where N may be any nucleotide), inhibits the binding of ribosomes to capped mRNAs and, correspondingly, their translation in cell-free systems. Similarly, antibodies against the cap-binding proteins inhibit the binding of ribosomes and the initiation of translation of capped mRNAs. It is noteworthy that procaryotic mRNAs devoid of caps may in some cases be translated by eucaryotic ribosomes, but at a low rate; the addition of the cap to their 5′-ends makes them highly effective templates in eucaryotic cell-free systems[43].

It should be pointed out, however, that the mRNAs of certain viruses do not possess caps at their 5′-ends. Even so, they serve as highly effective templates in eucaryotic systems. This is true, for example, of the RNAs of picornaviruses, e.g., poliovirus or encephalomyocarditis virus, as well as of the RNA of tobacco necrosis virus. Cap-binding proteins do not participate in the association of these RNA species with ribosomes. It is assumed that, in these exceptional cases, the 5′-end of mRNA has a special (maybe, unfolded) structure which confers high affinity to the 40S ribosomal subunit with initiation factors, and so the absence of the cap structure and cap-binding proteins is well compensated for.

As already mentioned, one of the functions assumed for the cap structure is the binding of the cap-binding proteins that provide for a necessary destabilization or rearrangement of secondary and/or tertiary structure of the mRNA 5′-terminal region. It is supposed that the unwinding effect of the cap-binding protein complex depends on ATP, and that ATP hydrolysis is required for the unwinding. Perhaps, only a specially prepared ("melted") conformation of the 5′-terminal region of mRNA is capable of effectively participating in the initial association with the eucaryotic ribosomal particle. Indeed, a number of observations indicate that the less structured is the 5′-terminal region of mRNA the less dependent is its association on the presence of the cap.

The monocistronic nature of eucaryotic mRNAs and the inability of eucaryotic ribosomes to perform "internal initiation" were first suggested from the early observations of Jacobson and Baltimore[44]. Later the need for the 5′-end of mRNA to initiate translation in eucaryotic systems has been demonstrated in the elegant experiments conducted by Kozak[45]. If the linear template polynucleotide, which is capable of being translated by both eucaryotic and procaryotic ribosomes, is converted into a circle by joining its 5′-end to its 3′-end using the RNA ligase, the resulting template can be translated only by procaryotic ribosomes; the eucaryotic ribosomes lose the ability to initiate translation on the circular template. A scission of the circle before the initiation codon AUG brings a recovery in the ability of the eucaryotic to translate this polynucleotide. Similarly, experiments performed by Rosenberg and Patterson[46] have demonstrated that if the procaryotic polycistronic mRNA is equipped with a 5′-end cap structure, which allows an effective translation by eucaryotic ribosomes, the translation is limited to the 5′-proximal cistron. Thus, in contrast to procaryotic ribosomes, eucaryotic ribosomes as a rule are incapable of initiating translation on the internal sections of the template polynucleotide.

An analysis of the primary structures of the eucaryotic ribosomal RNAs and mRNAs has shown no marked complementarity between the 3′-end of the RNA of the small ribosomal subunit and the preinitiation sequence of mRNA, in contrast to the situation in procaryotic organisms. Surprisingly, the 3′-terminal 50-nucleotide-long sequences of the ribosomal RNAs of the small ribosomal subunits, including both the 16S RNA of procaryotes and the 18S of eucaryotes, are very conservative in evolution and show high homology, forming a similar hairpin of the secondary structure; however, the polypyrimidine CCUCC block of the procaryotic 16S RNA is absent from the eucaryotic 18S RNA (fig. 121). This leads to the

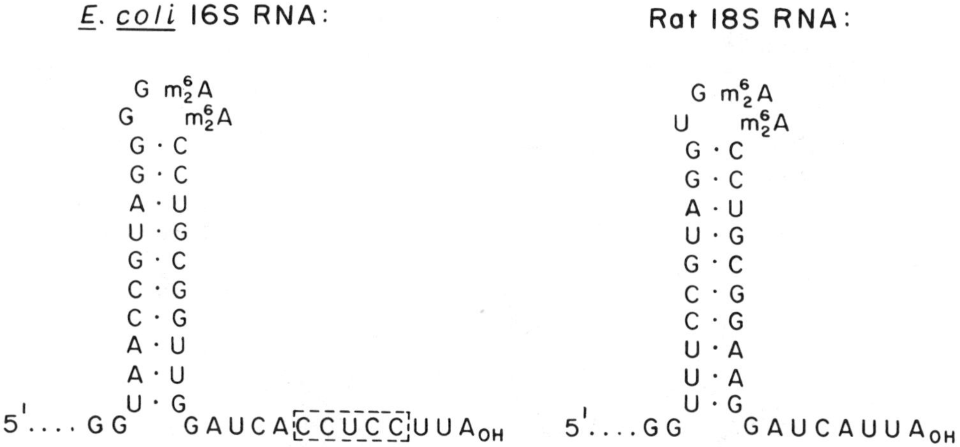

E. coli 16S RNA:

G m^6_2A
G m^6_2A
G · C
G · C
A · U
U · G
G · C
C · G
C · G
A · U
A · U
U · G
5′. . . . GG GAUCA[CCUCC]UUA$_{OH}$

Rat 18S RNA:

G m^6_2A
U m^6_2A
G · C
G · C
A · U
U · G
G · C
C · G
C · G
U · A
U · A
U · G
5′. . . . GG GAUCAUUA$_{OH}$

Figure 121 Comparison of the nucleotide sequences and secondary structures of the 3′-terminal regions of procaryotic 16S and eucaryotic 18S ribosomal RNAs[47]. The pyrimidine pentanucleotide insert into the procaryotic RNA is boxed by the broken line.

assumption that, for important reasons, eucaryotes had to dispose of the complementary recognition between the ribosomal RNA and mRNA in the initial association between ribosomes and the template and develop another way of recognizing the initiation sequence[47]. The new way appears to include the initial recognition of the 5′-end. (Of course, this reasoning may be turned upside down: recognizing the 5′-end may be a more ancient mechanism, and the evolution of procaryotes, particularly the evolution of operons and polycistronic templates, may have induced the appearance of a specialized mechanism for recognizing internal initiation regions; the pairing of the polypurine Shine–Dalgarno sequence of mRNA with the evolutionarily acquired CCUCC insert near the 3′-end of the ribosomal 16S RNA is part of such a new mechanism.)

Nevertheless, some complementary interaction between sequences located near the 5′-end of the capped mRNA and the 3′-end of the ribosomal 18S RNA still may take place in the course of eucaryotic initiation[48]; this pairing, however, is strictly dependent on the presence of initiation factors.

17-6 Recognition of the initiation codon

The association of the ribosomal particle with mRNA and recognition of the initiation codon are distinctly separated in space and time in the case of eucaryotes. The association of the 40S subunit with mRNA takes place at the 5′-end of the template polynucleotide. The initiation codon is often removed from the 5′-end by several dozens of nucleotides and, in some rare case, even by hundreds of nucleotides.

At the same time, a characteristic feature of many eucaryotic monocistronic mRNAs is the use of the first AUG triplet from the 5′-end as the initiation codon. On the basis of this and other evidence, Kozak has advanced an interesting hypothesis concerning the mechanism of the search by the 40S ribosomal subunit for the initiation region of mRNA[12,49]. According to this hypothesis, the native 40S ribosomal subunit loaded with initiation factors and the initiator methionyl-tRNA binds to the capped 5′-end of mRNA with the participation of the cap-binding proteins; upon its association with the 5′-end, the cap-binding proteins are released. Then, this 40S ribosomal subunit moves along the mRNA, scanning its sequence (it is postulated to consume ATP during this movement[50]), until it comes upon the AUG triplet, which is recognized by the anticodon of the bound initiator tRNA. It stops at this triplet, and the next step in initiation, the addition of the 60S ribosomal subunit, follows.

It should be pointed out, however, that the first AUG triplet from the 5′-end is not always used as an initiation codon. Apparently, the scanning 40S ribosomal subunit with initiator tRNA can skip some AUG triplets if they are not in a proper structural environment. Nearly all of the functional initiation AUG codons of eucaryotic mRNAs are preceded by the triplet beginning with the purine nucleotide, in most cases with A. It is assumed that when the 40S particle with the initiator tRNA scans the template, it preferentially recognizes the sequence $\genfrac{}{}{0pt}{}{\mathrm{A}}{\mathrm{G}}$NNAUG (where N may be any nucleotide residue) as the correct initiation site[51]. AUG triplets with the preceding $\genfrac{}{}{0pt}{}{\mathrm{C}}{\mathrm{U}}$NN triplet seem to be "weak" initiators and can be skipped without initiation. In addition, the functional initiation codons have G as a preferential neighbor at the 3′-side and the C-rich pentanucleotide sequence at the 5′-side (e.g., CCACC*AUG*G).

In some rare cases, the initiation of translation can take place at the AUG triplet which is removed far away from the 5′-end. For example, in the mRNA for the mouse epidermal growth factor the initiation codon is 350 nucleotide residues apart from the 5′-end; nevertheless, it is the first AUG triplet from the end. There are other rare cases where nontranslatable 5′-terminal sequences contain AUG triplets which are skipped by ribosomes without initiation; evidently they cannot serve as initiation codons due to unsuitable structural environment. Sometimes an AUG triplet in the 5′-terminal non-translatable sequence is followed, at a relatively short distance, by a termination codon which seems to prevent an occasional wrong initiation of synthesis of a long polypeptide. In cases of some virus-induced mRNAs, however, the phenomenon of "two initiations" may be observed: the first AUG triplet is recognized as an initiation codon only by a portion of scanning ribosomes, thus being just a "weak" initiator, while the other ribosomes skip it without initiation and initiate at the next "strong" AUG. In such cases, mRNA behaves as *functionally bicistronic*, since the synthesis of two different polypeptides is initiated and proceeds on the overlapping nucleotide sequences.

It is known that RNAs of many animal and plant viruses are exceptions from the rule of monocistronic nature of eucaryotic mRNA. In cases of polycistronic RNAs of plant and some animal viruses the eucaryotic ribosomes can initiate translation of only the first cistron, whereas the subsequent cistrons are read after specific fragmentation of RNA, opening new 5′-ends[12]. In other cases, such as that of poliovirus RNA, all the cistrons are translated continuously in the form of a single precursor polypeptide chain ("polyprotein") which is then cut into the chains of functionally active viral proteins[44].

Some rare cases when initiation takes place very far from the 5′-end of mRNA should be specially mentioned. For example, the initiation AUG codon of the poliovirus RNA is situated at a distance of 743 nucleotides from the 5′-end, and it is preceded by eight AUG triplets in all the three reading frames[52]. These preceding AUG triplets, however, seem to be not in a proper structural environment (see above); in addition, most of them are followed, at short distances, by termination triplets capable of interrupting translation if it were initiated. Hence, the scanning of a very long RNA sequence from the 5′-end is principally possible in order to find the right initiation codon inside the chain.

Nevertheless, there are not enough grounds yet to exclude completely a rare possibility of an internal initiation, i.e., of a direct recognition of the proper initiation codon by eucaryotic ribosomes without scanning a long preceding sequence.

17-7 Formation of the initiation ribosomal 80S complex[24,42]

After the 40S ribosomal subunit with the bound initiator tRNA has found the initiation codon, the complex binds the free native 60S subunit to yield the 80S ribosome where the initiator methionyl-tRNA is located at the P-site, while the A-site is vacant. In contrast to the situation with procaryotic ribosomes, this final stage of initiation requires the participation of a special large protein, the eIF-5. The association of the 60S ribosomal subunit, promoted by eIF-5, is accompanied by the hydrolysis of eIF-2-bound GTP, by the release of eIF-2 with GDP, and by the release of eIF-3 and other initiation factors:

$$\begin{aligned} &\text{40S·mRNA·Met-tRNA·eIF-2·GTP·eIF-3·eIF-4C·eIF-1} \\ &\qquad + \text{60S} \xrightarrow{\text{eIF-5}} \text{80S·mRNA·Met-tRNA} + \text{eIF-1} \\ &\qquad + \text{eIF-2·GDP} + \text{eIF-3} + \text{eIF-4C} + \text{P}_i. \end{aligned}$$

If a nonhydrolyzable GTP analog is present in the initiation 48S complex, the eIF-5-promoted addition of the 60S subunit is inhibited. It may be that eIF-5 induces GTP hydrolysis directly on the 40S ribosomal subunit, and contributes to the general destabilization of the 48S initiation complex, as a result of which the eIF-2 with GDP plus eIF-3 and other initiation factors leave the particle; the 40S ribosomal subunit remains in the complex with mRNA and the methionyl-tRNA and is then capable of associating with the 60S ribosomal subunit. In any case, it has been shown that if the 48S initiation complex is washed to remove eIF-2 and eIF-3, it can bind spontaneously the 60S ribosomal subunit, in the absence of eIF-5. There is some evidence that eIF-4C and eIF-4D are involved in the correct association of the subunits and in the formation of the eventual 80S initiation complex containing methionyl-tRNA in the P-site.

The sequence of events during the initiation of translation in eucaryotes is shown schematically in fig. 122.

17-8 Regulation of initiation

Although it is presumed that control at the level of translation is particularly important in eucaryotes and that this system of controlling protein synthesis must be well developed, there is surprisingly little direct experimental information and few clear examples of translational control mechanisms. Masking and unmasking mechanisms involved in storing mRNA in resting cells and its activation upon fertilization or germination, and the mechanisms for switching on and shutting down the translation of certain presynthesized mRNAs in the course of cell differentiation, are still completely unclear. The part played by numerous proteins associated with nontranslated and translated mRNA in the mRNPs of all species is also obscure. The information available is fragmentary and does not reveal the general regulatory system of the translational machinery of the eucaryotic cell.

The data available may be broken down into two groups of facts. First, there are many cases of selective discrimination of mRNA resulting from different rates ("strengths") of initiation with different mRNAs. This phenomenon is the result of the 5′-ends and the initiation region of templates having some unknown structural features which affect the initiation rate. Various cases of the inhibited translation of host mRNA and simultaneous, efficient translation of viral RNA in virus-infected eucaryotic cells also belong here. Second, the total control (total repression) of protein synthesis in the eucaryotic cell due to the modification of eIF-2, the key initiation factor, has been demonstrated clearly.

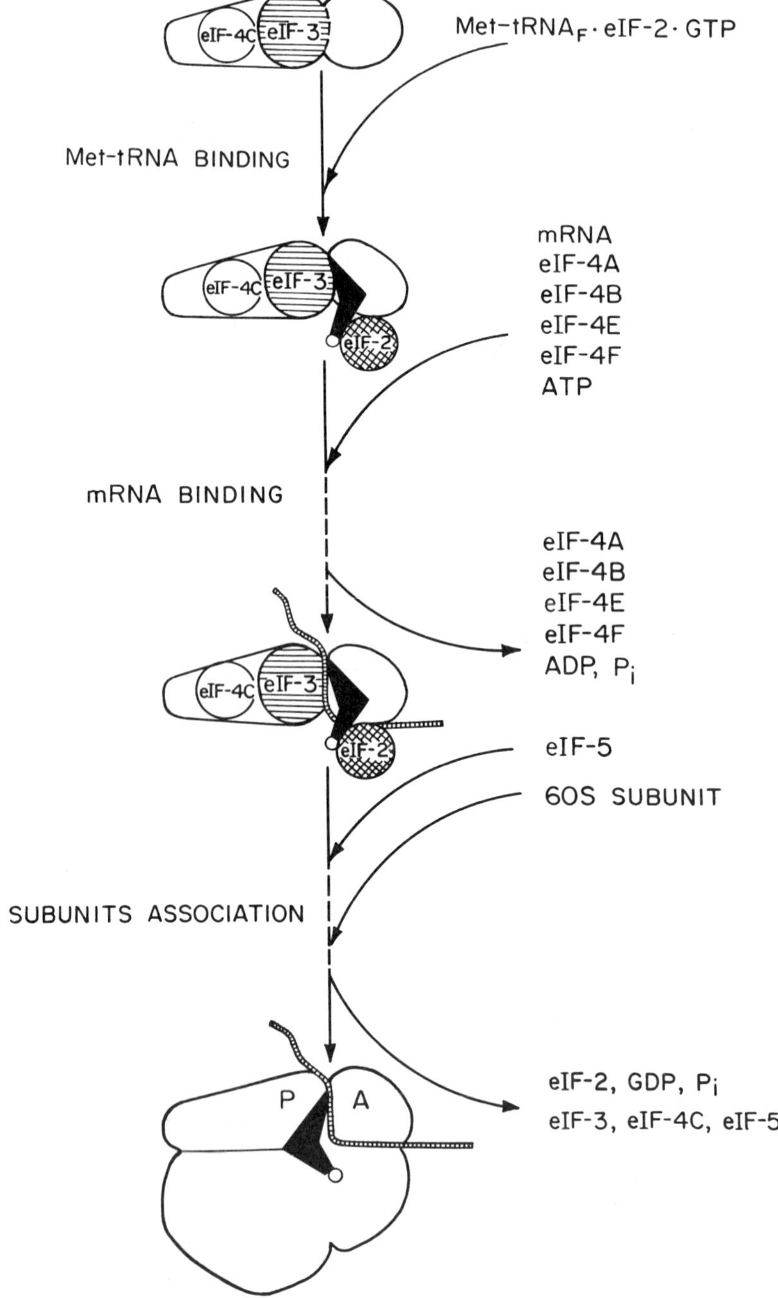

Figure 122 Sequence of events during initiation of translation in eucaryotes.

Selective discrimination of mRNA

Mammalian reticulocytes, and the cell-free systems derived from them, provide ideal models for studying control at the level of translation. The reticulocytes do not contain a cell nucleus and, therefore, there is no transcriptional effect. Furthermore, such a system synthesizes primarily one protein, hemoglobin. Formation of hemoglobin requires equimolar quantities of globin α- and β-chains. Two corresponding mRNA species present in the cytoplasm or in the extract are effectively translated by ribosomes. The elongation rates in polyribosomes for both these mRNAs have been shown to be identical, being about 5 to 10 amino acid residues per second at 37°C or one tenth of this value at 20°C[53–55]. The polyribosomes synthesizing β-chains, however, contain more ribosomes than those synthesizing α-chains, although the lengths of these two mRNA species are similar. It follows that initiation on mRNA coding for the β-chain takes place more effectively compared to the mRNA coding for the α-chain[54]. In other words, the β-chain mRNA is "stronger" in initiation. To compensate for this difference and to produce equimolar quantities of the α- and β-chains, the cell contains a correspondingly higher amount of mRNA for the α-chain than for the β-chain. It is thought that the 5′-terminal sequence of β-chain mRNA has some structural characteristics determining its higher affinity to initiation factors and perhaps to the 40S ribosomal subunits.

Another example of discrimination between two mRNAs is the different rates of initiation of the ovalbumin and the conalbumin mRNAs (mRNAs for hen egg proteins) in cell-free translation systems[56].

In the case of some viral infections, the inhibition of the host mRNA translation and the preferential synthesis of viral proteins are the results of a similar competition for initiation factors or ribosomes: viral mRNA species appear to be more effective than the host cell mRNA in the initiation of translation. This takes place, for example, when cells are infected with the encephalomyocarditis virus[57]. Here again, the affinity of the viral RNA 5′-terminal sequence to host initiation factors is far greater than that of the host mRNA. There is evidence that the affinity of the encephalomyocarditis virus RNA to eIF-4B is one order of magnitude greater than that of the cellular mRNA[58]; the depletion of this initiation factor, due to its binding with the viral mRNA, could explain the inhibition of the host protein synthesis initiation, although this mechanism has not been demonstrated directly. The initiation factors including the cap-binding proteins of the host cell do not undergo any modifications in this case, and their level in the cell does not change.

Another mechanism of selective mRNA discrimination operates when cells are infected with poliovirus. Here, the infection inactivates the cap-binding protein complex[59], resulting in a reduced rate of initiation on all capped mRNA, while initiation on the viral noncapped mRNA remains effective. The inactivation seems to be caused by virus-induced proteolysis of the 220,000-dalton subunit of the cap-binding protein complex (eIF-4F)[60].

Finally, there are cases where the selective discrimination of mRNAs in translation is achieved through changes in the ionic composition of the intracellular medium in response to viral infection[61]. Thus, infecting chicken fibroblasts with the Sindbis virus results in a decreased K^+ and an increased Na^+ concentration in the cell[62]. Under the changed ionic conditions, the affinity of the host mRNA to the components of the initiation complex lessens, while the viral mRNA retains its high initiation efficiency. Some special features in the structure of the viral RNA 5′-terminal sequence may explain the survival of its high affinity to corresponding initiation proteins under changed ionic conditions. Indeed, the optimal concentrations of monovalent cations for the translation of a number of viral RNAs *in vitro* have been shown to be higher than those required for the translation of most cellular mRNAs, the globin mRNA included[61]. It should be pointed out, incidentally, that the initiation rate for different cellular mRNAs also shows differential changes in response to the changed ionic conditions; for example, the proportion between synthesized α- and β-globin chains deviates from the normal toward an increased level of β-chain when KCl concentration in the cell-free reticulocyte system is increased[63]. It is possible that the differential effect of salt concentration on initiation with different mRNAs is directly related to the degree of secondary structure at the 5′-terminal region: the more developed is the secondary structure the stronger is its stabilization by increased ionic strength and, hence, the less is the unwinding action of the cap-binding proteins[29].

As for initiation factors whose affinity to the mRNA 5′-terminal sequence is decisive for the selective discrimination, the information available is somewhat controversial. The possible role of the eIF-4A, eIF-4B, and eIF-4E, particularly in connection with their involvement in the formation of the cap-binding protein 8–10S complex CBP-II (eIF-4F), is often discussed[64,65]. In several cases the different affinity of eIF-2 to different mRNAs, e.g. α- and β-globin mRNAs, has been described[63].

The results on the tissue specificity of eIF-3 and its possible role in the tissue-specific discrimination of mRNA are particularly interesting. Heywood and co-workers tried to demonstrate some time ago that eIF-3 preparations from different tissues possessed functional differences[66,67]. When an eIF-3 preparation from a muscle was added to the cell-free system containing both globin and myosin mRNA, the translation of myosin mRNA underwent selective stimulation. The purified eIF-3 from the muscle could be fractionated into the "core" and "discriminatory" subunits; the "core" eIF-3 stimulated the translation of both myosin and globin mRNAs, the myosin mRNA being translated less effectively; the addition of the discriminatory subunits resulted in the preferential translation of myosin mRNA[68]. This information in favor of a positive initiation control by eIF-3 deserves special consideration as evidence of a possible control mechanism at the level of translation in cell differentiation.

As already discussed (chapter 15), in eucaryotic cells the rate of elongation can be also controlled, and this control may be either selective or total. Changes in the elongation rate are important for the regulation of initiation, including selective discrimination between mRNAs: when the elongation is slowed down, the initiation ceases to be a rate-limiting stage of the whole translation process, so that the differences in initiation rates of different mRNAs become less displayed. In other words, the decrease of the elongation rate will result in a lower discrimination between mRNAs with a different strength of initiation. Indeed, it has been demonstrated directly, in the case of discrimination between mRNA coding for globin α- and β-chains, that slowing down the elongation rate by partial inhibition with some antibiotics results in a loss of preferential initiation on the β-chain mRNA[54].

Repression of initiation[42,69–71]

Hemin-controlled initiation of translation in reticulocytes. It has long been known that synthesis of proteins, particularly of hemoglobin, in rabbit and other mammalian reticulocytes, as well as in reticulocyte lysates and reticulocyte cell-free systems, requires hemin. In the absence of hemin, protein synthesis rapidly comes to a halt. A decrease in the protein synthesis rate in the absence of hemin is accompanied by a lower content of polyribosomes and by an accumulation of nontranslating 80S monoribosomes. This suggests that the initiation of translation becomes suppressed. Direct experiments have demonstrated that the amount of the initiation 43S complex, which contains initiator methionyl-tRNA$_F$, decreases when hemin is absent.

It has been shown that incubating reticulocyte extracts without hemin at physiological temperatures leads to an accumulation of a special inhibitor that is protein in nature. This inhibitor is called the *heme-controlled inhibitor* (HCI) or heme-controlled repressor (HCR). It is formed from the proinhibitor which preexists in the cell or, correspondingly, in the extract. The addition of an inhibitor or of an inhibitor-containing extract to a cell-free system results in the inhibition of globin synthesis even when hemin is present.

The activation of the proinhibitor and its conversion into the inhibitor in the absence of hemin, proceed in two phases. Initially, the proinhibitor without hemin is converted, within a few minutes, into the so-called reversible inhibitor; the addition of hemin returns the reversible inhibitor into the proinhibitor. The nature of this conversion, which requires a physiological temperature, is not clear since the physical characteristics of this protein appear to be similar whether it is in the proinhibitor or the reversible inhibitor state. It is assumed that an intramolecular rearrangement is responsible for the appearance of the inhibitory activity. Upon subsequent incubation, the reversible inhibitor is converted, within several hours, into an irreversible inhibitor. This conversion appears to be caused by the oxidation of the protein SH-groups and may be accelerated greatly by reagents such as N-ethylmaleimide or *o*-iodobenzoate blocking the SH-groups. In contrast, sulfhydryl reducing compounds, e.g. reduced glutathione, β-mercaptoethanol, or dithiotreitol, prevent the reversible inhibitor from transforming into an irreversible inhibitor. This transformation also involves self-phosphorylation of the inhibitor at the expense of ATP.

The inhibitor is a large protein. Its molecular mass ranges from 150,000 to 350,000 daltons according to different measurements, and the sedimentation coefficient is about 6S. Its subunit structure has not been determined accurately; the molecule appears to consist of either identical or different subunits, each with a molecular mass of about 80,000 to 100,000 daltons.

In an activated state the inhibitor possesses cAMP-independent protein kinase activity which specifically phosphorylates the eIF-2 α-subunit in an ATP-driven reaction. It is this activity that is responsible for the inhibition (repression) of initiation in reticulocytes as well as in their lysates, extracts, and cell-free translation systems when hemin is absent. The kinase activity of the reversible inhibitor is suppressed by hemin. The inhibitor is also capable of self-phosphorylation, which may play a part in controlling its activity.

The effect of eIF-2 phosphorylation catalyzed by the inhibitor is not simple. Phosphorylation of the eIF-2 α-subunit does not render it inactive. The pure preparations of the original eIF-2 and phosphorylated eIF-2 are equally active in the formation of the Met-$tRNA_F$·eIF-2·GTP ternary complex, in binding the ternary complex to the 40S ribosomal subunit, and in eventually forming the initiation 80S complex when eIF-2 is released from the ribosome.

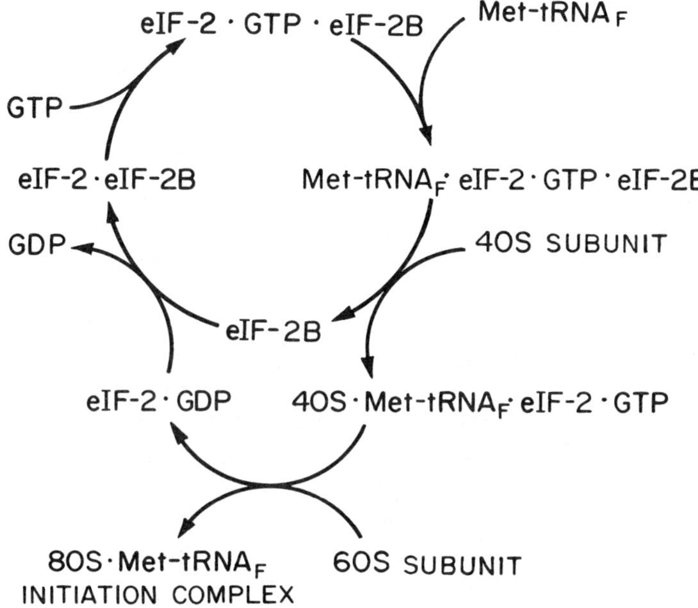

Figure 123 Scheme of the cyclic use of eucaryotic initiation factors eIF-2 and eIF-2B[74]. (Redrawn with some modifications from the scheme provided by Prof. H. O. Voorma, Department of Molecular Cell Biology, University of Utrecht)

It is now demonstrated that the phosphorylation of eIF-2 affects its interaction with a special high-molecular-mass protein (molecular mass of about 300,000 daltons) which is on the list of initiation factors and is designated as eIF-2B (or anti-HCI). According to the recent scheme[72–74] (fig. 123), eIF-2·GDP released from the ribosomal initiation complex at the stage of the 60S subunit joining interacts with eIF-2B (bottom of the scheme). In the eIF-2B·eIF-2 complex the eIF-2-bound GDP rapidly exchanges for GTP (left side of the scheme). Initiator Met-tRNA$_F$ associates directly with the eIF-2B·eIF-2·GTP complex, resulting in the quarternary complex eIF-2B·(Met-tRNA$_F$·eIF-2·GTP) being formed (right side of the scheme). During interaction with the 40S subunit, the eIF-2B dissociates from eIF-2, thus leaving Met-tRNA$_F$·eIF-2·GTP on the ribosomal particle. Hence, normally eIF-2B catalyzes the reutilization of eIF-2 after its release in the eIF-2·GDP form and, seemingly, the binding of the Met-tRNA·eIF-2·GTP to the 40S ribosomal subunit.

When eIF-2 is phosphorylated by the inhibitor (HCI), eIF-2B can still bind to eIF-2(P), but the resulting complex eIF-2(P)·eIF-2B is stable, and thus the eIF-2B becomes sequestered. It should be noted that in the cell or in a cell-free system the amount of eIF-2B is more limited as compared to eIF-2. Therefore, even partial phosphorylation of the eIF-2 pool (say, 30%) will be sufficient to sequester all eIF-2B in the stable inactive complexes. As a result, the cell or a cell-free system becomes depleted of eIF-2B, and eIF-2 accumulates in the form of eIF-2·GDP instead of being rapidly involved in the formation of the initiation 43S complex. In other words, in the absence of eIF-2B or after its sequestration on phosphorylated eIF-2, the catalytic use of eIF-2 (its recycling) becomes suppressed. That is why, when the action of pure eIF-2 in the original and the phosphorylated states are compared, these two molecular forms do not show any difference in intrinsic activity in the partial reactions of initiation. When, however, the additional protein eIF-2B is present, its stimulatory action can be seen only in combination with the nonphosphorylated form of eIF-2.

In conclusion, it should be said that eIF-2 phosphorylation by the heme-controlled inhibitor is partly counterbalanced by its dephosphorylation catalyzed by a specific phosphatase, which is present both in the cell and in the extract. Indeed, normally in the presence of hemin there is always a certain level of protein kinase activity, which phosphorylates eIF-2, and there is an equilibrium between eIF-2 phosphorylation and dephosphorylation:

$$\text{eIF-2} \underset{\text{phosphatase}}{\overset{\text{HCI + ATP}}{\rightleftarrows}} \text{eIF-2(P)}.$$

Hemin deficiency results in more HCI being activated and, consequently, the equilibrium toward phosphorylated eIF-2 being shifted.

Repression of translation by double-stranded RNA. Double-stranded RNA, including both double-stranded fragments of viral origin (polio- or reovirus RNA) and synthetic complexes such as poly(A)·poly(U) or poly(I)·poly(C), inhibit protein synthesis in reticulocyte lysates, even when hemin is present; all the characteristics are reminiscent of the repression caused by the lack of hemin. In order to inhibit translation, the double-stranded RNA should be at least 50 nucleotide pairs long. It has been demonstrated that in the presence of such a double-stranded RNA, another initiation inhibitor, designated as dsI, undergoes activation. This inhibitor is also a protein kinase phosphorylating the α-subunit of eIF-2. In contrast to HCI, however, the dsI, which is a protein with a molecular mass of about 67,000 daltons, is associated with ribosomes. ATP is required for activating the inhibitor, and this involves self-phosphorylation of the protein. This self-phosphorylation reaction is induced by the interaction between the protein and the double-stranded RNA. Apparently, the repression of initiation by activated dsI is analogous to that in the case of HCI and is also caused by a phosphorylation-induced change in the interaction between eIF-2 and the additional protein, eIF-2B (see above).

The dsI is of particular interest as far as the antiviral action of interferon upon animal cells is concerned. It is known that the interaction of cells with interferon results in several proteins being induced; this includes an increased level of cAMP-independent protein kinase phosphorylating the α-subunit of eIF-2. Like dsI, this protein kinase is inactive or latent until it comes in contact with the double-stranded RNA. It is possible that the protein kinase is identical to dsI, or that dsI is present in the enzyme of interferon-treated cells as one of its subunits.

It is noteworthy that double-stranded RNA induces a strong inhibition of protein synthesis in reticulocyte lysates without these cells being pretreated with interferon. In contrast, it affects only slightly translation in extracts of HeLa cells, L-fibroblasts, Ehrlich ascite carcinoma cells, and other cell types. Pretreating these cells with interferon increases markedly the sensitivity of their extracts to the inhibitory action of double-stranded RNA; this correlates directly with the phosphorylation of eIF-2 in the presence of the double-stranded RNA. It is now believed that interferon induces *de novo* synthesis of protein kinase and thereby prepares the cell for the "antiviral state." The penetration of the virus into the cell and the appearance of long double-stranded RNA regions in the cell activate the enzyme immediately. This leads to the phosphorylation of eIF-2 and, thus, the reduction of the initiation rate. This event may be significant to the inhibition of synthesis of viral components, among other events induced by interferon.

It should be pointed out that interferon also induces the formation of oligoisoadenylate synthetase E, which requires double-stranded RNA for its activity and which synthesizes the cofactor of preexisting RNAase F, (2′–5′)oligoadenylate; as a consequence, RNAase F begins to degrade mRNA.

References

1. A. S. Spirin (1966), "On 'masked' forms of messenger RNA in early embryogenesis and in other differentiating systems," Current topics in developmental biology, ed. A. A. Moscona and A. Monroy, vol. 1, pp. 1–38 (New York: Academic Press).

2. P. R. Gross (1967), "The control of protein synthesis in embryonic development and differentiation," Current topics in developmental biology, ed. A. A. Moscona and A. Monroy, vol. 2, pp. 1–47 (New York: Academic Press).

3. A. S. Spirin (1969), "Informosomes," *Eur. J. Biochem.* 10:20–35.

4. O. Civelli, A. Vincent, K. Maundrell, J.-F. Buri, and K. Scherrer (1980), "The translational repression of globin mRNA in free cytoplasmic ribonucleoprotein complexes," *Eur. J. Biochem.* 107:577–585.

5. A. J. Bester, D. S. Kennedy, and S. M. Heywood (1975), "Two classes of translational control RNA: Their role in the regulation of protein synthesis," *Proc. Nat. Acad. Sci. U.S.A.* 72:1523–1527.

6. M. E. Buckingham, A. Cohen, and F. Gros (1976), "Cytoplasmic distribution of pulse-labeled poly(A)-containing RNA, particularly 26S RNA, during myoblast growth and differentiation," *J. Mol. Biol.* 103:611–626.

7. A. A. Preobrazhensky and A. S. Spirin (1978), "Informosomes and their protein components: The present state of knowledge," Progress in nucleic acid research and molecular biology, ed. W. E. Cohn, vol. 21, pp. 1–38 (New York: Academic Press).

8. P. Pennequin, D. M. Robins, and R. T. Schimke (1978), "Regulation of translation of ovalbumin messenger RNA by estrogens and progesterone in oviduct of withdrawn chicks," *Eur. J. Biochem.* 90:51–58.

9. D. C. Beele and J. Piatigorsky (1981), "Translational regulation of δ-crystallin synthesis during lens development in the chicken embryo," *Dev. Biol.* 84:96–101.

10. M. Bienz and J. B. Gurdon (1982), "The heat-shock response in Xenopus oocytes is controlled at the translational level," *Cell* 29:811–819.

11. J. C. Brown and A. E. Smith (1980), "Initiation codons in eukaryotes," *Nature* 226:610–612.

12. M. Kozak (1978), "How do eukaryotic ribosomes select initiation regions of messenger RNA?" *Cell* 15:1109–1123.

13. A. E. Smith and K. A. Marcker (1970), "Cytoplasmic methionine transfer RNAs from eukaryotes," *Nature* 226:607–610.

14. N. K. Gupta, N. K. Chatterjee, K. K. Bose, S. Bhaduri, and A. Chung (1970), "Roles of methionine transfer RNAs in protein synthesis in rabbit reticulocytes," *J. Mol. Biol.* 54:145–154.

15. D. Houseman, M. Jacobs-Lorena, U. L. RajBhandary, and H. F. Lodish (1970), "Initiation of haemoglobin synthesis by methyonyl-tRNA," *Nature* 227:913–918.

16. M. Simsek and U. L. RajBhandary (1972), "The primary structure of yeast initiator transfer ribonucleic acid," *Biochem. Biophys. Res. Commun.* 49:508–515.

17. H. Gruhl and H. Feldman (1975), "The primary structure of a noninitiating methionine specific tRNA from brewer's yeast," *FEBS Letters* 57:145–148.

18. B. Safer, S. L. Adams, W. M. Kemper, K. W. Berry, M. Lloyd, and W. C. Merrick (1976), "Purification and characterization of two initiation factors required for maximal activity of a highly fractionated globin mRNA translation system," *Proc. Nat. Acad. Sci. U.S.A.* 73:2584–2588.

19. M. H. Schreier, B. Erni, and T. Staehelin (1977), "Initiation of mammalian protein synthesis. I. Purification and characterization of seven initiation factors," *J. Mol. Biol.* 116:727–753.

20. R. Benne, M. Brown-Luedi, and J. W. B. Hershey (1978), "Purification and characterization of protein synthesis initiation factors eIF-1, eIF-4C, eIF-4D, and eIF-5 from rabbit reticulocytes," *J. Biol. Chem.* 253:3070–3077.

21. R. Kaempfer (1974), "Identification and RNA-binding properties of an initiation factor capable of relieving translational inhibition induced by heme deprivation or double-stranded RNA," *Biochem. Biophys. Res. Commun.* 61:591–597.

22. L. P. Ovchinnikov, A. S. Spirin, B. Erni, and T. Staehelin (1978), "RNA-binding proteins of rabbit reticulocytes contain the two elongation factors and some of the initiation factors of translation," *FEBS Letters* 88:21–26.

23. T. N. Vlasik, S. P. Domogatsky, T. A. Bezlepkina, and L. P. Ovchinnikov (1980), "RNA-binding activity of eukaryotic initiation factors of translation," *FEBS Letters* 116:8–10.

24. A. A. M. Thomas, R. Benne, and H. O. Voorma (1981), "Initiation of eukaryotic protein synthesis," *FEBS Letters* 128:177–185.

25. M. A. Lloyd, J. C. Osborne, B. Safer, G. M. Powell, and W. C. Merrick (1980), "Characteristics of eukaryotic initiation factor 2 and its subunits", *J. Biol. Chem.* 255:1189–1193.

26. N. Sonenberg, M. A. Morgan, W. C. Merrick, and A. J. Shatkin (1978), "A polypeptide in eukaryotic initiation factors that crosslinks specifically to the 5′-terminal cap in mRNA," *Proc. Nat. Acad. Sci. U.S.A.* 45:4843–4847.

27. N. Sonenberg (1981), "ATP/Mg^{2+}-dependent crosslinking of cap binding proteins to the 5′-end of eukaryotic mRNA," *Nucleic Acids Res.* 9:1643–1656.

28. J. A. Grifo, S. M. Tahara, J. P. Leis, M. A. Morgan, A. J. Shatkin, and W. C. Merrick (1982), "Characterization of eukaryotic initiation factor 4A, a protein involved in ATP-dependent binding of globin mRNA," *J. Biol. Chem.* 257:5246–5252.

29. K. A. W. Lee, D. Guertin, and N. Sonenberg (1983), "mRNA secondary structure as a determinant in cap recognition and initiation complex formation", *J. Biol. Chem.* 258:707–710.

30. H. Amesz, H. Goumans, T. Haubrich-Morree, H. O. Voorma, and R. Benne (1979), "Purification and characterization of a protein factor that reverses the inhibition of protein synthesis by the heme-regulated translational inhibitor in rabbit reticulocyte lysates," *Eur. J. Biochem.* 98:513–520.

31. J. Sielierka, L. Mauser, and S. Ochoa (1982), "Mechanism of polypeptide

chain initiation in eukaryotes and its control by phosphorylation of the alpha-subunit of initiation factor 2," *Proc. Nat. Acad. Sci. U.S.A.* 79:2537–2540.

32. A. Konieczny and B. Safer (1983), "Purification of the eukaryotic initiation factor 2-eukaryotic initiation factor 2B complex and characterization of its guanine nucleotide exchange activity during protein synthesis initiation," *J. Biol. Chem.* 258:3402–3408.

33. R. Panniers and E. C. Henshaw (1983), "A GDP/GTP exchange factor essential for eukaryotic initiation factor 2 cycling in Ehrlich ascites tumor cells and its regulation by eukaryotic initiation factor 2 phosphorylation," *J. Biol. Chem.* 258:7928–7934.

34. R. L. Matts, D. H. Levin, and I. M. London (1983), "Effect of phosphorylation of the α-subunit of eukaryotic initiation factor 2 on the function of reversing factor in the initiation of protein synthesis," *Proc. Nat. Acad. Sci. U.S.A.* 80:2559–2563.

35. B. D. Davis (1971), "Role of subunits in the ribosome cycles," *Nature* 231:153–157.

36. H. A. Thompson, I. Sadnik, J. Scheinbuks, and K. Moldave (1977), "Studies on native ribosomal subunits from rat liver: Purification and characterization of a ribosome dissociation factor," *Biochemistry* 16:2221–2230.

37. W. J. van Venrooij, A. P. M. Janssen, J. H. Haeymakers, and B. M. DeMan (1976), "On the heterogeneity of native ribosomal subunits in Ehrlich-ascites-tumor cells cultured in vitro," *Eur. J. Biochem.* 64:429–435.

38. I. Sundkvist and T. Staehelin (1975), "Structure and function of free 40S ribosome subunits: Characterization of initiation factors," *J. Mol. Biol.* 99:401–418.

39. C. Freienstein and G. Blobel (1975), "Nonribosomal proteins associated with eukaryotic native small ribosomal subunits," *Proc. Nat. Acad. Sci. U.S.A.* 72:3392–3396.

40. N. K. Gupta, C. L. Woodley, Y. C. Chen, and K. K. Bose (1973), "Protein synthesis in rabbit reticulocytes: Assays, purification, and properties of different ribosomal factors and their roles in peptide chain initiation," *J. Biol. Chem.* 248:4500–4511.

41. B. Safer, S. L. Adams, W. F. Anderson, and W. C. Merrick (1975), "Binding of met-$tRNA_f$ and GTP to homogeneous initiation factor MP," *J. Biol. Chem.* 250:9076–9082.

42. R. Jagus, W. F. Anderson, and B. Safer (1981), "The regulation of initiation of mammalian protein synthesis," in Progress in nucleic acid research and molecular biology, ed. W. E. Cohn, vol. 25, pp. 127–185 (New York: Academic Press).

43. B. M. Patterson and M. Rosenberg (1979), "Efficient translation of prokaryotic mRNAs in a eukaryotic cell-free system require addition of a cap structure," *Nature* 279:692–696.

44. M. F. Jacobson and D. Baltimore (1968), "Polypeptide cleavage in the formation of poliovirus proteins," *Proc. Nat. Acad. Sci. U.S.A.* 61:77–84.

45. M. Kozak (1979), "Inability of circular mRNA to attach to eukaryotic ribosomes," *Nature* 280:82–85.

46. M. Rosenberg and B. M. Patterson (1979), "Efficient cap-dependent translation of polycistronic prokaryotic mRNAs is restricted to the first gene in the operon," *Nature* 279:696–701.

47. J. A. Steitz (1980), "RNA·RNA interactions during polypeptide chain initiation," in *Ribosomes: Structure, function, and genetics*, ed. G. Chambliss, G. R. Craven, J. Davies, K. Davis, L. Kahan, and M. Nomura, pp. 479–495 (Baltimore: University Park Press).

48. B. Erni and T. Staehelin (1982), "Base-pair formation between 18S ribosomal RNA and globin mRNA during initiation of protein synthesis in vitro," *FEBS Letters* 148:79–82.

49. M. Kozak (1980), "Evaluation of the 'scanning model' for initiation of protein synthesis in eukaryotes," *Cell* 22:7–8.

50. M. Kozak (1980), "Role of ATP in binding and migration of 40S ribosomal subunits," *Cell* 22:459–467.

51. M. Kozak (1981), "Possible role of flanking nucleotides in recognition of the AUG initiator codon by eukaryotic ribosomes," *Nucleic Acids Res.* 9:5233–5252.

52. A. J. Dorner, L. F. Dorner, G. R. Larsen, E. Wimmer, and C. W. Anderson (1982), "Identification of the initiation site of poliovirus polyprotein synthesis," *J. Virol.* 42:1017–1028.

53. T. Hunt, T. Hunter, and A. Munro (1969), "Control of haemoglobin synthesis: Rate of translation of the messenger RNA for the α and β chains," *J. Mol. Biol.* 43:123–133.

54. H. F. Lodish (1971), "Alpha and Beta globin messenger ribonucleic acid: Different amounts and rates of initiation of translation," *J. Biol. Chem.* 246:7131–7138.

55. H. F. Lodish and M. Jacobsen (1972), "Regulation of hemoglobin synthesis: Equal rates of translation and termination of α- and β-globin chains," *J. Biol. Chem.* 247:3622–3629.

56. R. D. Palmiter (1974), "Differential rates of initiation on conalbumin and ovalbumin messenger ribonucleic acid in reticulocyte lysates," *J. Biol. Chem.* 249:6779–6787.

57. Yu. V. Svitkin, V. A. Ginevskaya, T. Yu. Ugarova, and V. I. Agol (1978), "A cell-free model of the encephalomyocarditis virus-induced inhibition of host cell protein synthesis," *Virology* 87:199–203.

58. C. Baglioni, M. Simili, and D. A. Shafritz (1978), "Initiation activity of EMC virus RNA, binding to initiation factor eIF-4B, and shut-off of host cell protein synthesis," *Nature* 275:240–243.

59. K. A. W. Lee and N. Sonenberg (1982), "Inactivation of cap-binding proteins accompanies the shut-off of host protein synthesis by poliovirus," *Proc. Nat. Acad. Sci. U.S.A.* 79:3447–3451.

60. D. Etchison, S. C. Milburn, I. Edery, N. Sonenberg, and J. W. B. Hershey (1982), "Inhibition of HeLa cell protein synthesis following poliovirus infection correlates with the proteolysis of a 220,000 dalton polypeptide associated with eucaryotic initiation factor 3 and a cap binding protein complex," *J. Biol. Chem.* 257:14806–14810.

61. L. Carrasco and A. E. Smith (1976), "Sodium ions and the shut-off of host cell protein synthesis by picornaviruses," *Nature* 264:807–809.

62. R. F. Garry, J. M. Bishop, S. Parker, K. Westbrook, G. Lewis, and M. R. F. Waite (1979), "Na^+ and K^+ concentrations and the regulation of protein synthesis in Sindbis virus-infected chick cells," *Virology* 96:108–120.

63. G. Di Segni, H. Rosen, and R. Kaempfer (1979), "Competition between α- and β-globin messenger ribonucleic acids for eucaryotic initiation factor 2," *Biochemistry* 18:2847–2854.

64. H. van Steeg, M. van Grinsven, F. van Mansfeld, H. O. Voorma, and R. Benne (1981), "Initiation of protein synthesis in neuroblastoma cells infected by Semliki Forest virus: A decreased requirement of late viral mRNA for eIF-4B and cap binding protein," *FEBS Letters* 129:62–66.

65. A. Parets Soler, L. Reibel, and G. Schapira (1981), "Differential stimulation of α- and β-globin mRNA translation by M_r 50,000 and 28,000 polypeptide containing fractions isolated from reticulocyte polysomes," *FEBS Letters* 136:259–264.

66. D. S. Kennedy and S. M. Heywood (1976), "The role of muscle and reticulocyte initiation factor 3 on the translation of myosin and globin messenger RNA in a wheat germ cell-free system," *FEBS Letters* 72:314–318.

67. S. M. Heywood and D. S. Kennedy (1979), "Messenger RNA affinity column fractionation of eukaryotic initiation factor and the translation of myosin messenger RNA," *Arch. Biochem. Biophys.* 192:270–281.

68. W. R. Gette and S. M. Heywood (1979), "Translation of myosin heavy chain messenger ribonucleic acid in an eukaryotic initiation factor 3- and messenger-dependent muscle cell-free system," *J. Biol. Chem.* 254:9879–9885.

69. G. Kramer, A. B. Henderson, N. Grankowski, and B. Hardesty (1980), "Translational control in eukaryotes," in *Ribosomes: Structure, function, and genetics*, ed. G. Chambliss, G. R. Craven, J. Davies, K. Davis, L. Kahan, and M. Nomura, pp. 825–845 (Baltimore: University Park Press).

70. S. A. Austin and M. J. Clemens (1980), "Control of the initiation of protein synthesis in mammalian cells," *FEBS Letters* 110:1–7.

71. S. Ochoa (1983), "Regulation of protein synthesis initiation in eucaryotes," *Arch. Biochem. Biophys.* 223:325–349.

72. B. Safer (1983), "2B or not 2B: regulation of the catalytic utilization of eIF-2," *Cell* 33:7–8.

73. J. Siekierka, V. Manne, and S. Ochoa (1984), "Mechanism of translational control by partial phosphorylation of the α-subunit of eukaryotic initiation factor 2," *Proc. Nat. Acad. Sci. U.S.A.* 81:352–356.

74. M. Salimans, H. Goumans, H. Amesz, R. Benne, and H. O. Voorma (1984), "Regulation of protein synthesis in eukaryotes. Mode of action of eRF, and eIF-2-recycling factor from rabbit reticulocytes involved in GDP/GTP exchange," *Eur. J. Biochem.* 145:91–98.

Further reading

Anderson, W. F.; Bosch, L.; Cohn, W. E.; Lodish, H.; Merrick, W. C; Weissbach, H.; Wittmann, H. G.; and Wool, I. G. (1977). International symposium on protein synthesis: Summary of Fogarty Center-NIH Workshop held in Bethesda, Maryland, on 18–20 October, 1976. *FEBS Letters* 76:1–10.

Bloemendal, H. (1972). Mammalian messenger RNA. In *The mechanism of protein synthesis and its regulation* (L. Bosch, ed.), pp. 487–514. Amsterdam and London: North-Holland.

Bosch, L., and Voorma, H. O. (1972). Translation of viral RNA. In *The mechanism of protein synthesis and its regulation* (L. Bosch, ed.), pp. 395–440. Amsterdam and London: North-Holland.

Grunberg-Manago, M., and Gros, F. (1977). Initiation mechanisms of protein synthesis. Progress in nucleic acid research and molecular biology (W. E. Cohn, ed.), vol. 20, pp. 209–284. New York: Academic Press.

Kozak, M. (1983). Comparison of initiation of protein synthesis in procaryotes, eucaryotes, and organelles. *Microbiol. Rev.* 47:1–45.

Lodish, H. F. (1976). Translational control of protein synthesis. *Ann. Rev. Biochem.* 45:39–72.

mRNA: The relation of structure to function (1976). Progress in nucleic acid research and molecular biology (W. E. Cohn and E. Volkin, eds.), vol. 19. New York: Academic Press.

Revel, M. (1972). Polypeptide chain initiation: The role of ribosomal protein factors and ribosomal subunits. In *The mechanism of protein synthesis and its regulation* (L. Bosch, ed.), pp. 87–131. Amsterdam and London: North-Holland.

——— (1977). Initiation of messenger RNA translation into protein and some aspects of its regulation. In *Molecular mechanisms of protein biosynthesis* (H. Weissbach and S. Pestka, eds.), pp. 245–321. New York: Academic Press.

Revel, M., and Groner, Y. (1978). Post-transcriptional and translational controls of gene expression in eukaryotes. *Ann. Rev. Biochem.* 47:1079–1126.

Safer, B.; Grunberg-Manago, M.; Badman, D.; Bergman, F.; Freeman, C.; Galasso, G.; Jagus, R.; and Williams, B. (1982). International symposium on translational/transcriptional regulation of gene expression: Highlights of the Fogarty International Center Meeting held in Bethesda, Maryland, on 7–9 April, 1982. *FEBS Letters* 147:1–10.

Shafritz, D. A. (1977). Messenger RNA and its translation. In *Molecular mechanisms of protein biosynthesis* (H. Weissbach and S. Pestka, eds.), pp. 555–601. New York: Academic Press.

Sonenberg, N., and Trachsel, T. (1982). Probing the function of the eukaryotic 5′-cap structure using monoclonal antibodies to cap-binding proteins. Current topics in cellular regulation, vol. 21, pp. 65–88. New York: Academic Press.

Spirin, A. S. (1972). Non-ribosomal ribonucleoprotein particles (informosomes) of animal cells. In *The mechanism of protein synthesis and its regulation* (L. Bosch, ed.), pp. 515–537. Amsterdam and London: North Holland.

Chapter 18

Termination of Translation

18-1 Termination codons

After translation has started at the initiation triplet AUG (or GUG, UUG, etc.), the ribosome reads mRNA triplet by triplet and elongates the polypeptide chain until it comes across one of the following codons: UAA, UAG, or UGA. None of these triplets possess a cognate aminoacyl-tRNA; they all stop translation. Thus, they serve as *termination codons*[1–3].

The termination codon is always present at the end of the coding region of any naturally occurring mRNA. Termination codons are sometimes present in tandem, e.g., at the end of the MS2 bacteriophage coat protein cistron, where the termination codon UAA is followed by UAG (see chapter 2, fig. 6).

It should be noted that triplets UAA, UAG, and UGA are found far more often in wrong reading frames within mRNA coding regions than in the correct reading frame, where, as a rule, only one such triplet per coding sequence is present. Therefore an accidental frame shift in the course of elongation does not usually result in the synthesis of a long wrong polypeptide and generally leads to an early termination of this mistranslation. The frequency of termination triplets is also high in the noncoding mRNA regions, including the intercistronic regions of polycistronic mRNAs.

The termination triplet may appear in the reading frame of the mRNA coding region as a result of mutation. For example, a change of G to A in the tryptophan codon UGG results in the appearance of either UAG or UGA; a change of C to U in the glutamine codons CAA or CAG results in either UAA or UAG. Such mutations are referred to as *nonsense mutations*; the appearance of UAG is called the "*amber*" mutation, UAA the "*ochre*" mutation, and UGA the "*opal*" mutation. In contrast to the usual point mutations which result in the replacement of the amino acid in the synthesized polypeptide, these mutations lead to premature termination, which takes place at just the point where the nonsense codon appears. Another mutation changing the anticodon of some tRNA species in such a way that it becomes complementary to the nonsense codon may result in the *suppression* of the nonsense mutation; e.g., tyrosine tRNA, in which the anticodon GUA is changed into CUA, recognizes the termination codon UAG and thus suppresses the amber mutations.

Translation may proceed with errors, including ones in which the normal termination codon at the end of the mRNA coding sequence is recognized by the tRNA with an anticodon having partial complementarity to it; e.g., the normal tryptophan tRNA (anticodon CCA) can occasionally recognize the termination UGA codon (or, in rare cases, UAG codon). This results in a message being *read through* and a longer polypeptide being synthesized. The formation of such longer products, in addition to the normal translation product, has often been observed in studies of protein synthesis both *in vivo* and *in vitro*[4–6]. Similar mechanisms underlie the formation of small amounts of the normal product in nonsense mutants ("leakage").

The codon UGA is the "weakest" among the three termination codons: it can be read-through by a translating ribosome most frequently, seemingly due to its recognition by tryptophanyl-tRNA. In some cases this termination codon is specially used in nature, in order to form a small amount of a physiologically important protein from a read-through product, in addition to the main protein product whose translation is terminated by this codon. Such situation is observed in the case of Qβ phage RNA translation: coat protein cistron is ended by the termination codon UGA which is occasionally read-through by ribosomes thus resulting in the synthesis of a small amount of the polypeptide much longer than the coat protein; the read-through polypeptide is a necessary product of the Qβ RNA translation since it is required for the assembly of the normally infectious phage particle[7].

The mitochondrial genetic code should be mentioned specially. The codon UGA does not serve as a termination codon in mammalian and fungal mitochondria. Rather, it codes for tryptophan, just as UGG does. In addition, it has been reported that AGA and AGG codons in mammalian mitochondria do not code for arginine but serve as termination codons.

18-2 Termination protein factors

When the termination codon comes to the ribosomal A-site, it is recognized by a special soluble protein which binds to the ribosome and induces the hydrolysis of the ester bond between the tRNA and the polypeptide in the peptidyl-tRNA at the P-site. As a result, the polypeptide is released from the ribosome. Proteins recognizing the termination codons and inducing the release of polypeptide are called *termination* or *release factors* (RF)[8,9].

Procaryotic organisms have three proteins taking part in termination: RF-1, RF-2, and RF-3. These proteins have molecular masses ranging from 45,000 to 50,000 daltons. The codons UAA and UAG are recognized by RF-1; RF-2 is specific for the UAA and UGA codons; and RF-3 does not participate directly in codon recognition but is necessary for involving GTP/GDP in the termination process.

Eucaryotes have only one termination factor, referred to as the RF. The protein in the monomeric form has a molecular mass of about 50,000 to 60,000 daltons, but it is usually present as a dimer with a molecular mass of about 120,000 daltons. It is capable of recognizing all three termination codons, UAA, UAG, and UGA; it interacts with GTP and shows GTPase activity in the presence of ribosomes.

In the process of interacting with the termination codon and the ribosome, the termination factors RF-1 and RF-2 of procaryotes (or RF of eucaryotes) appear to simulate the binding of the aminoacyl-tRNA to the A-site of the translating ribosome. However, instead of an attack by the amino group of the aminoacyl-tRNA on the peptidyl-tRNA ester bond, this bond is attacked by a water molecule. The transfer of a peptidyl residue to the water molecule, like its transfer to the amino group in transpeptidation, is catalyzed by the ribosomal peptidyl transferase center (see below).

The occurrence of the protein factor-codon interaction, instead of the codon-anticodon interaction, is intriguing. Surprisingly, the protein recognizes the nucleotide triplet, and the recognition has a high specificity similar to that for the codon-anticodon recognition. Moreover, in the presence of a suppressor tRNA complementary to the termination codon, the aminoacyl-tRNA and the termination factor compete equally for binding to the ribosomal A-site. Experiments using various modified nucleotides in the termination codons have indicated that the specificity of RF in codon recognition resembles greatly the Watson–Crick base-pairing specificity including Crick's wobble pairing. Determining the protein "anticodon" structure is undoubtedly a fascinating and important problem pertinent to the general problem of protein-nucleic acid recognition.

The function of the termination factors, like that of the elongation factors and one of the initiation factors (IF-2), depends on GTP. This dependence has been demonstrated clearly for eucaryotic systems: GTP, as well as its nonhydrolyzable analog, stimulates the codon-dependent binding of RF to the ribosome; at the same time, the RF bound to the ribosome possesses GTPase activity which appears necessary for cleaving GTP in order to release the RF from the terminated ribosome.

The following observations concerning procaryotes have been made: (1) RF-3 possesses affinity to GTP and to GDP; (2) RF-3 in the presence of the ribosome shows GTPase activity; (3) RF-3 stimulates the codon-dependent binding of RF-1 and RF-2 with the ribosome; (4) RF-3 with GDP stimulates the release of RF-1 and RF-2 from the ribosome. Although it has not been demonstrated directly that the formation of a complex between RF-3 and GTP is necessary for the codon-dependent binding of RF-1 or RF-2, it is possible that this is indeed the case, particularly under *in vivo* conditions.

18-3 Ribosomal site for binding termination factors

As already mentioned, the termination factors recognize the termination codon positioned in the A-site of the ribosome[10]. Thus, the codon-dependent binding of RF is possible only after the peptidyl-tRNA has been translocated from the A-site to the P-site. Under *in vitro* conditions, the AUG triplet and the initiator F-Met-tRNA can be bound directly to the P-site of the vacant ribosomes, and this may be followed by the addition of RF and one of the termination triplets, UAA, UAG, or UGA; these in turn bind and induce hydrolysis of the F-Met-tRNA at the P-site, yielding free formyl-methionine. It is evident that, in this case, the termination triplet and RF come into the vacant A-site. The fact that the suppressor aminoacyl-tRNA in the complex with EF-T_u and GTP, which possesses an anticodon complementary to the termination codon, competes with RF in the course of translation also indicates that RF binds to the A-site. Such inhibitors of the ribosomal A-site as tetracycline and streptomycin inhibit the codon-dependent binding of the termination factors with the ribosome. The EF-G bound to the ribosome in a stable manner (e.g. in the presence of fusidic acid) interferes with the binding of RF to the ribosome, just as with the binding of the aminoacyl-tRNA·EF-T_u·GTP complex. The protein L7/L12 of the procaryotic ribosome forming the lateral stalk of the 50S ribosomal subunit contributes to the binding of RF, as well as of the elongation factors. The inhibition of codon-dependent RF binding by antibodies against individual ribosomal proteins confirms that the RF-binding site overlaps considerably both the A-site and the site responsible for binding elongation factors.

18-4 Hydrolysis of peptidyl-tRNA

The codon-dependent binding of the termination factor is necessary for the hydrolysis of the peptidyl-tRNA at the P-site and, consequently, for the release of the polypeptide from the ribosome. As has been demonstrated, the hydrolysis of the peptidyl-tRNA in the ribosome is suppressed by the inhibitors interfering with peptidyl transferase reaction[10]. Chloramphenicol, amicetin, lincomycin, gougerotin, sparsomycin, and other common inhibitors of the ribosomal peptidyl transferase, suppress the RF-induced release of the peptide from bacterial ribosomes. These antibiotics, however, do not inhibit the codon-dependent binding of RF to the bacterial ribosome. Hence, the terminating hydrolysis of the peptidyl-tRNA appears to be accomplished by the ribosomal peptidyl transferase center.

There is independent evidence that the ribosomal peptidyl transferase center is capable of catalyzing the formation of the ester bonds between the C-terminus of the peptide and the hydroxyanalogs of the aminoacyl-tRNA or puromycin[10]. Furthermore, in the presence of methyl or ethyl alcohols and deacylated tRNA, methanolysis or ethanolysis of the peptidyl-tRNA at the ribosomal P-site takes place; in other words, the peptide residue is transferred to the alcohol hydroxyl group. In the presence of other organic solvents, e.g. acetone, the peptide residue can be transferred to a water molecule, i.e. the ester bond of the peptidyl-tRNA is subjected to ribosome-catalyzed hydrolysis; the hydrolysis depends on the presence of deacylated tRNA or its 3′-terminal CCA sequence. In this case as well, the hydrolysis of the peptidyl-tRNA is prevented by inhibitors of ribosomal peptidyl transferase.

It is likely that the codon-dependent binding of the termination factor makes the ribosomal peptidyl transferase center more accessible to water, which is a good acceptor substrate; as a result, the peptide is transferred to water, particularly since under these conditions there is no competing amino group of an aminoacyl-tRNA. This accessibility can be achieved, for example, as a consequence of a certain unlocking (drawing apart) of the ribosomal subunits or a slight opening (loosening) of the large subunit which carries the peptidyl transferase center. Other alternatives, however, cannot be excluded: either some nucleophilic group of the RF protein may participate directly in an attack on the ester bond and accept transiently the peptide, or the RF-fixed water molecule may serve as a specific acceptor.

18-5 Sequence of events during termination

Thus the termination of translation appears to include several consecutive steps. The first step is the *recognition* of the A-site-bound termination codon by the protein factor of termination. Recognition ends in the *binding* of the termination factor, involving both the codon on the 30S ribosomal subunit and the factor-binding site on the 50S ribosomal subunit. There are reasons for assuming that the association with the factor-binding site (which seems to be located at the base of the L7/L12 stalk) requires GTP. In procaryotes this is accomplished by the two interacting proteins, RF-1 (or RF-2), which recognizes the codon, and RF-3, which binds GTP and probably is responsible for the interaction with the factor-binding site at the base of the L7/L12 stalk (by analogy with the aminoacyl-tRNA and EF-T_u, respectively). In eucaryotes, a single large RF participates in the codon recognition and in the interaction with GTP.

The second step involves *hydrolysis* of the ester bond of the peptidyl-tRNA at the P-site. This hydrolysis is catalyzed by the ribosomal peptidyl transferase center of the 50S ribosomal subunit. The binding of the termination factors both to the termination codon and to the ribosomal factor-binding site induces hydrolase activity of

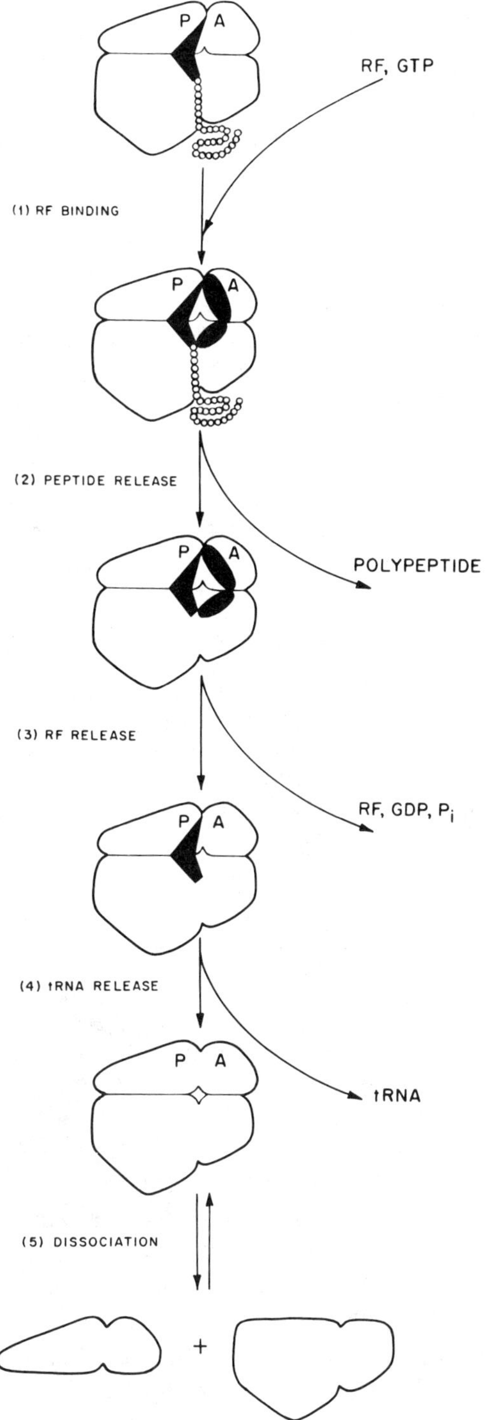

Figure 124 Sequence of events during termination of translation.

the peptidyl transferase center or somehow provides for the presence of the water molecule as an acceptor substrate there. As a result, peptide is released from the ribosome. The ribosome remains associated with the mRNA, the deacylated tRNA at the P-site, and the termination factor (or factors) with GTP at the A-site.

The third step is the *evacuation* of the ligands from the ribosome. The termination factors are probably the first to leave. By analogy with the EF-T_u, EF-G, and IF-2, it may be assumed that GTP hydrolysis is induced and, as a result, the affinity of the termination factor to the ribosome reduces. There is evidence that the release of deacylated tRNA from the ribosomal P-site is promoted by a special protein, called the *heat labile releasing factor*, or HLR[11].

Association between the ribosomal subunits in the vacant ribosome is far weaker than in the ribosome carrying the ligands. Therefore, after the ligands are released a reversible dissociation of the ribosome is facilitated. After the 50S subunit has gone, the 30S subunit is retained on mRNA less firmly and can dissociate easily from the template:

$$\text{mRNA·70S} \rightleftarrows \text{mRNA·30S} + \text{50S} \rightleftarrows \text{mRNA} + \text{30S} + \text{50S}.$$

As already indicated (see chapter 16), however, the 30S subunit should not necessarily leave the template after termination; it can slip along the mRNA for a while without translation and reinitiate translation at the next cistron of the same polycistronic mRNA.

The sequence of events in the course of termination is shown schematically in fig. 124.

References

1. S. Brenner, A. O. W. Stretton, and S. Kaplan (1965), "Genetic code: The 'nonsense' triplets for chain termination and their suppression," *Nature* 206:994–998.

2. S. Brenner, L. Barnett, E. R. Katz, and F. H. C. Crick (1967), "UGA: A third nonsense triplet in the genetic code," *Nature* 213:449–450.

3. A. Garen (1968), "Sense and nonsense in the genetic code," *Science* 160:149–159.

4. A. I. Geller and A. Rich (1980), "A UGA termination suppression $tRNA^{Trp}$ active in rabbit reticulocytes," *Nature* 283:41–46.

5. A. M. Weiner and K. Weber (1973), "A single UGA codon functions as a natural termination signal in the coliphage Q_β coat protein cistron," *J. Mol. Biol.* 80:837–855.

6. H. R. B. Pelham (1978), "Leaky UAG termination codon in tobacco mosaic virus RNA," *Nature* 272:469–471.

7. H. Hofstetter, H.-J. Monstein, and C. Weissman (1974), "The readthrough protein A_1 is essential for the formation of viable $Q\beta$ particles, *Biochim. Biophys. Acta* 374:238–251.

8. M. R. Capecchi and H. A. Klein (1969), "Characterization of three proteins involved in polypeptide chain termination," *Cold Spring Harbor Symp. Quant. Biol.* 34:469–477.

9. T. Caskey, E. Scolnick, R. Tomkins, J. Goldstein, and G. Milman (1969), "Peptide chain termination, codon, protein factor, and ribosomal requirements," *Cold Spring Harbor Symp. Quant. Biol.* 34:479–488.

10. C. T. Caskey (1977), "Peptide chain termination," in *Molecular mechanisms of protein biosynthesis*, ed. H. Weissbach and S. Pestka, pp. 443–465 (New York: Academic Press).

11. A. Kaji, K. Igarashi, and H. Ishitsuka (1969), "Interaction of tRNA with ribosomes: Binding and release of tRNA," *Cold Spring Harbor Symp. Quant. Biol.* 34:167–177.

Chapter 19

Cotranslational Folding, Compartmentalization, and Modifications of Proteins

19-1 Contribution of ribosomes to protein folding

The polypeptide chain on the ribosome elongates by consecutive growth from the N-terminus to the C-terminus. During the growth the C-terminus is always covalently fixed in the ribosomal peptidyl transferase center, whereas the N-terminus is free. Then, the free N-terminal region of the nascent peptide must acquire some conformation. This suggests that while the protein is being synthesized on the ribosome, it undergoes folding and that this folding begins from the N-terminal part[1].

Factors governing the folding of the N-terminal part of the growing peptide differ, however, depending on the distance from the fixed carboxyl end. Three different situations for amino acid residues of the nascent peptide can be considered.

1. In the ribosomal peptidyl transferase center two amino acid residues, the donor and the acceptor, should be positioned in a certain standard orientation with respect to each other (see chapter 13, section 13.4). The likely mutual orientation of the residues corresponds to the α-helical conformation of the polypeptide backbone.

2. Translocation moves the aminoacyl residue preceding the C-terminal residue out of the ribosomal peptidyl transferase center. The subsequent additions of new aminoacyl residues at the C-terminus moves the residue of interest as well as adjacent residues further and further away from the peptidyl transferase. However, the peptide section, which has a length of about 30 to 40 residues beginning from the peptidyl transferase center (i.e. from the growing carboxyl end), is still screened by the ribosome and is not exposed as a free chain to the solution[2–4]. The problem of conformation of this region and the effects of the ribosomal environment on it has yet to be solved. It seems unlikely that the peptide attached to the ribosome exists as an extended chain or a random coil. Each act of transpeptidation and subsequent translocation involves pushing the peptide by one residue through the ribosome; rigidity required for this vectorial process could be better provided by the α-helical conformation (i.e. by maintaining the *initial conformation* set up in the peptidyl transferase center). Another important argument should be added: the helical conformation of any polypeptide is preferred in a channel or another small space limiting the long-range mobility of parts of the macromolecule.

3. Beyond the section of 30 to 40 residues adjacent to the C-terminus, the N-terminal part of the nascent peptide hangs or protrudes from the ribosome and is accessible for the environment. Now, all the factors responsible for the spontaneous folding of a peptide into a conformation governed by environmental conditions are in action. However, three circumstances that make this situation different from that observed in the case of the spontaneous renaturation of unfolded protein *in vitro* should be taken into account. First, if the ribosome creates and maintains a certain universal conformation of the nascent peptide, e.g. the α-helix, then the protein folding may begin not from an extended or a random coiled state of the chain, but from the given initial conformation. Second, the search for a folding

pathway does not begin randomly from any region of the polypeptide chain but proceeds sequentially from the N-terminal part. Third, in the course of folding, the C-terminus is fixed at a particle of the large molecular mass and, therefore, its mobility is limited; this should result in the intermediate conformations having a higher stability than the analogous structures of the free polypeptide chain.

According to several pieces of evidence, the correct folding of the polypeptide into the protein takes place during its synthesis on the ribosomes, i.e. *cotranslationally*. Thus, the nascent peptide attached to the ribosome has been reported repeatedly to acquire the activities characteristic of the completed protein with a formed tertiary structure. The β-galactosidase synthesis is a well known example: the enzymatic activity of this protein requires, in addition to the folding of the polypeptide chain into a corresponding tertiary structure, that the four subunits be united into a quaternary structure; it has proved that the nascent chain, prior to its completion and when it is still attached to the ribosome, is capable of associating with the free subunits of this protein, and the ribosome-attached complex exhibits β-galactosidase activity[5,6].

Even more convincing are the experiments demonstrating that the nascent ribosome-attached peptide displays the immunological specificity characteristic of the native conformation of the mature protein. For example, ribosomes carrying nascent chains of β-galactosidase react with antibodies against the mature enzyme long before mRNA translation is completed and the protein is released[7]. The antibodies used were specific to the tertiary structure: denaturation of the ribosome-bound material by heating destroyed completely its capacity to react with antibodies.

It is possible that, in contrast to renaturation of the free unfolded polypeptide *in vitro*, attaining the correct final conformation of protein on the ribosome proceeds more directionally and is therefore quicker and more reliable. In other words, the ribosome may contribute to a certain *folding pathway*. This contribution from the ribosome could include at least such factors as the determination of the folding sequence from the N-terminus to the C-terminus; the setting of a certain initial conformation, specifically φ and ψ angles, for each amino acid residue; and the stabilization of intermediate local conformations due to the fixation of the C-terminus.

Unfortunately, the experimental solution of the protein-folding problem taking into account the ribosome contribution has not even begun.

19-2 Interaction of the ribosome and the nascent peptide with the membrane: Cotranslational transmembrane transport

Synthesis of proteins by free and membrane-bound polyribosomes

When the ribosome has synthesized a peptide more than 30 to 40 aminoacyl residues in length, the N-terminal part of this peptide protrudes beyond the ribosome. This extraribosomal part begins to fold into a certain structure depending on its amino acid sequence and the existing environment.

At the same time, in both procaryotic and eucaryotic cells, some of the ribosomes organized into polyribosomes are "free," i.e. not bound to membranes (although in eucaryotes they appear to still be associated with a cytoskeleton), but other polyribosomes are attached to membranes. In procaryotes the polyribosomes may be attached to the inner surface of the cell plasma membrane, whereas in eucaryotes the membrane-bound ribosomes are localized on the so-called rough endoplasmic reticulum (REF) of the cytoplasm. The attached ribosomes donate the nascent peptide directly to the membrane. Correspondingly, depending on ribosomal localization, the cotranslational extraribosomal folding of the nascent polypeptide proceeds either in the aqueous medium of the cytoplasm or in the hydrophobic environment of the membrane lipid bilayer.

As was noted years ago, free polyribosomes synthesize primarily water-soluble proteins for house-keeping use of the cytoplasm, whereas the membrane-bound particles synthesize either the proteins for incorporation into membranes or the secretory proteins which are transported out of the cell through the membranes[8–14]. Apparently, the soluble cytoplasmic proteins synthesized on free polyribosomes are folded in the aqueous medium as they emerge from the ribosomes. As a consequence, they form a typically globular structure, with a more or less polar surface and a hydrophobic core. At the same time, protein synthesis on membrane-bound ribosomes causes the growing peptide to come into contact with the hydrophobic surrounding of the membrane lipid bilayer. This implies that, at least initially, such a peptide should undergo folding in the hydrophobic environment. In the case of proteins destined to become components of a given membrane (the endoplasmic reticulum membrane of eucaryotes or the plasma membrane of bacteria), the hydrophobic environment probably should dictate the final definitive folding type, with numerous hydrophobic residues being exposed outside the molecule; the transmembrane hydrophobic sequences of such proteins often exist in the α-helical conformation (consider, for example, bacteriorhodopsin[15]).

With proteins passing through the membrane into the aqueous phase, however, the picture appears to be more complex. In this instance the aqueous phase may be represented either by the intermembrane lumen of the endoplasmic reticulum in eucaryotes, or by the periplasmic space of gram-negative bacteria, or simply by the external space in other bacteria. Such protein transport appears to be accompanied by multistage protein folding with the cotranslational and posttranslational processing and enzymatic covalent modifications of the polypeptide chain. In one way or another, in the case of water-soluble secretory proteins the polypeptide chain is found initially in the hydrophobic surrounding of the membrane lipid bilayer, and thus the folding should proceed without a compact hydrophobic core being formed; after this, the polypeptide, as it emerges from the membrane, should rearrange from this intermediate conformation and become a water-soluble globule with a hydrophobic core and a polar surface.

Interaction between the ribosome and the membrane

The idea that protein synthesis on membrane-bound ribosomes is coupled with transmembrane protein transport emerged from observations on the intimate association of nascent polypeptide chains with membranes of a rough endoplasmic reticulum in eucaryotic cells[16–18] and with the plasma membrane in bacteria[4,19]. The translating ribosomes were found to be anchored firmly on the membrane by the growing peptide. Only puromycin treatment, which resulted in the abortion of the peptide from the ribosomes, allowed the complex dissociation into free ribosomes and membranes, leaving the peptide in the membrane. Thus, it became clear that the growing peptide significantly contributes to the association between the translating ribosome and the membrane. A rupture of this anchor by puromycin in the bacteria results in the immediate release of the ribosomes from the membranes. On the basis of this observation, it has been concluded that the nascent peptide provides for the sole attachment of polyribosomes to the plasma membrane in bacteria[19].

In contrast, with eucaryotic cells, after the peptide anchor is broken the ribosomes still show a marked affinity to endoplasmic reticulum membranes. The complete dissociation of the ribosomes from endoplasmic reticulum membranes *in vitro* may be achieved only by combining the treatment of microsomes with puromycin and high ionic strength solutions[17,18]. It also can be demonstrated that non-translating ribosomes, or ribosomes that have just started translation and contain only a short peptide which does not protrude from the particle, have a certain affinity to endoplasmic reticulum membranes. From all this, it is assumed that the membranes of the rough endoplasmic reticulum contain specialized *receptors* which are responsible for the reversible association of ribosomes with a membrane; these receptors are also thought to help the membrane accept the nascent peptide when it emerges from the ribosome (hypothetical membrane "pores" for the polypeptide chain).

Two glycoproteins, referred to as *ribophorins* I and II, have been found in the rough endoplasmic reticulum membranes of eucaryotic cells. These proteins always accompany membrane-bound ribosomes and are located near them; the latter follows from the observation that ribophorins can be crosslinked chemically with the ribosomes by bifunctional reagents[20,21]. Ribophorins I and II have molecular masses of 65,000 and 63,000 daltons, respectively. These proteins span the membrane i.e. are transmembrane proteins. In experiments on membrane reconstitution from detergent-solubilized mixtures of microsomal proteins and phospholipids, the incorporation of ribophorins into the membrane correlates with the recovery of the affinity of the membrane for the ribosomes.

It was demonstrated long ago that eucaryotic ribosomes attach to the membrane through their large (60S) subunit[22]. The 60S subunit appears to have a special site with affinity to the membrane of the endoplasmic reticulum. Therefore, all ribosomes attach to the membrane through a strictly fixed point in the same orientation[23]. In this orientation the axis connecting the large and the small subunits is roughly parallel to the membrane surface (fig. 125). Electron microscopy of negatively stained eucaryotic ribosomes[23] demonstrates that

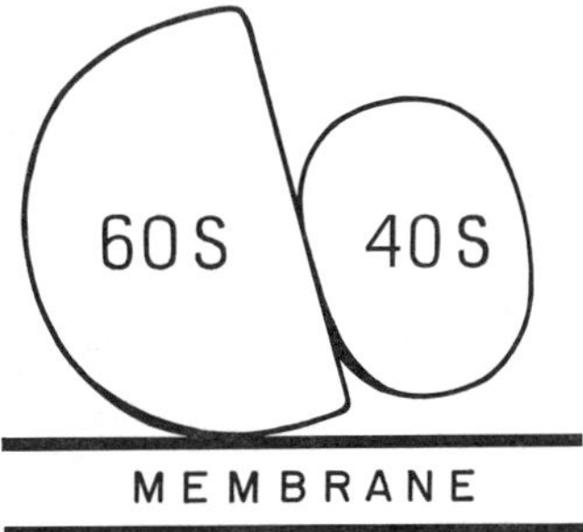

Figure 125 Scheme illustrating the orientation of the 80S ribosomal subunits relative to the surface of the endoplasmic reticulum membrane[23].

the long axis of the small subunit should be roughly parallel to the membrane surface. This leads to the assumption that the attachment of the ribosome to the membrane probably takes place at the side of the lateral protuberances of the subunits (equivalent to the L1 ridge of the 50S subunit and the platform of the 30S subunit of *E. coli* ribosomes; see chapter 6, section 6.2); the region of the presumed "pocket" for tRNA (chapter 11, fig. 85) and the stalk of the large subunit (chapter 6, fig. 39, bottom) should be at the side turned away from the membrane.

Apparently, the nascent peptide emerges from the ribosome somewhere at the side contacting the membrane. Direct immuno-electron microscopic attempts at detecting the site of exit of the synthesized protein (β-galactosidase on *E. coli* 70S ribosomes[24] or ribulose-1,5-diphosphate carboxylase on plant *Lemna gibba* 80S ribosomes[25]) suggest that the region of the large subunit located well away from the head (from the central protuberance) and probably near the cleft dividing the subunit body, is involved (fig. 126).

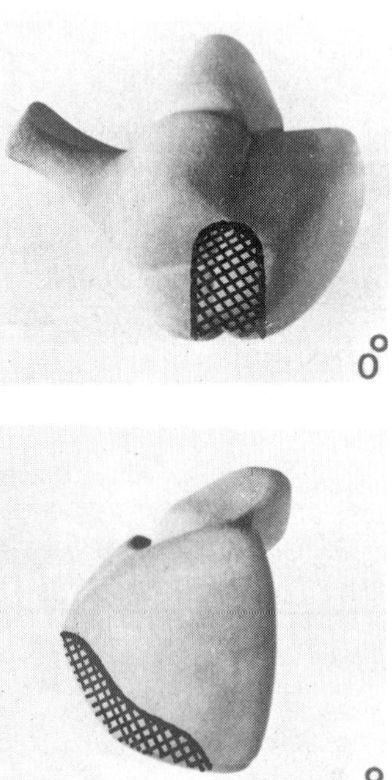

Figure 126 Proposed region of the nascent peptide exit[24,25] shown as a crosshatched area on the 60S subunit model (viewed in two projections).

N-terminal signal sequence of the nascent polypeptide

Despite a thorough search no significant differences have been found either in the composition and structure of the free and membrane-bound ribosomes, or in the functional characteristics of the isolated particles. In every experimental test, both the ribosomes isolated from the free polyribosomes and the ribosomes isolated from the microsomal fraction of the cell homogenate have been found to be equal and interchangeable.

Moreover, although synthesis of intrinsic membrane proteins and secretory proteins proceeds on membrane-bound ribosomes, translation is initiated on the free particles: the "native" 40S subunits used for initiation are always free, and the 80S particles resulting from initiation also are not yet attached to the membrane. Attachment to membranes takes place some time after the beginning of elongation. This is understandable since the growing peptide plays a decisive role in anchoring ribosomes to the membrane.

In 1971, on the basis of several known facts and their own observations, Blobel and Sabatini proposed that mRNAs, which are to be translated by membrane-bound ribosomes, have a special sequence immediately following the initiation codon. This sequence codes for the characteristic N-terminal sequence of the nascent polypeptide, which provides a signal for recognition either by the membrane or by a factor mediating the association between the ribosome and the membrane[26]. Thus, if a ribosome translates mRNA coding for the secretory or membrane protein, some time after initiation the N-terminal part of the growing peptide will protrude from the ribosome and signal the attachment to the membrane. In this instance the ribosome becomes membrane bound. If the N-terminus of the growing peptide does not carry such a signal sequence, as is the case with water-soluble cytoplasmic proteins, the corresponding ribosomes (polyribosomes) will remain free throughout elongation.

Independent experimental evidence for the existence of special N-terminal sequences in newly synthesized polypeptide chains of certain secretory proteins of eucaryotic cells appeared at almost the same time. These sequences were found to be temporary: they were cleaved off in the membrane even before synthesis of the corresponding protein had been completed. For example, the translation of mRNA coding for the light chain of immunoglobulin in the cell-free system on free ribosomes yielded a polypeptide which had an additional sequence at the N-terminus compared to the authentic light chain of the immunoglobulin; translation of the same mRNA in the presence of microsomes yielded a normal product[27]. It has been concluded that translation of the first mRNA codons results in the synthesis of the N-terminal signal sequence which determines the attachment of the ribosome to the membrane; in the course of further elongation this sequence is cleaved off on the membrane, and the final product does not contain it. Subsequently, it has been demonstrated that such an N-terminal sequence is rich in hydrophobic aminoacyl residues[28]; therefore, it was natural to conclude that the hydrophobic nature of the signal sequence of the nascent polypeptide chain contributes to the interaction of the polypeptide with the membrane.

Since then, numerous secretory proteins of eucaryotes, e.g. the light and heavy chains of immunoglobulin, serum albumin, procollagen, uteroglobin, caseins, α-lactalbumin and β-lactalbumin of milk, ovomucoid, lysozyme and conalbumin of egg white, trypsinogen, as well as proinsulin and other peptide hormones, have been reported to be synthesized with an additional N-terminal sequence of between 15 and 30 amino acid residues in length (see table 3). This sequence is later cleaved off on the membrane during protein synthesis. If the corresponding mRNAs are translated in the cell-free system in the absence of membranes, then elongated products referred to as *pre-proteins* are formed. Cleavage of the N-terminal signal sequence usually does not take place when the membranes (the microsomal fraction) are added to the system after synthesis of the pre-protein is completed or even at the late stages of elongation. Only if membranes are present in the system from the beginning, or if they are added soon after the emergence of the N-terminal section of the polypeptide from the ribosome, does the ribosome attach to the membrane, the nascent peptide enter the membrane, and cleaving off of the signal sequence take place. Elongation of the peptide in an aqueous medium above a certain size seems to result in peptide folding, which screens the hydrophobic N-terminal region and thus prevents its interaction with the membrane.

Similar additional N-terminal sequences are also characteristic of the nascent chains of a number of bacterial proteins, exported from the cytoplasm[29] (see table 3). In gram-negative bacteria the export of proteins takes place either into the periplasmic space (as for alkaline phosphatase, maltose-binding protein, arabinose-binding protein, and penicillinase) or into the outer membrane (as for outer membrane lipoprotein and λ-receptor). The beginning of the synthesis of proteins destined for exportation from the cell seems to be followed by their hydrophobic N-terminal sequences interacting with the inner cytoplasmic membrane of the bacterial cell, so that the subsequent steps of their synthesis proceed on the membrane-bound ribosomes[30]. Cleavage of the N-terminal sequences may take place during or, in some cases, following elongation. After synthesis has been completed

Table 3 N-terminal signal sequence of selected proteins

Protein		−25		−20		−15	N-terminal sequence	−10		−5	Cleavage point ↓	+1	
Pre-K-immunoglobulin, L-chain (mouse)			Met	Glu	ThrAspThr	Leu	LeuLeuTrpVal	Leu	LeuLeuTrpVal	Pro	GlySerThrGly	Asp	IleVal
Pre-proalbumin (rat)					MetLysTrp	Val	ThrPheLeuLeu	Leu	LeuPheIleSer	Gly	SerAlaPheSer	Arg	GlyVal
Pre-α-s1-casein (ovine)						Met	LysLeuIle	Leu	ThrCysLeuVal	Ala	ValAlaLeuAla	Arg	ProLys
Pre-conalbumin (chicken)					MetLysLeuIle	Leu	CysThrValLeu	Ser	LeuGlyIleAla	Ala	ValCysPheAla	Ala	ProPro
Pre-lysozyme (chicken)					MetArgSer	Leu	LeuIleLeuVal	Leu	CysPheLeuPro	Leu	AlaAlaLeuGLy	Lys	ValPhe
Pre-ovomucoid (chicken)			MetAlaMetAla	Gly	ValPheValLeu	Phe	SerPheValLeu	Cys	GlyPheLeuPro	Asp	AlaAlaPheGly	Ala	GluVal
Pre-interferon (human)			MetAlaLeu	Thr	PheAlaLeuLeu	Val	AlaLeuLeuVal	Leu	SerCysLysSer	Ser	CysSerValGly	Cys	AspLeu
Pre-melittin (bee)			Met	Lys	PheLeuValAsn	Val	AlaLeuValPhe	Met	ValValTyrIle	Ser	TyrIle Tyr Ala	Ala	ProGlu
Pre-proinsulin I (rat)			MetAlaLeuTrp	Met	ArgPheLeuPro	Leu	LeuAlaLeuLeu	Val	LeuTrpGluPro	Lys	ProAlaGlnAla	Phe	ValLys
Pre-growth hormone (rat)	Met	Ala	AlaAspSerGln	Thr	ProTrpLeuLeu	Thr	PheSerLeuLeu	Cys	LeuLeuTrpPro	Gln	GluAlaGlyAla	Leu	ProAla
Pre-proparathyroid hormone (bovine)		Met	MetSerAlaLys	Asp	MetValLysVal	Met	IleValMetLeu	Ala	IleCysPheLeu	Ala	ArgSerAspGly	Lys	SerVal
Pre-prosomatostatin (rat)			MetLeuSerCys	Arg	LeuGlnCysAla	Leu	AlaAlaLeuCys	Ile	ValLeuAlaLeu	Gly	GlyValThrGly	Ala	ProSer
Pre-glycoprotein (VS virus)					Met	Lys	CysLeuLeuTyr	Leu	AlaPheLeuPhe	Ile	HisValAsnCys	Lys	PheThr
Pre-haemagglutinin (influenza virus)						Met	AlaIleIleTyr	Leu	IleLeuLeuPhe	Thr	AlaValArgGly	Asp	GlnIle
Pre-phosphatase, alkaline (*E. coli*)			Met	Lys	GlnSerThrIle	Ala	LeuAlaLeuLeu	Pro	LeuLeuPheThr	Pro	ValThrLysAla	Arg	Thr
Pre-penicillinase (*E. coli*)			MetSerIle	Gln	HisPheArgVal	Ala	LeuIleProPhe	Phe	AlaAlaPheCys	Leu	ProValPheAla	His	ProGlu
Pre-lipoprotein (*E. coli*)				Met	LysAlaThrLys	Leu	ValLeuGlyAla	Val	IleLeuGlySer	Thr	LeuLeuAlaGly	Cys	SerSer
	Nonsplittable												
Ovalbumin (chicken)	MetGlySerIleGlyAlaAlaSerMetGluPheCysPheAspValPheLysGluLeuLysValHisHisAlaAsnGluAsnIlePheTyrCys												

and after the termination of translation, the protein passes into the periplasmic space and then, depending on the hydrophobicity or hydrophilicity of its surface, either remains in the periplasmic space as a water-soluble protein or is integrated into the outer membrane. All this is very similar to the situation with secreted proteins in eucaryotic cells.

Moreover, there are reasons to believe that the most fundamental mechanisms of signal peptide-membrane recognition and subsequent intramembrane peptide cleavage are common to both eucaryotes and procaryotes. Indeed, bacteria transformed by recombinant plasmids with the genes of eucaryotic secreted proteins can synthesize these proteins and effectively export them through the plasma membrane, specifically cleaving off the amino-terminal signal segment. Rat pre-proinsulin synthesized in *E. coli*, for example, is correctly processed into proinsulin, which is then secreted from the bacterial cytoplasm into the periplasmic space[31,32].

In addition to secreted proteins, the nascent polypeptide chains of many membrane-incorporated proteins possess transient signal N-terminal sequences as well. One of the first examples to be studied was vesicular stomatitis virus glycoprotein; this protein, along with the host cell membrane, is involved in making the viral envelope. It is synthesized with the N-terminal signal sequence, which is similar to that of secretory pre-proteins. The signal sequence is necessary for the attachment of the translating ribosome to the endoplasmic reticulum membrane; further synthesis of this protein thus takes place on membrane-bound ribosomes. In the course of elongation the N-terminal sequence, consisting of 16 amino acid residues, is cleaved off in the membrane[33]. In other words, the situation is no different than in the case of water-soluble secreted proteins. In contrast to the situation with secreted proteins, however, the final product after termination of translation does not pass through the membrane but remains inserted in it as a transmembrane protein. Further examples of integral membrane proteins synthesized with cleavable N-terminal signal sequences can be given (hemagglutinine of the influenza virus, the heavy chain of histocompatibility antigens A and B, glycophorin A of red cells, cytochrome P-448, etc.). Apparently, the synthesis of both secretory and integral membrane proteins involves exactly the same mechanisms for signal peptide-membrane recognition, peptide penetration into the membrane, and cleavage of the N-terminal signal fragment; however, termination of translation results either in the final product passing through the membrane, as is the case with water-soluble secretory proteins, or in its sticking within the membrane, as is the case with more hydrophobic proteins which are destined to become membrane components. Proteins stuck in the endoplasmic reticulum membrane may then be subjected to post-translational transport into the membrane structures of other types, including the cell plasma membrane.

The scheme of the membrane-bound polyribosome demonstrating initiation on free particles, attachment of particles to the membrane through the signal N-terminal sequence, removal of the signal peptide, etc., is presented in fig. 127.

Examples of the cleavable N-terminal sequences of secretory and membrane pre-proteins are given in table 3. As already mentioned, all these sequences are hydrophobic. The region of particularly high hydrophobicity (most often consisting of only hydrophobic amino acids) is located roughly in the middle of the sequence to be cleaved; this hydrophobic "core" usually consists of about a dozen amino acid residues. The last amino acid residue of the cleaved sequence (the cleavage site) is generally characterized by the small size of the side group; it is either Gly, Ala, Ser, or Cys; this residue is often, but not always, followed by a charged amino acid. However, there is no apparent similarity or homology between cleaved signal sequences other than that they all have high hydrophobicity. In connection with this, it is possible that the main factor underlying signal peptide-membrane recognition and the entry of the peptide into the membrane is the common conformation of the hydrophobic N-terminal peptide protruding from the ribosome. The α-helix is perhaps a conformation of the hydrophobic sequence most favorable for interaction with the hydrophobic environment[34].

Despite numerous examples of secretory and membrane proteins being synthesized with the *cleavable* signal N-terminal sequence, this is not a general rule. Some of the proteins are synthesized on membrane-bound ribosomes but *without cleavage* of the N-terminal or any other signal sequence. Such proteins include ovalbumin, a typical secretory protein[35], as well as the membrane proteins cytochrome P-450, epoxide hydratase, retinal opsin, and Sindbis virus glycoprotein PE_2. In these cases the growing polypeptide chain possesses a signal sequence which induces the translating ribosome to attach to the endoplasmic reticulum membrane and the cotranslational transport of the protein into the membrane, but the accompanying processing is absent. In cytochrome P-450, the N-terminal sequence is hydrophobic and resembles the cleavable signal sequence; however, this region is retained in the mature protein[36]. The N-terminal sequences of ovalbumin[35] and opsin[37] do not resemble the usual signal sequence. It may well be that, in these cases, either the signal role is played by the special inner hydrophobic regions of the growing polypeptide[38], or the less hydrophobic N-terminal sequence may also provide a signal for membrane attachment. As in other cases,

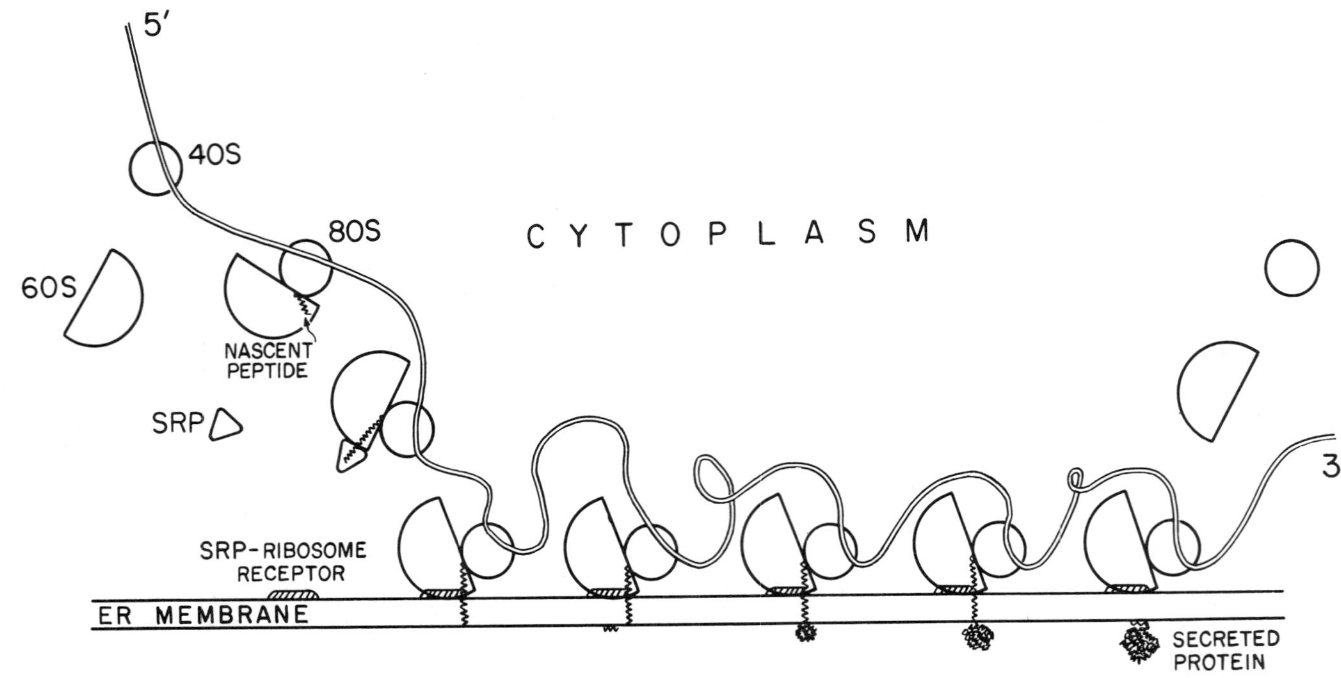

Figure 127 Schematic representation of a membrane-bound polyribosome.

interaction with the membrane is possible only during translation, not after it has been completed; the folding of the completed polypeptide chain seems somehow to block or screen the signal function of the corresponding region.

Not all membrane proteins are synthesized on membrane-bound ribosomes and enter the membrane cotranslationally. Cytochrome b_5 and NADH-cytochrome b_5-reductase, for example, are synthesized on free ribosomes and incorporated into the membranes post-translationally[39,40]. After they are released from the ribosomes into the cytoplasm as completed chains, these proteins show high affinity to the membranes (probably because of their hydrophobic C-terminal domains). Hence, they interact easily with membranes in the folded state, independently of ribosomes.

Signal recognition particles and their membrane receptors

The appearance of the hydrophobic sequence in the growing peptide protruding from the ribosome results in the ribosome being attached to the membrane. There is, however, the question of whether the hydrophobic sequence is recognized directly by the membrane or whether the attachment of the nascent peptide protruding from the ribosome to the membrane is mediated through some auxiliary recognition component.

In eucaryotic cells, the ribosome with the exposed signal sequence of the nascent peptide has been shown to interact initially with the special small particle called the *signal recognition particle* (SRP)[41,42]. This particle is a 11S ribonucleoprotein containing 7S RNA with a length of about 300 nucleotides[42] and six polypeptides with molecular masses of 72,000, 68,000, 54,000, 19,000, 14,000, and 9000 daltons[41]; all these components are present in the particle in equimolar amounts, as a single copy per particle. The particles appear to be universal, at least among eucaryotes[42]. The particles show an affinity to endoplasmic reticulum membranes and can be in labile and reversible association with them, although a large proportion of SRPs are found in the cell as a soluble cytoplasmic component. SRPs show a definite, although relatively low, affinity for the ribosomes as well. The presence of the signal peptide on the ribosome increases, by several orders of magnitude, the affinity of SRPs to the ribosome[43].

The recognition of the signal peptide and the attachment of the 11S particle to the ribosome result in two events. First, the elongation of a peptide on the ribosome is arrested[44]. Second, the attachment of the translating ribosome to the membrane is induced.

On the other hand, the endoplasmic reticulum membrane of eucaryotic cells possesses a special receptor which accepts the signal recognition particle complexed with the ribosome. The receptor is a protein with a molecular mass of 72,000 daltons. It is partially submerged into the membrane, while its main domain is turned toward the cytoplasm and seems to provide a direct "dock" for the signal recognition particle[45–48]. This protein is called the *docking protein*. The interaction of the ribosome-associated signal recognition particle with the membrane docking protein restores the previously arrested elongation, and polypeptide synthesis proceeds further. However, the nascent peptide then protrudes not into the aqueous phase but directly into the membrane; further elongation of the polypeptide results in it submerging and entering the membrane directly from the ribosome, without passing through the aqueous cytoplasmic environment. This process is referred to as the cotranslational translocation of the polypeptide through the membrane. More detailed mechanisms responsible for the polypeptide entering the membrane and for the polypeptide passing through the membrane in the case of secretory proteins have yet to be discovered.

The peptidase cleaving off the signal peptide is not found on the cytoplasmic side of the membrane. The enzyme appears to be localized on the opposite membrane surface; as for the eucaryotic cell endoplasmic reticulum, it occurs at the side of the intermembrane lumen. Apparently, the signal peptidase comes into play at relatively late stages of elongation when the peptide "pierces" through the membrane and comes into contact with its opposite (luminal) surface. Cleavage of the signal peptide is probably necessary for the N-terminal part of the protein molecule to enter the aqueous phase and for a corresponding reorganization of protein folding into the globular outer extracytoplasmic domain, as is the case with transmembrane proteins, or into the globular structure as a whole, as is the case with secretory proteins.

Thus this sequence of events may be visualized as follows[46] (see fig. 128). Synthesis of a protein predestined for secretion or for incorporation into the membrane begins on free ribosomes. In the course of elongation, as the first 30 to 40 N-terminal amino acid residues are being added, the peptide does not yet emerge from the ribosome. Later, the hydrophobic signal sequence appears from the ribosome synthesizing a secreted or integral membrane protein. When the signal sequence protruding from the ribosomes reaches 15 to 30 amino acid residues in length, i.e. when the nascent peptide as a whole reaches a length of about 50 to 70 residues, the signal recognition 11S ribonucleoprotein particle (SRP) binds to the ribosome and arrests elongation. This translational arrest continues until contact is made

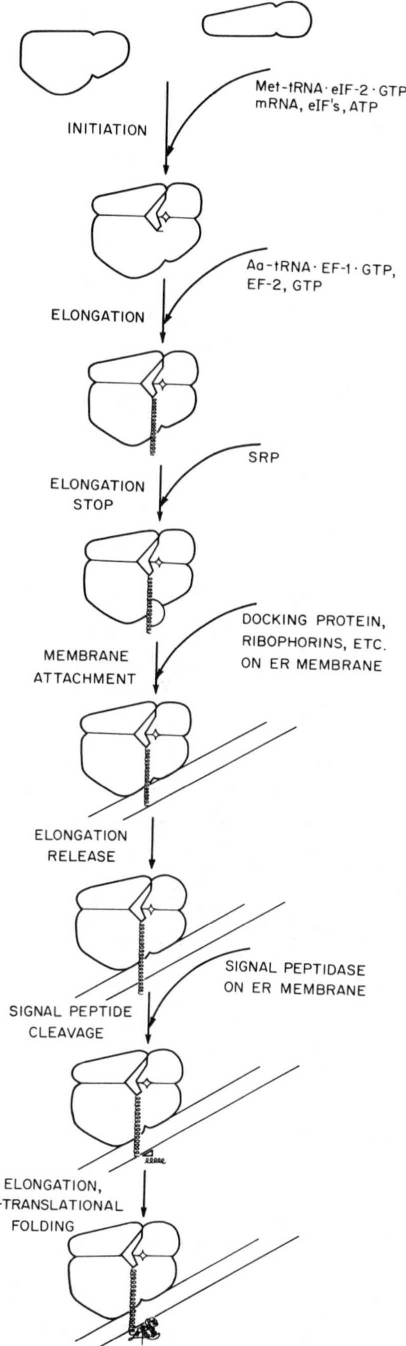

Figure 128 Sequence of events during the attachment of the eucaryotic ribosome to the membrane.

with the docking protein of the endoplasmic reticulum membrane. Then, elongation is restored and the growing peptide directly enters the membrane; the so-called cotranslational translocation of the peptide through the membrane takes place.

The arrest of elongation after the signal sequence has emerged from the ribosome is an extremely important step toward normal secretion of the synthesized protein or its integration into the membrane. Indeed, if there were no such an arrest, the elongation would continue and could result in the additional folding of the protruding nascent peptide in the aqueous phase of the cell and, consequently, in the signal sequence being screened. This would make it impossible for the membrane to recognize the signal peptide, and eventual secretion would, thus, be prevented. Hence, an arrest of elongation provides the signal sequence in the proper conformation with the necessary "waiting time" for the "docking" to the membrane.

Cleavage of the signal sequence at the luminal side of the endoplasmic reticulum membrane reduces the hydrophobicity of the growing peptide, and the continued presence of the polypeptide in the lipid bilayer may become less favorable than the transition into the aqueous phase of the intramembrane cistern. Correspondingly, depending on the amino acid composition and sequence, either only the water-soluble part, e.g. the N-terminal domain, is emerged into the aqueous phase of the lumen (this is the case with many transmembrane proteins), or the entire completed protein is pushed out (as is the case for numerous secreted proteins). It is logical that transition into the aqueous phase should be accompanied by the three-dimensional structure being rearranged and acquiring a globular conformation (hydrophobic residues become buried inside the globule or the globular domain, while the hydrophilic residues are exposed outside).

Secretory proteins accumulating in the intermembrane lumen (cistern space) of the endoplasmic reticulum are then transported to the Golgi complex, concentrated in secretory granules, and eventually secreted from the cell by exocytosis (for a review, see Palade (1975)).

19-3 Cotranslational modifications of proteins

The emergence of the nascent peptide out of the ribosome and its cotranslational folding in the aqueous medium, or its passing through the membrane, may be accompanied by enzymatic modifications of the chain.

In procaryotes, as well as in the mitochondria and chloroplasts of eucaryotic cells, the synthesis of all peptides on the ribosome begins from the formylmethionine residue (chapter 16). The formyl group, however, is not found in mature proteins, and methionine is found as an N-terminal residue in only some proteins. In procaryotic cells, as in mitochondria and chloroplasts, a special *deformylase* hydrolytically cleaves the formyl-group from the N-terminal methionine soon after the N-terminus of the nascent peptide appears from the ribosome[49,50]. Procaryotic, as well as eucaryotic, cells also contain a special methionine-specific *aminopeptidase*, which is capable of cleaving the N-terminal methionine[49,50]. Its action, however, appears to depend on the cotranslational folding of the nascent peptide. In any case, the N-terminal methionine is retained in some proteins[51]; this is thought to be the result of its inaccessibility because of some conformational features in the growing peptide.

The cotranslational proteolytic cleavage of the signal hydrophobic sequence in a number of secretory and transmembrane proteins of eucaryotes has already been considered. The *signal peptidase* is localized in the membrane apparently on the surface turned away from the cytoplasm, i.e. on the luminal side of the endoplasmic reticulum membrane[52,53]. This enzyme is an endopeptidase. The characteristic site of polypeptide cleavage by the signal peptidase is the peptide bond at a small amino acid residue, such as Gly or Ala, less often Ser or Cys, from its C-side (often, but by no means always, such a residue is followed by a charged residue, e.g. Arg, Lys, His, or Asp). In addition, the region of cleavage should somehow be marked by a more open conformation. Cleavage of the signal peptide seems in most cases to be a prerequisite for nascent peptide entry into the water-containing transmembrane space and for the cotranslational folding of the peptide there.

The formation of disulfide linkages between cysteine residues may well play an important part in cotranslational protein folding. Disulfide bridges locking the tertiary structure are particularly common for the secretory proteins of eucaryotes. In contrast, the intracellular proteins are often characterized by the presence of free sulfhydryl groups of cysteine residues. Indeed, the extracellular medium is more oxidative than the intracellular environment. Disulfide bridges may be formed between cysteine residues in the nascent polypeptide chain as it passes through the membrane into the luminal space. If the oxidative conditions of the medium are appropriate, such bridges may arise spontaneously. Two points, however, should be kept in mind. First, the rate of spontaneous disulfide linkage formation in proteins is low compared to the rate of protein synthesis and folding. Second, in the course of protein folding, disulfide bridges may be occasionally formed between the wrong cysteine residues and not between those that should participate in forming bridges in the completely folded mature protein molecule. More than 20 years ago it was discovered that the endoplasmic reticulum of animal cells contains an enzyme that catalyzes the formation of disulfide linkages in proteins[54–56]. The enzyme accelerates both the forward and the reverse reaction and, in fact, catalyzes thiol-disulfide exchange. This results not only in disulfide bridges being formed rapidly, but also in wrong or intermediate (nonnative) linkages being exchanged for correct ones during protein folding. Thus, the enzyme catalyzes the folding of disulfide crosslinked proteins. This enzyme has been termed the *protein disulfide isomerase*[57]. It appears to be connected very loosely with the luminal surface of the endoplasmic reticulum membrane, and it is possible that it is present in the lumen in a soluble form. It may be assumed that as a protein is synthesized and the growing polypeptide chain extends to the lumen, the disulfide isomerase catalyzes the cotranslational formation of intradomain disulfide bridges and their run-through while the correct tertiary structure of the domain is being formed.

One important and interesting type of cotranslational modification is the primary or *"core" N-glycosylation* of certain asparagine residues in the nascent chains of a number of transmembrane and secretory proteins. This reaction takes place in the endoplasmic reticulum membranes of eucaryotic cells[58,59]. N-glycosylation is the transfer of a carbohydrate (oligosaccharide) group from the membrane-bound polyprenol lipid, dolichyl phosphate, to the amide group of the asparagine residue of the nascent peptide. The process is as follows. Dolichyl phosphate is a lipid acceptor, upon which an oligosaccharide group is eventually built up. Initially, dolichyl phosphate reacts with uridine diphosphate-N-acetylglucosamine (UDP-GlcNAc) and accepts the residue of N-acetylglucosamine phosphate:

$$\text{Dol-Ⓟ} + \text{UDP-GlcNAc} \rightarrow \text{Dol-Ⓟ-Ⓟ-GlcNAc} + \text{UMP}.$$

In the next reaction, another residue of N-acetylglucosamine is added:

$$\text{Dol-Ⓟ-Ⓟ-GlcNAc} + \text{UDP-GlcNAc} \rightarrow \text{Dol-Ⓟ-Ⓟ-GlcNAc}_2 + \text{UDP}.$$

Then, in a series of sequential reactions, the glycosylated dolichyl diphosphate reacts with guanosine diphosphate mannose (GDP-Man) and, as a result, a branched polymannoside is synthesized on the first carbohydrate residues:

$$\text{Dol-Ⓟ-Ⓟ-GlcNAc}_2 + n\ \text{GDP-Man} \rightarrow \text{Dol-Ⓟ-Ⓟ-GlcNAc}_2\text{-Man}_n + n\ \text{GDP}.$$

Finally, glucose residues are transferred from uridine diphosphate glucose (UDP-Glc):

$$\text{Dol-Ⓟ-Ⓟ-GlcNAc}_2\text{-Man}_n + 3\,\text{UDP-Glc} \xrightarrow{\text{Dol-Ⓟ}} \text{Dol-Ⓟ-Ⓟ-GlcNAc}_2\text{-Man}_n\text{-Glc}_3 + 3\,\text{UDP}.$$

It is this oligosaccharide that is transferred by a corresponding glycosyltransferase to the asparagine residue of the nascent peptide:

$$\text{Dol-Ⓟ-Ⓟ-GlcNAc}_2\text{-Man}_n\text{-Glc}_3 + \text{(Asn)protein} \rightarrow \text{Glc}_3\text{-Man}_n\text{-GlcNAc}_2\text{-(Asn)protein} + \text{Dol-Ⓟ-Ⓟ}.$$

Soon afterward the glucose residues are partially removed by glycosidases.

However, such an oligosaccharide structure is not generally found in mature proteins. Rather, this structure is an intermediate in the overall process of protein N-glycosylation. The cotranslational step described above is followed by a posttranslational step which usually proceeds when the synthesized protein comes to the Golgi complex. The posttranslational step results in the removal of glucose and the consecutive, partial removal of mannose residues. This is often followed by the addition of N-acetylglucosamine, galactose, and sialic acid residues and sometimes, of fucose and xylose residues as well[58,59]. The result of this addition is the formation of a branched hetero-oligosaccharide residue attached to the amide group of an asparagine in the corresponding mature protein.

In fungi, cotranslational glycosylation on the membrane involving dolichyl phosphate may involve not only asparagine residues (N-glycosylation) but also serine and threonine residues (O-glycosylation)[60]. In this case, dolichyl phosphate initially accepts the mannose residue from guanosine diphosphate mannose:

$$\text{Dol-Ⓟ} + \text{GDP-Man} \rightarrow \text{Dol-Ⓟ-Man} + \text{GDP}.$$

The mannose residue is then transferred to the serine or threonine hydroxygroup by the corresponding glycosyltransferase:

$$\text{Dol-Ⓟ-Man} + \text{(Ser/Thr)protein} \rightarrow \text{Man} - \text{(Ser/Thr)protein} + \text{Dol-Ⓟ}.$$

After translation is completed, another glycosyltransferase in the Golgi complex extends the carbohydrate residue of the peptide chain by several additional mannose residues; these are transferred from guanosine diphosphate mannose:

$$\text{GDP-Man} + \text{Man-(Ser/Thr)protein} \rightarrow \text{Man}_{1\text{–}3}\text{-Man-(Ser/Thr)protein} + \text{GDP}.$$

In animal and plant cell proteins, O-glycosylation, involving serine and threonine residues, proceeds via another mechanism. In these cases, this is a posttranslational process taking place in the Golgi complex.

It is important to emphasize that the cotranslational glycosylation of nascent chains is not necessarily a prerequisite for this chain to enter the endoplasmic reticulum membrane. Moreover, this glycosylation appears to take place *after* the distal part of the growing peptide passes across the membrane, on the extracytoplasmic (luminal) membrane side. Correspondingly, each act of glycosylation (if there are several glycosylation points along the polypeptide chain) proceeds only after a certain length of the corresponding chain section is reached; this length seems to be necessary for the acceptor Asn to reach the luminal side of the membrane[61].

Hydroxylation of proline and lysine residues is another type of specific modification of polypeptide chains. It is characteristic of collagen synthesis in the connective tissue cells (fibroblasts) of animals. This process is also a cotranslational event, at least in part[62,63]. Prolyl-4-hydroxylase, prolyl-3-hydroxylase, and lysylhydroxylase are membrane-bound enzymes capable of converting the proline and lysine residues of nascent collagen polypeptide chains into 4-*trans*-hydroxyproline, 3-*trans*-hydroxyproline, and 5-hydroxylysine residues, respectively[64]. Further modification of collagen chains involves, in particular, the O-glycosylation of hydroxylysine residues.

References

1. D. C. Phillips (1967), "The hen egg-white lysozyme molecule," *Proc. Nat. Acad. Sci. U.S.A.* 57:484–495.
2. L. I. Malkin and A. Rich (1967), "Partial resistance of nascent polypeptide chains to proteolytic digestion due to ribosomal shielding," *J. Mol. Biol.* 26:329–346.
3. G. Blobel and D. D. Sabatini (1970), "Controlled proteolysis of nascent polypeptides in rat liver cell fractions. I. Location of the polypeptides within ribosomes," *J. Cell Biol.* 45:130–145.
4. W. P. Smith, P. C. Tai, and B. D. Davis (1978), "Interaction of secreted nascent chains with surrounding membrane in *Bacillus subtilis*," *Proc. Nat. Acad. Sci. U.S.A.* 75:5922–5925.
5. D. Zipser and D. Perrin (1963), "Complementation on ribosomes," *Cold Spring Harbor Symp. Quant. Biol.* 28:533–537.

6. Y. Kiho and A. Rich (1964), "Induced enzyme formed on bacterial polyribosomes," *Proc. Nat. Acad. Sci. U.S.A.* 51:111–118.

7. J. Hamlin and I. Zabin (1972), "β-Galactosidase: Immunological activity of ribosome-bound, growing polypeptide chains," *Proc. Nat. Acad. Sci. U.S.A.* 69:412–416.

8. P. Siekevitz and G. E. Palade (1960), "A cytochemical study on the pancreas of the guinea pig. V. *In vivo* incorporation of leucine-1-C^{14} into the chymotrypsinogen of various cell fractions," *J. Biophys. Biochem. Cytol.* 7:619–630.

9. M. Takagi and K. Ogata (1968), "Direct evidence for albumin biosynthesis by membrane-bound polysomes in rat liver," *Biochem. Biophys. Res. Commun.* 33:55–60.

10. C. M. Redman (1969), "Biosynthesis of serum proteins and ferritin by free and attached ribosomes of rat liver," *J. Biol. Chem.* 244:4308–4315.

11. M. C. Ganoza and C. A. Williams (1969), "*In vitro* synthesis of different categories of specific protein by membrane-bound and free ribosomes," *Proc. Nat. Acad. Sci. U.S.A.* 63:1370–1376.

12. S. J. Hicks, J. W. Drysdale, and H. N. Munro (1969), "Preferential synthesis of ferritin and albumin by different populations of liver polysomes," *Science* 164:584–585.

13. R. Cancedda and M. J. Schlessinger (1974), "Localization of polyribosomes containing alkaline phosphatase nascent polypeptides on membranes of *Escherichia coli*," *J. Bacteriol.* 117:290–301.

14. T. G. Morrison and H. F. Lodish (1975), "Site of synthesis of membrane and nonmembrane proteins of vesicular stomatitis virus," *J. Biol. Chem.* 250:6955–6962.

15. R. Henderson and P. N. T. Unwin (1975), "Three-dimensional model of purple membrane obtained by electron microscopy," *Nature* 257:28–32.

16. D. D. Sabatini and G. Blobel (1970), "Controlled proteolysis of nascent polypeptides in rat liver cell fractions. II. Location of the polypeptides in rough microsomes," *J. Cell. Biol.* 45:146–157.

17. M. R. Adelman, D. D. Sabatini, and G. Blobel (1973), "Ribosome-membrane interaction: Nondestructive disassembly of rat liver rough microsomes into ribosomal and membranous components," *J. Cell Biol.* 56:206–229.

18. T. M. Harrison, G. G. Brownlee, and C. Milstein (1974), "Studies on polysome-membrane interactions in mouse myeloma cells," *Eur. J. Biochem.* 47:613–620.

19. W. P. Smith, P.-C. Tai, and B. D. Davis (1978), "Nascent peptide as sole attachment of polysomes to membranes in bacteria," *Proc. Nat. Acad. Sci. U.S.A.* 75:814–817.

20. G. Kreibich, B. L. Ulrich, and D. D. Sabatini (1978), "Proteins of rough microsomal membranes related to ribosome binding. I. Identification of

ribophorins I and II, membrane proteins characteristic of rough microsomes," *J. Cell Biol.* 77:464–487.

21. G. Kreibich, C. M. Freienstein, B. N. Pereyra, B. L. Ulrich, and D. D. Sabatini (1978), "Proteins of rough microsomal membranes related to ribosomal binding. II. Crosslinking of bound ribosomes to specific membrane proteins exposed at the binding sites," *J. Cell Biol.* 77:488–506.
22. D. D. Sabatini, Y. Tashiro, and G. E. Palade (1966), "On the attachment of ribosomes to microsomal membranes," *J. Mol. Biol.* 19:503–524.
23. P. N. T. Unwin (1979), "Attachment of ribosome crystals to intracellular membranes," *J. Mol. Biol.* 132:69–84.
24. C. Bernabeu and J. A. Lake (1982), "Nascent polypeptide chains emerge from the exit domain of the large ribosomal subunit: Immune mapping of the nascent chain," *Proc. Nat. Acad. Sci. U.S.A.* 79:3111–3115.
25. C. Bernabeu, E. M. Tobin, A. Fowler, I. Zabin, and J. A. Lake (1983), "Nascent polypeptide chains exit the ribosome in the same relative position in both eucaryotes and procaryotes," *J. Cell Biol.* 96:1471–1474.
26. G. Blobel and D. D. Sabatini (1971), "Ribosome-membrane interaction in eukaryotic cells," Biomembranes, ed. L. A. Manson, vol. 2, pp. 193–195 (New York and London: Plenum Press).
27. C. Milstein, G. G. Brownlee, T. H. Harrison, and M. B. Mathews (1972), "A possible precursor of immunoglobulin light chains," *Nature New Biol.* 239:117–120.
28. I. Schechter, D. J. McKean, R. Guyer, and W. Terry (1975), "Partial amino acid sequence of the precursor of immunoglobulin light chain programmed by messenger RNA *in vitro*," *Science* 188:160–162.
29. H. Inouye and J. Beckwith (1977), "Synthesis and processing of an *Escherichia coli* alkaline phosphatase precursor *in vitro*," *Proc. Nat. Acad. Sci. U.S.A.* 74:1440–1444.
30. W. P. Smith, P.-C. Tai, R. C. Thompson, and B. D. Davis (1977), "Extracellular labeling of nascent polypeptides traversing the membrane of *Escherichia coli*," *Proc. Nat. Acad. Sci. U.S.A.* 74:2830–2834.
31. K. Talmadge, S. Stahl, and W. Gilbert (1980), "Eukaryotic signal sequence transports insulin antigen in *Escherichia coli*," *Proc. Nat. Acad. Sci. U.S.A.* 77:3363–3373.
32. K. Talmadge, J. Kaufman, and W. Gilbert (1980), "Bacteria mature preproinsulin to proinsulin," *Proc. Nat. Acad. Sci. U.S.A.* 77:3988–3992.
33. V. R. Lingappa, F. N. Katz, H. F. Lodish, and G. Blobel (1978), "A signal sequence for the insertion of a transmembrane glycoprotein: Similarities to the signals of secretory proteins in primary structure and function," *J. Biol. Chem.* 253:8667–8670.
34. A. V. Finkelstein, P. Bendzko, and T. A. Rapoport (1983), "Recognition of signal sequences," *FEBS Letters* 161:176–179.

35. R. D. Palmiter, J. Gagnon, and K. A. Walsh (1978), "Ovalbumin: A secreted protein without a transient hydrophobic leader sequence," *Proc. Nat. Acad. Sci. U.S.A.* 75:94–98.

36. S. Bar-Nun, G. Kreibich, M. Adesnik, L. Alterman, M. Negishi, and D. D. Sabatini (1980), "Synthesis and insertion of cytochrome P-450 into endoplasmic reticulum membranes," *Proc. Nat. Acad. Sci. U.S.A.* 77:965–969.

37. I. Schechter, G. Burstein, R. Zemell, E. Ziv, F. Kantor, and D. S. Papermaster (1979), "Messenger RNA of opsin from bovine retina: Isolation and partial sequence of the *in vitro* translation product," *Proc. Nat. Acad. Sci. U.S.A.* 76:2654–2658.

38. V. R. Lingappa, J. R. Lingappa, and G. Blobel (1979), "Chicken ovalbumin contains an internal signal sequence," *Nature* 281:117–121.

39. R. A. Rachubinski, D. P. S. Verma, and J. J. M. Bergeron (1980), "Synthesis of rat liver microsomal cytochrome b_5 by free ribosomes," *J. Cell Biol.* 84:705–716.

40. N. Borgese and S. Gaetani (1980), "Site of synthesis of rat liver NADH-cytochrome-b_5 reductase, an integral membrane protein," *FEBS Letters* 112:216–220.

41. P. Walter and G. Blobel (1980), "Purification of a membrane-associated protein complex required for protein translation across the endoplasmic reticulum," *Proc. Nat. Acad. Sci. U.S.A.* 77:7112–7116.

42. P. Walter and G. Blobel (1982), "Signal recognition particle contains a 7S RNA essential for protein translation across the endoplasmic reticulum," *Nature* 209:691–698.

43. P. Walter, I. Ibrahimi, and G. Blobel (1981), "Translocation of proteins across the endoplasmic reticulum. I. Signal recognition protein (SRP) binds to in-vitro-assembled polysomes synthesizing secretory protein," *J. Cell Biol.* 91:545–550.

44. P. Walter and G. Blobel (1981), "Translocation of proteins across the endoplasmic reticulum. III. Signal recognition protein (SRP) causes signal sequence-dependent and site-specific arrest of chain elongation that is released by microsomal membranes," *J. Cell Biol.* 91:557–561.

45. D. I. Meyer, D. Lonvard, and B. Dobberstein (1982), "Characterization of molecules involved in protein translocation using a specific antibody," *J. Cell Biol.* 92:579–583.

46. D. I. Meyer, E. Krause, and B. Dobberstein (1982), "Secretory protein translocation across membranes: The role of the 'docking protein,'" *Nature* 297:647–650.

47. R. Gilmore, G. Blobel, and P. Walter (1982), "Protein translocation across the endoplasmic reticulum. I. Detection in the microsomal membrane of a receptor for signal recognition particle," *J. Cell Biol.* 95:463–469.

48. R. Gilmore, P. Walter, and G. Blobel (1982), "Protein translocation across

the endoplasmic reticulum. II. Isolation and characterization of the signal recognition particle receptor," *J. Cell Biol.* 95:470–477.

49. M. R. Capecchi (1966), "Initiation of *E. coli* proteins," *Proc. Nat. Acad. Sci. U.S.A.* 55:1517–1524.
50. J. M. Adams (1968), "On the release of the formyl group from nascent protein," *J. Mol. Biol.* 33:571–589.
51. J.-P. Waller (1963), "The NH_2 terminal residues of the proteins from cell-free extracts of *E. coli*," *J. Mol. Biol.* 7:483–496.
52. R. C. Jackson and G. Blobel (1977), "Post-translational cleavage of presecretory proteins with an extract of rough microsomes from dog pancreas containing signal peptidase activity," *Proc. Nat. Acad. Sci. U.S.A.* 74:5598–5602.
53. P. Walter, R. C. Jackson, M. M. Marcus, V. R. Lingappa, and G. Blobel (1979), "Tryptic dissection and reconstitution of translocation activity for nascent presecretory proteins across microsomal membranes," *Proc. Nat. Acad. Sci. U.S.A.* 76:1795–1799.
54. R. F. Goldberger, C. J. Epstein, and C. B. Anfinsen (1963), "Acceleration of reactivation of reduced bovine pancreatic ribonuclease by a microsomal system from rat liver," *J. Biol. Chem.* 238:628–635.
55. F. De Lorenzo, R. F. Goldberger, E. Steers, D. Givol, and C. B. Anfinsen (1966), "Purification and properties of an enzyme from beef liver which catalyses sulfhydryldisulfide interchange in proteins," *J. Biol. Chem.* 241:1562–1567.
56. P. Venetianer and F. B. Straub (1963), "The enzymic reactivation of reduced ribonuclease," *Biochim. Biophys. Acta* 67:166–168.
57. R. B. Freedman and D. A. Hillson (1980), "Formation of disulfide bonds," The enzymology of post-translational modifications of proteins, ed. R. B. Freedman and H. C. Hawkins, vol. 1, pp. 157–212 (London and New York: Academic Press).
58. C. F. Phelps (1980), "Glycosylation," The enzymology of post-translational modifications of proteins, ed. R. B. Freedman and H. C. Hawkins, vol. 1, pp. 105–155 (London and New York: Academic Press).
59. S. C. Hubbard and R. J. Ivatt (1981), "Synthesis and processing of asparagine-linked oligosaccharides," *Ann. Rev. Biochem.* 50:555–583.
60. A. Haselbeck and W. Tanner (1983), "O-glycosylation in *Saccharomyces cerevisiae* is initiated at the endoplasmic reticulum," *FEBS Letters* 158:335–338.
61. J. E. Rothman and H. F. Lodish (1977), "Synchronised transmembrane insertion and glycosylation of a nascent membrane protein," *Nature* 269:775–780.
62. E. L. Lazarides, L. N. Lukens, and A. A. Infante (1971), "Collagen polysomes: Site of hydroxylation of proline residues," *J. Mol. Biol.* 58:831–846.

63. R. F. Diegelmann, L. Bernstein, and B. Peterkofsky (1973), "Cell-free collagen synthesis on membrane-bound polysomes of chick embryo connective tissue and the localization of prolyl hydroxylase on the polysome-membrane complex," *J. Biol. Chem.* 248:6514–6521.

64. K. I. Kivirikko and R. Myllylä (1980), "The hydroxylation of prolyl and lysyl residues," The enzymology of post-translational modifications of proteins, ed. R. B. Freedman and H. C. Hawkins, vol. 1, pp. 53–104 (London and New York: Academic Press).

Further reading

Blobel, G., and Dobberstein, B. (1975). Transfer of protein across membranes. I. Presence of proteolytically processed and unprocessed nascent immunoglobulin light chains on membrane-bound ribosomes by murine myeloma; II. Reconstitution of functional rough microsomes from heterologous components. *J. Cell Biol.* 67:835–851, 852–862.

Hand, A. R., and Oliver, C., eds. (1981). *Basic mechanisms of cellular secretion.* "Methods in Cell Biology," Acad. Press, New York, London. Vol. 23.

Harwood, R. (1980). Protein transfer across membranes: The role of signal sequences and signal peptidase activity. The enzymology of post-translational modifications of proteins (R. B. Freedman and H. C. Hawkins, eds.), vol. 1, pp. 3–52. London and New York: Academic Press.

Inouye, M., and Halegoua, S. (1980). Secretion and membrane localization of proteins in *Escherichia coli*. *CRC Crit. Rev. Biochem.* 7:339–371.

Palade, G. (1975). Intracellular aspects of the process of protein synthesis. *Science* 189:347–358.

Rothman, J. E., and Lenard, J. (1977). Membrane asymmetry. *Science* 195:743–753.

Sabatini, D. D.; Kreibich, G.; Morimoto, T.; and Adesnik, M. (1982). Mechanisms for the incorporation of proteins in membranes and organelles. *J. Cell Biol.* 92:1–22.

Strauss, A. W., and Boime, I. (1982). Compartmentation of newly synthesized proteins. *CRC Crit. Rev. Biochem.* 9:205–235.

Wickner, W. (1979). The assembly of proteins into biological membranes: The membrane trigger hypothesis. *Ann. Rev. Biochem.* 48:23–45.

Zimmerman, M.; Mumford, R. A.; and Steiner, D. F., eds. (1980). *Precursor processing in the biosynthesis of proteins.* Annals of the New York Academy of Sciences, vol. 343.

Index